Spinors and Calibrations

PERSPECTIVES IN MATHEMATICS, Vol. 9

J. Coates and S. Helgason, editors

Spinors and Calibrations

F. Reese Harvey

Department of Mathematics
Rice University
Houston, Texas

ACADEMIC PRESS, INC.
Harcourt Brace Jovanovich, Publishers

Boston   San Diego   New York
Berkeley   London   Sydney
Tokyo   Toronto

ACADEMIC PRESS, INC.
1250 Sixth Avenue, San Diego, CA 92101

United Kingdom Edition published by
ACADEMIC PRESS LIMITED
24–28 Oval Road, London NW1 7DX

Library of Congress Cataloging-in-Publication Data

Harvey, F. Reese.
    Spinors and calibrations/F. Reese Harvey.
      p.   cm. – (Perspectives in mathematics: vol. 9)
    Bibliography: p.
    Includes index.
    ISBN 0-12-329650-1
    1 Spinor analysis.   2. Matrix groups.   I. Title.   II. Series.
QA433.H327 1990
515'.63–dc19   89-74
                                                        CIP

Printed in the United States of America
90 91 92   9 8 7 6 5 4 3 2 1

This book is dedicated to my wife, Linda

TABLE OF CONTENTS

vii

PREFACE

This book is intended to be a collection of examples. The (simple) Lie groups, the spin groups for general signature, normed algebras for general signature, the exceptional groups G_2 and F_4, the orbit structure of the simpler representations of these groups, and the special Lagrangian and associative calibrations are all discussed in some detail. The underlying and not always mentioned motivation for these examples is differential geometry: Riemannian, symplectic, Kähler, hyperKähler, as well as complex and quaternionic.

The book is divided into two parts. Readers who are primarily interested in the spin groups are encouraged to start with Part II—Spinors.

Part I begins with an introduction to certain specific matrix groups, with the entries real, complex, or quaternionic. Some of these groups are defined by requiring that the matrix fix one of several types of (generalized) inner products. This leads to a discussion in Chapter 2 of eight types of inner products.

The groups introduced in Chapter 1 are also the subjects of Chapter 3. A brief discussion of group representations and orbits is followed by the computation of orbit structures for some of the classical representations of Chapter 1. Lie Algebras are also introduced in this chapter.

Chapter 4 is devoted to a study of inner product spaces in the usual sense, i.e., a real vector space equipped with a real nondegenerate symmetric bilinear form. This is just one of the eight types discussed in Chapter 2. The topics vary in intensity, beginning with a very elementary but fairly complete discussion of the Cauchy–Schwarz "equality."

The reader is assumed to have Euclidean intuition. A discussion of Lorenzian intuition is presented based on both special relativity and the analogue between complex numbers $\mathbf{C}$ and Lorentz numbers $\mathbf{L}$. Some readers will prefer to skip the long section on the Cartan–Dieudonné Theorem until after reading Chapter 9, where additional motivation is provided.

The next chapter, titled "Differential Geometry," is quite different from the first four chapters. Chapter 5 is intended to provide two things: first, motivation (but not prerequisites) for other material in this book; second, a skeleton or bare outline of some of the various types of geometry.

Hopefully the reader will be inclined to build on this outline by further study outside the book.

In contrast, Chapter 6, on normed algebras, provides a fairly complete discussion. In fact, some of the information presented here is not available elsewhere in the literature. The chapter naturally includes a discussion of the exceptional Lie group G_2 as the automorphism group of the octonians.

Various examples of calibrations are examined in Chapter 7. The group that fixes a particular calibration provides an important ingredient in discussing the geometry of submanifolds determined by the calibration. The beautiful Angle Theorem of Lawlor and Nance is discussed. Both halves of the proof involve introducing an interesting class of examples: a) the Nance calibrations; and b) the Lawlor special Lagrangian submanifolds.

Chapter 8 is a brief, elementary, and somewhat tedious discussion of matrix algebras, but the material is essential for understanding Clifford algebras. Algebraists already familiar with this material are encouraged to skip to Part II.

Part II begins with a presentation of Clifford algebras that is independent of Part I. The reader may wish to start the book here. The exterior algebra is assumed to be a familiar object, although a brief definition is provided. Since the exterior algebra and the Clifford algebra are canonically isomorphic as vector spaces (but not as algebras), a considerable amount of intuition about the exterior algebra carries over to the Clifford algebra. Our presentation emphasizes this relationship between the Clifford algebra and the exterior algebra. For example, the natural inner product induced on the exterior algebra immediately provides a natural inner product on the Clifford algebra. This norm on the Clifford algebra usually appears somewhat myteriously as a norm only defined on those Clifford elements in the Clifford group.

Chapter 10 describes the Spin and Pin groups. Again the discussion emphasizes the connection with the exterior algebra, via the grassmannians as subsets of the exterior algebra. Each plane through the origin, once it is oriented in one of the two possible ways, can be considered a Clifford element. By utilizing Clifford multiplication, both of these two oriented planes are associated with the orthogonal transformation "reflection along the plane," providing the core of the double cover of the orthogonal group by the Pin group. For the reader willing to accept the Cartan–Dieudonné Theorem, this Chapter is also independent of Part I.

As algebras—forgetting the more subtle extra structure—the Clifford algebras are just matrix algebras. The matrices act on the vector space of pinors. Although not canonical, this pinor space is unique up to a scalar multiple of the identity. Chapter 11 describes the spaces of pinors and spinors.

Each space of pinors or of spinors can, in a natural way, be given the extra structure of one of the types of (generalized) inner products discussed in Part I. This inner product is unique up to a change of scale, and it plays an important role in understanding the spin groups and the space of spinors. Chapter 12 analyzes the case of split signature and the complex case in even dimensions. This provides a basis for determining the spinor inner product for arbitrary signature in Chapter 13. Moreover, this spinor inner product determines a classical (companion) group containing the spin group that is, probably, the smallest such group.

Chapter 14 consists of just a few of the many interesting applications of, as well as some very explicit models for, the various Clifford algebras in low dimensions.

The importance of spinors in geometry (including, of course, general relativity) is not universally accepted. Two recent texts should play an important role in correcting any previous oversight. The two-volume book by Penrose and Rindler on general relativity contains many interesting topics on general relativity and Twistor Theory. *Spin Geometry*, by Lawson and Michelson, is a very exciting presentation of some of the most beautiful topics in geometry.

I wish to thank Robert Bryant, Blaine Lawson, and Roger Penrose for very valuable discussions and frequent encouragement concerning this book. In addition, Jack Mealy deserves special appreciation for his careful reading of the manuscript. Finally, I wish to thank typists Janie McBane, Anita Poley, and Janice Want.

F. Reese Harvey
Houston, Texas

PART I. CLASSICAL GROUPS
AND
NORMED ALGEBRAS

1. Classical Groups I

The elements of the groups defined in this chapter are matrices with entries in one of the three fields:

R the field of *real numbers*,
C the field of *complex numbers*,
H the field of *quaternions*.

Note that H, the field of quaternions (or hamiltonians), is *not* commutative. The quaternions will be examined in great detail, along with the *octonions* O (or *Cayley numbers*), in Chapter 6 on normed algebras. For the purposes of this chapter, a rudimentary knowledge of H is all that is presupposed. Consult Problem 6 for the multiplication rules for quaternions.

Let $M_n(\mathbf{R})$, $M_n(\mathbf{C})$, and $M_n(\mathbf{H})$ denote the algebras of $n \times n$ matrices with entries in R, C, and H respectively. Represent elements x of $\mathbf{R}^n, \mathbf{C}^n$, and $\mathbf{H}^n$ as column n-tuples. Then each matrix A determines a linear transformation or endomorphism $x \mapsto Ax$ by letting the matrix A act on the left of the column vector x, at least in the real and complex case. Special consideration is necessary for the quaternionic case since H is not commutative. In order for the map $A : \mathbf{H}^n \to \mathbf{H}^n$ (defined by A acting on x on left) to be H-linear, we are forced to let the scalars H act on the H-vector space $\mathbf{H}^n$ on the right!

Although it will be convenient to consider both *right* H-*vector spaces* (where the scalars H act on the right of the vectors) and *left* H-*vector spaces* (where the scalars H act on the left of the vectors), the space $\mathbf{H}^n$ of

3

column n-tuples will always be considered as a right **H**-vector space. Then we have

(1.1) $M_n(\mathbf{R}) \cong \mathrm{End}_\mathbf{R}(\mathbf{R}^n)$,

(1.2) $M_n(\mathbf{C}) \cong \mathrm{End}_\mathbf{C}(\mathbf{C}^n)$,

(1.3) $M_n(\mathbf{H}) \cong \mathrm{End}_\mathbf{H}(\mathbf{H}^n)$.

Here $\mathrm{End}_F(V)$ denotes the F-linear maps from a vector space V, with scalar field F, into itself. If V is a real vector space, then $\mathrm{End}_\mathbf{R} V$ is naturally a real algebra (associative and with unit). If V is a complex vector space, then $\mathrm{End}_\mathbf{C} V$ is naturally a complex algebra (associative and with unit) but may also be considered as a real algebra, which is convenient for some immediate purposes. Finally, if V is a right quaternionic vector space, then $\mathrm{End}_\mathbf{H}(V)$ is naturally a *real* algebra—in fact, a real subalgebra of the algebra $\mathrm{End}_\mathbf{R}(V)$. There is no canonical way to make $\mathrm{End}_\mathbf{H}(V)$ into even a quaternionic vector space (right or left), much less a "quaternionic" algebra (see Problem 7).

THE GENERAL LINEAR GROUPS

The group of units, or invertible elements, in the matrix algebra $M_n(F)$ is called the *F-general linear group* for $F \equiv \mathbf{R}, \mathbf{C}$, or $\mathbf{H}$ and is denoted by $\mathrm{GL}(n, \mathbf{R}), \mathrm{GL}(n, \mathbf{C})$, or $\mathrm{GL}(n, \mathbf{H})$, respectively. If the group of units, or invertible elements, in $\mathrm{End}_F(V)$ is denoted by $\mathrm{GL}_F(V)$, then

(1.1') $\mathrm{GL}(n, \mathbf{R}) \cong \mathrm{GL}_\mathbf{R}(\mathbf{R}^n)$,

(1.2') $\mathrm{GL}(n, \mathbf{C}) \cong \mathrm{GL}_\mathbf{C}(\mathbf{R}^n)$,

(1.3') $\mathrm{GL}(n, \mathbf{H}) \cong \mathrm{GL}_\mathbf{H}(\mathbf{H}^n)$.

In the quaternion case, there is another important group, larger than the **H**-general linear group $\mathrm{GL}(n, \mathbf{H})$, which we will call the *enhanced* **H**-*general linear group*. First note that the **H**-general linear group $\mathrm{GL}(n, \mathbf{H})$ (which acts on the left) consists entirely of **H**-linear maps. However, right multiplication by a scalar $\lambda \in \mathbf{H}$, denoted R_λ, is not necessarily **H**-linear. In fact, R_λ is **H**-linear if and only if λ commutes with all scalars $\mu \in \mathbf{H}$

because $R_\lambda(x\mu) = x\mu\lambda$, while $R_\lambda(x)\mu = x\lambda\mu$. The reader should confirm that λ commutes with all $\mu \in \mathbf{H}$ if and only if $\lambda \in \mathbf{R} \subset \mathbf{H}$. Thus, R_λ is $\mathbf{H}$-linear if and only if $\lambda \in \mathbf{R} \subset \mathbf{H}$. Let $\mathbf{H}^*$ denote the group of right multiplications by nonzero scalars. Then $\mathbf{H}^*$ is not a subgroup of $GL(n, \mathbf{H})$. However, both are contained in the algebra $\text{End}_{\mathbf{R}}(\mathbf{H}^n)$ of $\mathbf{R}$-linear maps. As noted above, the intersection $GL(n, \mathbf{H}) \cap \mathbf{H}^*$ equals $\mathbf{R}^*$ the group of real nonzero multiples of the identity.

The *enhanced* $\mathbf{H}$-*general linear group*, denoted $GL(n, \mathbf{H}) \cdot \mathbf{H}^*$, is defined to be the image of $GL(n, \mathbf{H}) \times \mathbf{H}^*$ in $\text{End}_{\mathbf{R}}(\mathbf{H}^n)$ via the map sending the pair (A, λ) to $L_A \cdot R_\lambda$, where $\cdot$ denotes multiplication in the algebra $\text{End}_{\mathbf{R}}(\mathbf{H}^n)$, i.e., composition. Thus, the following sequence of groups is exact:

$$(1.4) \qquad 1 \to \mathbf{R}^* \to GL(n, \mathbf{H}) \times \mathbf{H}^* \to GL(n, \mathbf{H}) \cdot \mathbf{H}^* \to 1$$

with $GL(n, \mathbf{H}) \cdot \mathbf{H}^* \subset \text{End}_{\mathbf{R}}(\mathbf{H}^n)$. Note that the larger group $GL(n, \mathbf{H}) \cdot \mathbf{H}^*$, as well as the smaller group $GL(n, \mathbf{H})$, maps quaternion lines to quaternion lines.

Given $A \in M_n(\mathbf{R})$, the real determinant of A will be denoted $\det_{\mathbf{R}} A$. Similarly, $\det_{\mathbf{C}} A$ denotes the complex determinant of $A \in M_n(\mathbf{C})$. The lack of commutativity for $\mathbf{H}$ eliminates the possibility of any useful notion of "quaternionic determinant." Of course,

$$GL(n, \mathbf{R}) = \{A \in M_n(\mathbf{R}) : \det_{\mathbf{R}} A \neq 0\},$$

(1.5)   and

$$GL(n, \mathbf{C}) = \{A \in M_n(\mathbf{C}) : \det_{\mathbf{C}} A \neq 0\}.$$

The group

$$(1.6) \qquad GL^+(n, \mathbf{R}) = \{A \in M_n(\mathbf{R}) : \det_{\mathbf{R}} A > 0\}$$

is called the *orientation-preserving general linear group*.

In both the real and the complex case, we have a *special linear group*, defined by

$$(1.7) \qquad SL(n, \mathbf{R}) \equiv \{A \in M_n(\mathbf{R}) : \det_{\mathbf{R}} A = 1\},$$

$$(1.8) \qquad SL(n, \mathbf{C}) \equiv \{A \in M_n(\mathbf{C}) : \det_{\mathbf{C}} A = 1\}.$$

Since there is no quaternion determinant, if we proceed in exact analogy with the real or the complex case, the special quaternion linear group does not exist. However, it is useful to retain the notation $\mathrm{SL}(n, \mathbf{H})$ by employing the real determinant. Let

(1.9) $$\mathrm{SL}(n, \mathbf{H}) \equiv \{A \in \mathrm{GL}(n, \mathbf{H}) : \det_{\mathbf{R}} A = 1\}$$

denote the *special quaternion linear group*.

GROUPS DEFINED BY BILINEAR FORMS

Some very interesting groups are best defined as subgroups of the groups defined above that fix a certain bilinear form.

R-*symmetric*

The *orthogonal group* $O(p, q)$ *with signature* p, q is defined to be the subgroup of $\mathrm{GL}(n, \mathbf{R})$ $(n \equiv p + q)$ that fixes the *standard* R-*symmetric form*

(1.10) $$\varepsilon(x, y) \equiv x_1 y_1 + \cdots + x_p y_p - x_{p+1} y_{p+1} - \cdots - x_n y_n.$$

That is,

$$O(p, q) \equiv \{A \in \mathrm{GL}(n, \mathbf{R}) : \varepsilon(Ax, Ay) = \varepsilon(x, y) \text{ for all } x, y \in \mathbf{R}^n\}.$$

R-*skew (or symplectic)*

The *real symplectic group* $\mathrm{Sp}(n, \mathbf{R})$ is defined to be the subgroup of $\mathrm{GL}(2n, \mathbf{R})$ that fixes the *standard* R-*symplectic* (or R-*skew*) *form*

(1.11) $$\varepsilon \equiv dx_1 \wedge dx_2 + \cdots + dx_{2n-1} \wedge dx_{2n},$$

or

(1.11′) $$\varepsilon(x, y) \equiv x_1 y_2 - x_2 y_1 + \cdots + x_{2n-1} y_{2n} - x_{2n} y_{2n-1}.$$

That is,

$$\mathrm{Sp}(n, \mathbf{R}) \equiv \left\{A \in \mathrm{GL}(2n, \mathbf{R}) : \varepsilon(Ax, Ay) = \varepsilon(x, y) \text{ for all } x, y \in \mathbf{R}^{2n}\right\}.$$

C-*symmetric*

The *complex orthogonal group* $O(n, \mathbf{C})$ is defined to be the subgroup of $GL(n, \mathbf{C})$ that fixes the standard **C**-*symmetric form*

$$(1.12) \qquad \varepsilon(z, w) \equiv z_1 w_1 + \cdots + z_n w_n.$$

C-*skew* (*or symplectic*)

The *complex symplectic group* $Sp(n, \mathbf{C})$ is defined to be the subgroup of $GL(2n, \mathbf{C})$ that fixes the *standard* **C**-*symplectic* (or **C**-*skew*) *form*

$$(1.13) \qquad \varepsilon \equiv dz_1 \wedge dz_2 + \cdots + dz_{2n-1} \wedge dz_{2n},$$

or

$$(1.13') \qquad \varepsilon(z, w) \equiv z_1 w_2 - z_2 w_1 + \cdots + z_{2n-1} w_{2n} - z_{2n} w_{2n-1}.$$

C-*hermitian* (*symmetric*)

The *complex unitary group* $U(p, q)$ *with signature* p, q is defined to be the subgroup of $GL(n, \mathbf{C})$ $(n \equiv p + q)$ that fixes the *standard* **C**-*hermitian symmetric form*

$$(1.14) \qquad \varepsilon(z, w) \equiv z_1 \overline{w}_1 + \cdots + z_p \, \overline{w}_p - z_{p+1} \, \overline{w}_{p+1} - \cdots - z_n \, \overline{w}_n.$$

Remark 1.15. $i\varepsilon(z, w)$ is called the *standard* **C**-*hermitian skew form*. Note that the group that fixes $i\varepsilon$ is just the same group $U(p, q)$ that fixes ε. This contrasts sharply with the quaternion case.

H-*hermitian symmetric*

The *hyper-unitary group* $HU(p, q)$ *with signature* p, q is defined to be the subgroup of $GL(n, \mathbf{H})$ $(n \equiv p + q)$ that fixes the *standard* **H**-*hermitian symmetric form*

$$(1.16) \qquad \varepsilon(x, y) \equiv \overline{x}_1 y_1 + \cdots + \overline{x}_p y_p - \overline{x}_{p+1} y_{p+1} - \cdots - \overline{x}_n y_n.$$

Note: $\varepsilon(x, y)$ is **H**-*hermitian*. This means that ε is additive in both variables x and y, and $\varepsilon(x\lambda, y) = \overline{\lambda}\varepsilon(x, y)$, $\varepsilon(x, y\lambda) = \varepsilon(x, y)\lambda$ for all scalars $\lambda \in \mathbf{H}$. Also note that xy is not **H**-linear in x. In fact, there is no standard **H**-symmetric or **H**-skew form (see Problem 8).

Remark. The group $HU(p, q)$ is usually denoted "$Sp(p, q)$" and called the "symplectic group."

H-*hermitian skew*

The *skew* **H**-*unitary group* $SK(n, \mathbf{H})$, or $SK(n)$, is defined to be the subgroup of $GL(n, \mathbf{H})$ that fixes the *standard* **H**-*hermitian skew form*

(1.17) $\varepsilon(x, y) \equiv \overline{x}_1 i y_1 + \cdots + \overline{x}_n i y_n.$

Remark. This i is the quaternion i (see Problem 6). In Chapter 2, we shall see that if the i occurring in (1.17) is replaced by any unit imaginary quaternion $u \in S^2 \subset \operatorname{Im} \mathbf{H}$, then the new form ε' differs from the old form ε by a coordinate change, i.e., an element of $GL(n, \mathbf{H})$.

Table 1.18. The groups defined by bilinear forms

	symmetric ϵ	skew ϵ	hermitian symmetric ϵ	hermitian skew ϵ
R	$O(p, q)$	$Sp(n, \mathbf{R})$		
C	$O(n, \mathbf{C})$	$Sp(n, \mathbf{C})$	$U(p, q)$	$U(p, q)$
H			$HU(p, q)$	$SK(n, \mathbf{H})$

OTHER MISCELLANEOUS GROUPS

The subgroups defined by requiring either $\det_{\mathbf{R}}$ or $\det_{\mathbf{C}}$ to be equal to one can also be defined by requiring that an n-form be fixed. The skew n-form

(1.19) $dx \equiv dx_1 \wedge \cdots \wedge dx_n$

on $\mathbf{R}^n$ is called the *standard volume form on* $\mathbf{R}^n$, while the skew n-form

(1.19') $dz \equiv dz_1 \wedge \cdots \wedge dz_n$

on $\mathbf{C}^n$ is called the *standard complex volume form* on $\mathbf{C}^n$. The volume form transforms, under a coordinate change, by multiplication by the determinant:

$$A^* dx = (\det_{\mathbf{R}} A)\, dx \quad \text{for all } A \in \operatorname{End}_{\mathbf{R}} V$$

and
$$B^* dz = (\det_{\mathbf{C}} B) \, dz \quad \text{for all } B \in \operatorname{End}_{\mathbf{C}} V.$$

Here A^* denotes the *dual* (or *pull back*) *map* associated with A, which is defined by $(A^*\alpha)(u) = \alpha(Au)$ if α is a form of degree one and by $A^*(\alpha^1 \wedge \cdots \wedge \alpha^k) = A^*\alpha^1 \wedge \cdots \wedge A^*\alpha^k$ if $\alpha = \alpha^1 \wedge \cdots \wedge \alpha^k$ is the simple product of degree one forms. This provides the most elegant definition of the determinant. Frequently, this is also the most useful. For example, see Problem 4. This definition gives

(1.20) $$\operatorname{SL}(n, \mathbf{R}) = \{A \in \operatorname{GL}(n, \mathbf{R}) : A^* dx = dx\},$$

(1.20′) $$\operatorname{SL}(n, \mathbf{C}) = \{A \in \operatorname{GL}(n, \mathbf{C}) : A^* dz = dz\}.$$

The *special orthogonal group with signature* p, q is defined by

(1.21) $$\operatorname{SO}(p, q) \equiv \{A \in O(p, q) : \det_{\mathbf{R}} A = 1\}.$$

The *special complex orthogonal group* is defined by

(1.22) $$\operatorname{SO}(n, \mathbf{C}) \equiv \{A \in O(n, \mathbf{C}) : \det_{\mathbf{C}} A = 1\}.$$

The *special unitary group* is defined by

(1.23) $$\operatorname{SU}(p, q) \equiv \{A \in U(p, q) : \det_{\mathbf{C}} A = 1\}.$$

The various other possibilities do not lead to new groups. This is a consequence of the facts presented below—see (1.24), (1.25), (1.26), Lemma 1.28, (1.29), and (1.30).

Consult Problem 5 for proofs of the following:

(1.24) $$\text{if } A \in \operatorname{Sp}(n, \mathbf{R}), \text{ then } \det_{\mathbf{R}} A = 1;$$
and
(1.25) $$\text{if } A \in \operatorname{Sp}(n, \mathbf{C}), \text{ then } \det_{\mathbf{C}} A = 1.$$

Forgetting the complex structure on $\mathbf{C}^n$, the complex vector space $\mathbf{C}^n$ becomes a real vector space of dimension $2n$. This embeds the algebra $\operatorname{End}_{\mathbf{C}}(\mathbf{C}^n)$ of complex linear maps into the algebra $\operatorname{End}_{\mathbf{R}}(\mathbf{C}^n)$ of all real linear maps. Thus, for $a \in M_n(\mathbf{C})$, the real determinant $\det_{\mathbf{R}} A$ has meaning as well as $\det_{\mathbf{C}} A$. See Problem 4 for a proof of the result:

(1.26) $$\text{if } A \in M_n(\mathbf{C}), \text{ then } \det_{\mathbf{R}} A = |\det_{\mathbf{C}} A|^2.$$

The quaternion vector space $\mathbf{H}^n$ can be considered as a complex vector space in a variety of natural ways (more precisely, a 2-sphere S^2 of natural ways). Let $\operatorname{Im}\mathbf{H}$ denote the real hyperplane in $\mathbf{H}$ with normal $1 \in \mathbf{H}$. Let S^2 denote the unit sphere in $\operatorname{Im}\mathbf{H}$. Then, for each $u \in S^2, u^2 = -u\,\overline{u} = -|u|^2 = -1$. Therefore, right multiplication by u, defined by

$$R_u x \equiv xu \quad \text{for all } x \in \mathbf{H}^n,$$

is a *complex structure* on $\mathbf{H}^n$; that is, $R_u^2 = -1$. This property enables one to define a complex scalar multiplication on $\mathbf{H}^n$ by $(a + bi)x \equiv (a + bR_u)(x)$ for all $a, b \in R$ and all $x \in \mathbf{H}^n$, where $i^2 = -1$. Note that $\operatorname{End}_{\mathbf{H}}(\mathbf{H}^n) \subset \operatorname{End}_{\mathbf{C}}(\mathbf{H}^n)$ for each of the complex structures R_u on $\mathbf{H}^n$, where $u \in S^2 \subset \operatorname{Im}\mathbf{H}$. Choosing a complex basis for $\mathbf{H}^n$ provides a complex linear isomorphism $\mathbf{H}^n \cong \mathbf{C}^{2n}$. Sometimes it is convenient to select this complex basis as follows. Let $\mathbf{C}(u)$ denote the complex line containing 1 in each of the axis subspaces $\mathbf{H} \subset \mathbf{H}^n$. Thus, $\mathbf{C}(u)$ is the real span of 1 and u. Let $\mathbf{C}(u)^\perp$ denote the complex line orthogonal to $\mathbf{C}(u)$ in $\mathbf{H} \subset \mathbf{H}^n$. Then

(1.27) $$\mathbf{H}^n \cong \left[\mathbf{C}(u) \oplus \mathbf{C}(u)^\perp \right]^n \cong \mathbf{C}^{2n}.$$

Assume the complex structure on $\mathbf{H}^n$ has been fixed, say R_i, then as noted above $\operatorname{End}_{\mathbf{H}}(\mathbf{H}^n) \subset \operatorname{End}_{\mathbf{C}}(\mathbf{C}^{2n})$. Moreover, given $A \in \operatorname{End}_{\mathbf{C}}(\mathbf{C}^{2n})$, one can show that

$$A \in \operatorname{End}_{\mathbf{H}}(\mathbf{H}^n) \text{ if and only if } AR_j = R_j A.$$

This is a useful characterization of the subspace $\operatorname{End}_{\mathbf{H}}(\mathbf{H}^n)$ of $\operatorname{End}_{\mathbf{C}}(\mathbf{C}^{2n})$.

Lemma 1.28. *For each complex structure R_u on $\mathbf{H}^n$ (determined by right multiplication by a unit imaginary quaternion $u \in S^2 \subset \operatorname{Im}\mathbf{H}$) and for each $A \in M_n(\mathbf{H})$ the complex determinant $\det_{\mathbf{C}} A$ is the positive square root of $\det_{\mathbf{R}} A$, independent of the complex structure R_u.*

Proof: First, we show that the complex determinant of $A \in M_n(\mathbf{H})$ is real for all $A \in M_n(\mathbf{H})$. We will give the proof for the particular complex structure R_i. Consider the case $n = 1$. Let $e_0 \equiv 1, e_1 \equiv i, e_2 \equiv j$, and $e_3 \equiv k$ denote the standard real basis for the quaternions $\mathbf{H}$. Let $\omega^0, \omega^1, \omega^2, \omega^3$ denote the standard dual basis. Then

$$dz^1 = \omega^0 + i\omega^1, \qquad dz^2 = \omega^2 - i\omega^3$$

is a basis for the complex forms of type $1, 0$ on $\mathbf{H} \cong \mathbf{C}^2$ (with complex structure R_i). Note that $R_j^*(dz^1) = -d\overline{z}^2$ and $R_j^*(dz^2) = d\overline{z}^1$, so that $R_j(dz^1 \wedge dz^2) = d\overline{z}^1 \wedge d\overline{z}^2$.

Now

$$R_j^* A^*(dz^1 \wedge dz^2) = R_j^*(\det_{\mathbf{C}} A \ dz^1 \wedge dz^2) = \det_{\mathbf{C}} A \ d\bar{z}^1 \wedge d\bar{z}^2,$$

while

$$A^* R_j^*(dz^1 \wedge dz^2) = A^*(d\bar{z}^1 \wedge d\bar{z}^2) = \overline{\det_{\mathbf{C}} A} \ d\bar{z}^1 \wedge d\bar{z}^2.$$

Therefore, $\det_{\mathbf{C}} A \in \mathbf{R}$ is real since $AR_j = R_j A$. The proof, for $n > 1$, that $\det_{\mathbf{C}} A \in \mathbf{R}$ for all $A \in M_n(\mathbf{H})$ is similar and omitted. Because of (1.26), it remains to show that

$$\det_{\mathbf{C}} A > 0 \quad \text{if } A \in \mathrm{GL}(n, \mathbf{H}).$$

Since $\det_{\mathbf{C}} I = 1$ and $\mathrm{GL}(n, \mathbf{H})$ is connected (Problem 3), the set $\{\det_{\mathbf{C}} A : A \in \mathrm{GL}(n, \mathbf{H})\}$ is a connected subset of $\mathbf{R} - \{0\}$ containing 1, and hence it is contained in $\mathbf{R}^+$. ∎

For elements of the subgroups $\mathrm{HU}(p, q)$ and $\mathrm{SK}(n, \mathbf{H})$ of $\mathrm{GL}(n, \mathbf{H})$, the real determinant is already equal to one (and hence by Lemma 1.28 all the various complex determinants are also equal to one). That is,

(1.29) $\det_{\mathbf{R}} A = 1$ if $A \in \mathrm{HU}(p, q)$;

(1.30) $\det_{\mathbf{R}} A = 1$ if $A \in \mathrm{SK}(n, \mathbf{H})$.

Both of these facts follow from (1.24), since both $\mathrm{HU}(p, q)$ and $\mathrm{SK}(n, \mathbf{H})$ are contained in $\mathrm{Sp}(2n, \mathbf{R})$ for a suitable choice of coordinates. For example, if A fixes the ε defined by (1.16), i.e., $A \in \mathrm{HU}(p, q)$, then A fixes the real valued skew form $\mathrm{Re} \, i\varepsilon(x, y)$, which under a coordinate change is the symplectic form given by (1.11'). The details are provided in the next chapter—see Lemma 2.80 and Equation (2.91).

In the quaternion case, there is always the option of enlarging the group by utilizing right scalar multiplications. Recall (1.4) how the group $\mathrm{GL}(n, \mathbf{H}) \cdot \mathbf{H}^*$ is an enhancement of the quaternionic general linear group $\mathrm{GL}(n, \mathbf{H})$. For another example, consider the *enhanced hyper-unitary group* (perhaps a better name is the *quaternionic unitary group*). This group is denoted $\mathrm{HU}(p, q) \cdot \mathrm{HU}(1)$ and defined to be the subgroup of $\mathrm{End}_{\mathbf{R}}(\mathbf{H}^n)$ generated by letting $\mathrm{HU}(p, q)$ act on $\mathbf{H}^n$ on the left and the unit scalars $\mathrm{HU}(1) \cong S^3$ act on $\mathbf{H}^n$ on the right. Since

(1.31) $1 \longrightarrow \mathbf{Z}_2 \longrightarrow \mathrm{HU}(p, q) \times \mathrm{HU}(1) \overset{\chi}{\longrightarrow} \mathrm{End}_{\mathbf{R}}(\mathbf{H}^n)$

is exact, where $\mathbf{Z}_2 = \{1, -1\}$ and where $\mathrm{HU}(1) \cong \{R_y : y \in S^3 \subset \mathbf{H}\}$, it follows that

$$(1.32) \qquad \mathrm{HU}(p, q) \cdot \mathrm{HU}(1) \cong (\mathrm{HU}(p, q) \times \mathrm{HU}(1))/\mathbf{Z}_2.$$

See Problem 3.15 for more information about the quaternionic unitary group $\mathrm{HU}(p, q) \cdot \mathrm{HU}(1)$. For example, this group fixes a 4-form $\Phi \in \Lambda^4(\mathbf{H}^n)^*$.

Remark 1.33. In the special case of $n = p = 1$, and $q = 0, \mathrm{HU}(1)$ acting on the left equals $\{L_a : |a| = 1\}$, while $\mathrm{HU}(1)$ acting on the right equals $\{R_b : |b| = 1\}$. In fact, the quaternionic unitary group is just the special orthogonal group. That is,

$$(1.34) \qquad \mathrm{HU}(1) \cdot \mathrm{HU}(1) = \mathrm{SO}(4),$$

or equivalently,

$$(1.34') \qquad \chi : \mathrm{HU}(1) \times \mathrm{HU}(1) \longrightarrow \mathrm{SO}(4),$$

is a surjective group homomorphism with kernel $\mathbf{Z}_2 = \{-1, 1\}$, where the map χ is defined by

$$\chi_{a,b}(x) \equiv a x \overline{b} \quad \text{for all } x \in \mathbf{H}.$$

To prove $(1.34')$, first note that by (1.29), or more directly, by Problem 6(b), $\det_{\mathbf{R}} L_a = 1$ if $|a| = 1$ (similarly $\det_{\mathbf{R}} R_b = 1$ if $|b| = 1$). Second, one can show that $|ax| = |a| \, |x|$ under quaternion multiplication. Thus, $L_a \in O(4)$ if $|a| = 1$ (similarly $R_b \in O(4)$ if $|b| = 1$). This proves that $\mathrm{HU}(1) \cdot \mathrm{HU}(1) \equiv \chi(\mathrm{HU}(1) \times \mathrm{HU}(1)) \subset \mathrm{SO}(4)$. The surjectivity of χ can be demonstrated with a topological argument based on dimension, once it is known that $\mathrm{SO}(4)$ is connected (see Corollary 3.31). A nontopological proof that χ is surjective is provided by Problem 4.9.

If $\mathrm{HU}(1)$ denotes the diagonal copy of $S^3 \equiv \{a \in \mathbf{H} : |a| = 1\}$ embedded in $\mathrm{HU}(1) \times \mathrm{HU}(1)$ and χ is restricted to $\mathrm{HU}(1)$, then

$$(1.35) \qquad \frac{\mathrm{HU}(1)}{\mathbf{Z}_2} \overset{\chi}{\cong} \mathrm{SO}(3).$$

To prove (1.35), it suffices to note that the subgroup of $\mathrm{SO}(\mathbf{H})$ that fixes $1 \in \mathbf{H}$ is just $\mathrm{SO}(\mathrm{Im}\,\mathbf{H})$ and that the subgroup of $\mathrm{HU}(1) \times \mathrm{HU}(1)$ that maps into $\mathrm{SO}(\mathrm{Im}\,\mathbf{H})$ equals $\{(a, b) \in \mathrm{HU}(1) \times \mathrm{HU}(1) : a\overline{b} = 1\} = \mathrm{HU}(1)$.

The quaternionic enhancements are summarized as follows.

Group		Enhanced Group	
general linear	$GL(n,\mathbf{H})$	$GL(n,\mathbf{H})\cdot\mathbf{H}^*$	enhanced general linear
special linear	$SL(n,\mathbf{H})$	$SL(n,\mathbf{H})\cdot HU(1)$	enhanced special linear
hyper-unitary	$HU(p,q)$	$HU(p,q)\cdot HU(1)$	enhanced hyperunitary
skew-unitary	$SK(n,\mathbf{H})$	$SK(n,\mathbf{H})\cdot HU(1)$	enhanced skew unitary

Of course, one can always enhance a group G with $\mathbf{R}^+$, or $\mathbf{R}^* \equiv \mathbf{R} - \{0\}$, if the nonzero multiples of the identity do not already belong to G. The groups $G \cdot \mathbf{R}^+$ are usually referred to as *conformal groups*. For example,

$$(1.36) \qquad CO(p,q) \equiv O(p,q) \cdot \mathbf{R}^+ \equiv O(p,q) \times \mathbf{R}^+$$

is called the *conformal (orthogonal) group of signature* p,q. This group, perhaps the most important conformal group, is usually defined by requiring that the inner product ε (see (1.10)) be fixed up to a positive scalar multiple (or conformal factor):

$$(1.36') \qquad \begin{aligned} CO(p,q) \equiv \{A \in GL(n,\mathbf{R}) : &\text{ for some } \lambda \in \mathbf{R}^+, \varepsilon(Ax, Ay) \\ &= \lambda\varepsilon(x,y) \text{ for all } x,y \in \mathbf{R}^n \}. \end{aligned}$$

Similarly,

$$(1.37) \qquad \begin{aligned} CSO(p,q) &\equiv SO(p,q) \cdot \mathbf{R}^+ \\ &\equiv \{A \in GL^+(n,\mathbf{R}) : A^*\varepsilon = \lambda\varepsilon \text{ for some } \lambda \in \mathbf{R}^+\} \end{aligned}$$

is called the *special conformal group of signature* p,q.

If both $p,q \geq 1$, then (see Chapter 3) $SO(p,q)$ has two connected components. The connected component of the identity, denoted by $SO^\uparrow(p,q)$, is, of course, a subgroup of $SO(p,q)$. This subgroup $SO^\uparrow(p,q)$ of $SO(p,q)$ is called the *reduced special orthogonal group*. Later, in Chapter 4, additional subgroups of $O(p,q)$, denoted $O^+(p,q)$, and $O^-(p,q)$ will be discussed in some detail. Briefly, if $p,q \geq 1$, then $O(p,q)$ has four connected components. Adding any one of the remaining three components to $SO^\uparrow(p,q)$ yields three additional subgroups of $O(p,q)$, denoted $SO(p,q), O^+(p,q)$, and $O^-(p,q)$. Thus, the intersection of any two of these three is always $SO^\uparrow(p,q)$. See Chapter 4 for the details.

ISOMORPHISMS

The unit circle

$$(1.38) \qquad S^1 \equiv \{z \in \mathbf{C} : |z| = 1\}$$

in the complex plane $\mathbf{C}$ is a group under complex multiplication. By definition, the groups $U(1)$ and S^1 are the same. Of course, $S^1 \equiv \{e^{i\theta} : \theta \in \mathbf{R}\} \cong \mathbf{R}/2\pi\mathbf{Z}$. The group of nonzero complex numbers under complex multiplication is denoted by $\mathbf{C}^*$, and by definition, $GL(1, \mathbf{C}) \equiv \mathbf{C}^*$.

The set of unit quaternions

$$(1.39) \qquad\qquad S^3 \equiv \{x \in \mathbf{H} : |x| = 1\}$$

also forms a group under quaternionic multiplication. Again, by definition, the groups $HU(1)$ and S^3 are the same.

Also, by definitions (1.20) and (1.20'), we have $SL(2, \mathbf{R}) = Sp(1, \mathbf{R})$ and $SL(2, \mathbf{C}) = Sp(1, \mathbf{C})$. The more difficult equality $HU(1) \cdot HU(1) = SO(4)$ has already been discussed. These and other coincidences are listed in the next proposition.

Proposition 1.40. *The following isomorphisms hold*

$$(1.41) \qquad\qquad SO(2) \cong U(1) \cong SK(1) \cong S^1,$$

$$(1.42) \qquad\qquad CSO(2) \cong GL(1, \mathbf{C}) \cong \mathbf{C}^* \cong SO(2, \mathbf{C}),$$

$$(1.43) \qquad\qquad SU(2) \cong HU(1) \cong SL(1, \mathbf{H}) \cong S^3,$$

$$(1.44) \qquad\qquad Sp(1, \mathbf{R}) = SL(2, \mathbf{R}) \cong SU(1, 1),$$

$$(1.45) \qquad\qquad Sp(1, \mathbf{C}) = SL(2, \mathbf{C}),$$

$$(1.46) \qquad HU(1) \cdot HU(1) \cong SO(4) \quad \text{and} \quad GL(1, \mathbf{H}) \cdot \mathbf{H}^* \cong CSO(4),$$

$$(1.47) \qquad\qquad SO^\uparrow(3, 1) \cong SO(3, \mathbf{C}).$$

The last isomorphism (1.47) will be verified in the section on special relativity in Chapter 3.

The proofs of all of the other isomorphisms in Proposition 1.40 are left as an exercise (see Problems 9, 10, and 11). One of these isomorphisms, $SU(2) \cong HU(1)$, warrants the following discussion.

Let $\mathbf{H}$ have the complex structure $R_i x \equiv xi$ (right multiplication by i). Thus, $\mathbf{H} \cong \mathbf{C}^2$, where each $p \in \mathbf{H}$ can be expressed as $p = z + jw$ with

$z, w \in \mathbf{C} \subset \mathbf{H}$. Now each $A \in M_1(\mathbf{H}) \cong \mathrm{End}_{\mathbf{H}}(\mathbf{H})$ can be considered as acting on $\mathbf{H}$ on the left, hence $A \in \mathrm{End}_{\mathbf{C}}(\mathbf{C}^2) \cong M_2(\mathbf{C})$ is a complex linear transformation of $\mathbf{H} \cong \mathbf{C}^2$. Using the coordinates $p = z + jw = (z, w)$ for $p \in \mathbf{H} \cong \mathbf{C}^2$, the complex linear map A expressed as a complex matrix is given by

$$A = \begin{pmatrix} a & -\bar{b} \\ b & \bar{a} \end{pmatrix},$$

where $A = a + jb$, $a, b \in \mathbf{C} \subset \mathbf{H}$. This is because

$$Aj = (A \cdot 1)j = (a + jb)j = -\bar{b} + j\bar{a}.$$

This proves

(1.48) $\qquad M_1(\mathbf{H}) \cong \left\{ \begin{pmatrix} a & -\bar{b} \\ b & \bar{a} \end{pmatrix} \in M_2(\mathbf{C}) : a, b \in \mathbf{C} \right\}.$

The isomorphism $\mathrm{HU}(1) \cong \mathrm{SU}(2)$ is derived from (1.48) (see Problem 10).

Remark. In the standard reference (Helgason [10]), $\mathrm{SL}(n, \mathbf{H})$ is denoted by $\mathrm{SU}^*(2n)$, $\mathrm{SK}(n, \mathbf{H})$ is denoted by $\mathrm{SO}^*(2n)$, and $\mathrm{HU}(p, q)$ is denoted by $\mathrm{Sp}(p, q)$.

SUMMARY

The three general linear groups $\mathrm{GL}(n, \mathbf{R}), \mathrm{GL}(n, \mathbf{C})$, and $\mathrm{GL}(n, \mathbf{H})$ and the seven groups described in Table 1.18 can be changed by imposing restrictions on determinants and/or by enhancing with scalar multiplication. The connected component of the identity in $\mathrm{SO}(p, q)$ with $p, q \geq 1$ is also a group. All the groups introduced in this chapter can be obtained in this way.

In low dimension, some of these groups coincide. One of the most interesting isomorphisms is $\mathrm{SU}(2) \cong \mathrm{HU}(1)$. The topic of special isomorphisms in low dimensions will be discussed again in Chapter 14.

PROBLEMS

1. Establish $M_n(F) \cong \mathrm{End}_F(F^n)$ and $\mathrm{GL}(n, F) \cong \mathrm{GL}_F(F^n)$ for $F \equiv \mathbf{R}, \mathbf{C}, \mathbf{H}$.

2. If $A \in M_n(\mathbf{H}) \cong \mathrm{End}_{\mathbf{H}}(\mathbf{H}^n)$ is injective, then A^{-1} is $\mathbf{H}$-linear.

3. (a) Let e denote the column vector $(1, 0, \ldots, 0) \in F^n$. Show that for each $x \in F^n - \{0\}$ there exist $A \in \mathrm{GL}(n, F)$ with $Ae = x$. Let K denote the subgroup of $\mathrm{GL}(n, F)$ that fixes e. Show that K has the same number of connected components as $\mathrm{GL}(n-1, F)$.

(b) Use part (a) and induction to show that $\mathrm{GL}(n, \mathbf{C})$ and $\mathrm{GL}(n, \mathbf{H})$ are connected, while $\mathrm{GL}(n, \mathbf{R})$ has exactly two components.

(c) Show that $\mathrm{SL}(n, \mathbf{R}), \mathrm{SL}(n, \mathbf{C})$, and $\mathrm{SL}(n, \mathbf{H})$ are all connected.

(d) Show that $O(n), U(n)$, and $\mathrm{HU}(n)$ are compact.

(e) Show that each of the groups defined in Table 1.18 is the level set of a vector-valued polynomial, and hence is a closed set.

4. Suppose $A \in \mathrm{End}_{\mathbf{C}}(\mathbf{C}^n) \subset \mathrm{End}_{\mathbf{R}}(\mathbf{R}^{2n})$. Let dt denote the standard volume form on $\mathbf{R}^{2n}$. Use the facts that:

(a) $A^* dt = (\det_{\mathbf{R}} A) dt$,

(b) $A^* dz = (\det_{\mathbf{C}} A) dz$,

(c) $dt = k \, dz \wedge d\bar{z}$ for some constant k,

to show that $\det_{\mathbf{R}} A = |\det_{\mathbf{C}} A|^2$.

5. Let ε denote the standard symplectic form on $\mathbf{R}^{2n}$, and dx the standard volume form on $\mathbf{R}^{2n}$. Show that they are related by

$$\frac{1}{n!} \varepsilon \wedge \cdots \wedge \varepsilon = dx.$$

6. The quaternions $\mathbf{H}$ can be defined as the vector space $\mathbf{R}^4$ with

$$1 \equiv (1,0,0,0), \quad i \equiv (0,1,0,0), \quad j \equiv (0,0,1,0), \quad k \equiv (0,0,0,1),$$

and multiplication defined by

$$i^2 = j^2 = k^2 = -1,$$

$ij = k$ and all cyclic permutations of this equation are valid,

and i, j, k skew commute.

Given $x = x_0 + x_1 i + x_2 j + x_3 k$, define conjugation by $\bar{x} \equiv x_0 - x_1 i - x_2 j - x_3 k$.

(a) Show that $\overline{xy} = \bar{y}\,\bar{x}$, $x\bar{x} = |x|^2$, $2\langle x, y \rangle = \bar{x}y + \bar{y}x$.

(b) Given $a \in \mathbf{H}$, left multiplication by a, denoted L_a, belongs to $\mathrm{End}_{\mathbf{H}}(\mathbf{H}) \subset \mathrm{End}_{\mathbf{R}}(\mathbf{H})$. In terms of the standard basis for $\mathbf{H} \cong \mathbf{R}^4$, compute the 4×4 matrix for L_a, and then show that $\det_{\mathbf{R}} L_a = |a|^4$.

7. Suppose V is a right $\mathbf{H}$-vector space. Let $\mathrm{Hom}_{\mathbf{H}}(V, \mathbf{H})$ denote the real vector space of right $\mathbf{H}$-linear maps from V to $\mathbf{H}$. Given $f \in \mathrm{Hom}_{\mathbf{H}}(V, \mathbf{H})$ and $\lambda \in \mathbf{H}$, define $f\lambda \in \mathrm{Hom}_{\mathbf{H}}(V, \mathbf{H})$ by

(1.49) $$(f\lambda)(v) \equiv \overline{\lambda} f(v) \quad \text{for all } v \in V.$$

(a) Show that (1.49) exhibits a right $\mathbf{H}$-structure on $\mathrm{Hom}_{\mathbf{H}}(V, \mathbf{H})$. The $\mathbf{H}$-*dual* of V, denoted V^*, is defined to be the vector space $\mathrm{Hom}_{\mathbf{H}}(V, \mathbf{H})$ equipped with this right $\mathbf{H}$-structure (1.49).

(b) Exhibit a (canonical) right $\mathbf{H}$-linear isomorphism $(V^*)^* \cong V$.

(c) Given $f \in \mathrm{Hom}_{\mathbf{H}}(V, W)$, let f^* (the *dual map*) be defined by

(1.50) $$(f^*(w^*))(v) \equiv w^*(f(v)) \text{ for all } w^* \in W^* \text{ and } v \in V.$$

Show that f^* is right $\mathbf{H}$-linear, i.e., $f \in \mathrm{Hom}_{\mathbf{H}}(W^*, V^*)$.

(The real vector space $\mathrm{Hom}_{\mathbf{H}}(V, W)$ cannot be (canonically) given the extra structure of either a right or a left $\mathbf{H}$-vector space.)

8. Suppose $\varepsilon : V \times V \longrightarrow \mathbf{H}$ is $\mathbf{H}$ symmetric, i.e., $\varepsilon(x, y) = \varepsilon(y, x)$ and $\varepsilon(x, y\lambda) = \varepsilon(x, y)\lambda$. Prove that $\varepsilon \equiv 0$.

9. Show that
   (a) $U(1) \cong S^1$,
   (b) $\mathrm{SO}(2) \cong S^1$,
   (c) $\mathrm{SK}(1) \cong S^1$,
   (d) $\mathrm{Sp}(1, \mathbf{R}) = \mathrm{SL}(2, \mathbf{R})$,
   (e) $\mathrm{Sp}(1, \mathbf{C}) = \mathrm{SL}(2, \mathbf{C})$,
   (f) $\mathrm{SO}(2, \mathbf{C}) \cong \mathbf{C}^*$,
   (g) $\mathrm{SL}(1, \mathbf{H}) \cong S^3$.

10. Show that
    (a) $\mathrm{HU}(1) \cong S^3$,
    (b) $U(2) \cong \left\{ \begin{pmatrix} a & -e^{i\theta}\overline{b} \\ b & e^{i\theta}\overline{a} \end{pmatrix} : a, b \in \mathbf{C}, \ |a|^2 + |b|^2 = 1, \text{ and } \theta \in \mathbf{R} \right\}$,
    (c) $\mathrm{SU}(2) \cong \left\{ \begin{pmatrix} a & -\overline{b} \\ b & \overline{a} \end{pmatrix} : a, b \in \mathbf{C} \text{ and } |a|^2 + |b|^2 = 1 \right\}$,
    (d) $\mathrm{SU}(2) \cong \mathrm{HU}(1)$.

11. Prove the following.

(a) $U(1,1) \cong \left\{ \begin{pmatrix} a & e^{i\theta}\overline{c} \\ c & e^{i\theta}\overline{a} \end{pmatrix} : a, c \in \mathbf{C}, \ \theta \in \mathbf{R}, \ \text{and} \ |a|^2 - |c|^2 = 1 \right\}.$

(b) $SU(1,1) \cong \left\{ \begin{pmatrix} a & \overline{c} \\ c & \overline{a} \end{pmatrix} : a, c \in \mathbf{C} \ \text{and} \ |a|^2 - |c|^2 = 1 \right\}.$

(c) $SU(1,1)$ maps N into itself and M into itself, where $N = \{(z, \overline{z}) : z \in \mathbf{C}\}$ and $M = \{(z, -\overline{z}) \in \mathbf{C}^2 : z \in \mathbf{C}\}$ are both two-dimensional real subspaces of $\mathbf{C}^2$.

(d) $C^{-1}SU(1,1)C \cong SL(2, \mathbf{R})$, where $C = \begin{pmatrix} 1 & i \\ 1 & -i \end{pmatrix}$ is the *Cayley Transform*.

2. The Eight Types of Inner Product Spaces

The "inner products" (or "bilinear forms"), which we denoted by ε in the previous chapter, will be examined in more detail in this chapter.

Even if one is *only* interested in the orthogonal groups $O(p,q)$, all types appear (in a natural way), as will be seen in a later chapter on spinor inner products.

Suppose V is a vector space over $F \equiv \mathbf{R}, \mathbf{C}$, or $\mathbf{H}$ and ε is a biadditive map

$$(2.1) \qquad \varepsilon : V \times V \to F.$$

That is, $\varepsilon(x+y,z) = \varepsilon(x,z) + \varepsilon(y,z)$ and $\varepsilon(z,x+y) = \varepsilon(z,x) + \varepsilon(z,y)$. The biadditive map ε is said to be (*pure*) *bilinear* if

$$(2.2) \qquad \begin{array}{c} \varepsilon(\lambda x, y) = \lambda \varepsilon(x,y) \quad \text{and} \quad \varepsilon(x, \lambda y) = \lambda \varepsilon(x,y) \\ \text{for all scalars } \lambda \in F. \end{array}$$

The biadditive map ε is said to be (*hermitian*) *bilinear* if

$$(2.3) \qquad \begin{array}{c} \varepsilon(x\lambda, y) = \overline{\lambda}\varepsilon(x,y) \quad \text{and} \quad \varepsilon(x, y\lambda) = \varepsilon(x,y)\lambda \\ \text{for all scalars } \lambda \in F. \end{array}$$

19

Frequently, when it is clear from the context, the adjective "pure" or the adjective "hermitian" will be dropped. For example, if $F \equiv \mathbf{R}$ there is only one kind of bilinear; the two notions, pure bilinear and hermitian bilinear, agree. If $F \equiv \mathbf{H}$ and ε is pure bilinear, then $\varepsilon = 0$ because $\varepsilon(x, y)\lambda\mu = \varepsilon(x\lambda, y\mu) = \varepsilon(x, y)\mu\lambda$, while $\mu, \lambda \in \mathbf{H}$ may be chosen so that $[\mu, \lambda] \equiv \mu\lambda - \lambda\mu \neq 0$. Thus, for $F \equiv \mathbf{H}$ there is only one kind of bilinear, namely, hermitian bilinear.

Let $\mathrm{BIL}_{\mathrm{pure}}(V), \mathrm{BIL}_{\mathrm{herm}}(V)$—or just $\mathrm{BIL}(V)$ when the correct adjective pure or hermitian is clear from the context—denote the space of bilinear forms on V. Note that these spaces are always real vector spaces. Suppose $\varepsilon \in \mathrm{BIL}_{\mathrm{pure}}(V)$. The form ε is said to be F-symmetric, or F-pure symmetric if

(2.4) $\varepsilon(x, y) = \varepsilon(y, x)$ for all x and $y \in V$.

The form ε is said to be F-*skew* or F-*pure skew* if

(2.5) $\varepsilon(x, y) = -\varepsilon(y, x)$ for all $x, y \in V$.

Suppose $\varepsilon \in \mathrm{BIL}_{\mathrm{herm}}(V)$. The form ε is said to be F-(*hermitian*) *symmetric* if

(2.6) $\overline{\varepsilon(x, y)} = \varepsilon(y, x)$ for all $x, y \in V$.

Finally, the form ε is said to be F-(*hermitian*) *skew* if

(2.7) $\overline{\varepsilon(x, y)} = -\varepsilon(y, x)$ for all $x, y \in V$.

Note that for $F \equiv \mathbf{C}$ (but *not* for $F \equiv \mathbf{H}$) a *hermitian* bilinear form $\varepsilon \in \mathrm{BIL}(V)$ is skew if and only if $i\varepsilon$ is symmetric.

The notations $\mathrm{SYM}_{\mathrm{pure}}(V), \mathrm{SYM}_{\mathrm{herm}}(V), \mathrm{SK}_{\mathrm{pure}}(V)$, and $\mathrm{SK}_{\mathrm{herm}}(V)$ should be self-explanatory. Each $\varepsilon \in \mathrm{BIL}(V)$ has a unique decomposition $\varepsilon = \varepsilon_1 + \varepsilon_2$ into the sum of a symmetric part ε_1 and a skew part ε_2. If ε is pure, then

$$\varepsilon_1(x, y) = \frac{1}{2}\,\varepsilon(x, y) + \frac{1}{2}\,\varepsilon(y, x) \quad \text{and} \quad \varepsilon_2(x, y) = \frac{1}{2}\,\varepsilon(x, y) - \frac{1}{2}\,\varepsilon(y, x);$$

while if ε is hermitian then

$$\varepsilon_1(x, y) = \frac{1}{2}\,\varepsilon(x, y) + \frac{1}{2}\,\overline{\varepsilon(y, x)} \quad \text{and} \quad \varepsilon_2(x, y) = \frac{1}{2}\,\varepsilon(x, y) - \frac{1}{2}\,\overline{\varepsilon(y, x)}.$$

A bilinear form $\varepsilon \in \mathrm{BIL}(V)$ is said to be *nondegenerate* if

(2.8) $\varepsilon(x, y) = 0$ for all $y \in V$ implies $x = 0$,

and

(2.8′) $\quad\quad\quad \varepsilon(x,y) = 0$ for all $x \in V$ implies $y = 0$.

Definition 2.9 (Inner Product Space). *Suppose V is a finite dimensional vector space of F, with F one of the three fields $F \equiv \mathbf{R}, \mathbf{C}$, or $\mathbf{H}$. An inner product ε on V is a nondegenerate bilinear form on V that is either symmetric or skew. If $F \equiv \mathbf{C}$, then there are two types of symmetric and two types of skew: pure and hermitian.*

The eight *types* of inner products are

1. **R**-*-symmetric*

$$\varepsilon \text{ is } \mathbf{R}\text{-bilinear and } \varepsilon(x,y) = \varepsilon(y,x),$$

2. **R**-*skew* or **R**-*symplectic*

$$\varepsilon \text{ is } \mathbf{R}\text{-bilinear and } \varepsilon(x,y) = -\varepsilon(y,x),$$

3. **C**-*symmetric*

$$\varepsilon \text{ is } \mathbf{C}\text{-bilinear and } \varepsilon(x,y) = \varepsilon(y,x),$$

4. **C**-*skew* or **C**-*symplectic*

$$\varepsilon \text{ is } \mathbf{C}\text{-bilinear and } \varepsilon(x,y) = -\varepsilon(y,x),$$

5. **C**-*hermitian symmetric*

$$\varepsilon \text{ is } \mathbf{C}\text{-hermitian bilinear and } \overline{\varepsilon(x,y)} = \varepsilon(y,x),$$

6. **C**-*hermitian skew*

$$\varepsilon \text{ is } \mathbf{C}\text{-hermitian bilinear and } \overline{\varepsilon(x,y)} = -\varepsilon(y,x),$$

7. **H**-*hermitian symmetric*

$$\varepsilon \text{ is } \mathbf{H}\text{-hermitian symmetric and } \overline{\varepsilon(x,y)} = \varepsilon(y,x),$$

8. **H**-*hermitian skew*

$$\varepsilon \text{ is } \mathbf{H}\text{-hermitian and } \overline{\varepsilon(x,y)} = -\varepsilon(y,x).$$

In each of the eight cases the pair V, ε is then called an inner product space (of one of the eight types).

THE STANDARD MODELS

The *standard models* of inner product spaces are listed below. The reader may recall the names of the groups that fix these bilinear forms from Chapter 1. Please note that the conjugation in the C-hermitian symmetric case has been changed from the second variable w (as in (1.14)) to the first variable z in order to be consistent with the quaternion case.

1. **R**-*symmetric*: The vector space is $\mathbf{R}^n$, denoted by $\mathbf{R}(p, q)(n \equiv p + q)$, with

$$(2.10) \qquad \varepsilon(x, y) \equiv x_1 y_1 + \cdots + x_p y_p - \cdots - x_n y_n.$$

2. **R**-*skew* or **R**-*symplectic*: The vector space is $\mathbf{R}^{2n}$ with

$$(2.11) \qquad \varepsilon(x, y) \equiv x_1 y_2 - x_2 y_1 + \cdots + x_{2n-1} y_{2n} - x_{2n} y_{2n-1}.$$

3. **C**-*symmetric*: The vector space is $\mathbf{C}^n$ with

$$(2.12) \qquad \varepsilon(z, w) \equiv z_1 w_1 + \cdots + z_n w_n.$$

4. **C**-*skew* or **C**-*symplectic*: The vector space is $\mathbf{C}^{2n}$ with

$$(2.13) \qquad \varepsilon(z, w) \equiv z_1 w_2 - z_2 w_1 + \cdots + z_{2n-1} w_{2n} - z_{2n} w_{2n-1}.$$

5. **C**-*hermitian* (*symmetric*): The vector space is $\mathbf{C}^n$, denoted by $\mathbf{C}(p, q)$ $(n \equiv p + q)$ with

$$(2.14) \qquad \varepsilon(z, w) \equiv \bar{z}_1 w_1 + \cdots + \bar{z}_p w_p - \cdots - \bar{z}_n w_n.$$

6. **C**-*hermitian* (*skew*): The vector space is $\mathbf{C}^n (n \equiv p + q)$ with

$$(2.14') \qquad \varepsilon(z, w) = i\bar{z}_1 w_1 + \cdots + i\bar{z}_p w_p - \cdots - i\bar{z}_n w_n.$$

7. **H**-*hermitian symmetric*: The vector space is $\mathbf{H}^n$, denoted by $\mathbf{H}(p, q)$ $(n = p + q)$ with

$$(2.15) \qquad \varepsilon(x, y) \equiv \bar{x}_1 y_1 + \cdots + \bar{x}_p y_p - \cdots - \bar{x}_n y_n.$$

8. **H**-*hermitian skew*: The vector space is $\mathbf{H}^n$ with

(2.16) $\varepsilon(x, y) \equiv \bar{x}_i i y_1 + \cdots + \bar{x}_n i y_n.$

Some of the **R**-symmetric special cases are very important. The positive definite case $\mathbf{R}(n, 0)$ is called *euclidean space*, while the case of a single minus, $\mathbf{R}(n - 1, 1)$, is called *Lorentzian space*. If $n = 4$, then Lorentzian space is also called *Minkowski space* or *Minokowski space–time*. The special case $\mathbf{R}(p, p)$ will be referred to as the *split case*.

ORTHOGONALITY

Definition 2.17. *Suppose* $f : V \to \tilde{V}$ *is a F-linear map from an inner product space* V, ε *to another inner product space* $\tilde{V}, \tilde{\varepsilon}$ *of the same type. The map* f *is said to preserve inner products if*

(2.18) $\tilde{\varepsilon}(f(x), f(y)) = \varepsilon(x, y)$ *for all* $x, y \in V.$

Since ε *is nondegenerate such a map must be one to one. If, in addition, such a map is onto then* f *is called an isometry. The inner product spaces* V, ε *and* $\tilde{V}, \tilde{\varepsilon}$ *are said to be isometric if there exists an isometry between them.*

Remark 2.19. Suppose f is an isometry from V, ε to $\tilde{V}, \tilde{\varepsilon}$. Let G denote the ε-isometry subgroup of $\mathrm{GL}(V, F)$, and let $\tilde{G}$ denote the $\tilde{\varepsilon}$-isometry subgroup of $\mathrm{GL}(\tilde{V}, F)$. Then the groups G and $\tilde{G}$ are isomorphic:

(2.20) $\tilde{G} = f \circ G \circ f^{-1}.$

For example, the model $\mathbf{C}^2, \tilde{\varepsilon}$ with $\tilde{\varepsilon}(z, w) \equiv z_1 w_2 + z_2 w_1$ is isometric to the standard model $\mathbf{C}^2, \varepsilon$ with $\varepsilon(\xi, \eta) \equiv \xi_1 \eta_1 + \xi_2 \eta_2$. Using the model $\mathbf{C}^2, \tilde{\varepsilon}$ it is almost immediate that $\mathrm{SO}(2, \mathbf{C}) \cong \mathbf{C}^*$ (cf. Problem 1.9(f)).

Given a vector subspace W of the inner product space V, ε, let $\varepsilon|_W$ denote the restriction of ε to W. The restriction $\varepsilon|_W$ is *positive definite* if $\varepsilon(x, x) > 0$ for all nonzero $x \in W$. Similarly, one defines *negative definite*. A subspace W of V is said to be

(2.21)   *positive* (or *spacelike*)   if $\varepsilon|_W$ is positive definite,

(2.22)   *null* (or *degenerate* or *lightlike*)   if $\varepsilon|_W$ is degenerate,

(2.23)   *negative*   if $\varepsilon|_W$ is negative definite.

These three possibilities are by no means exhaustive.

Positive (negative) subspaces can only occur for three types of inner product spaces:

R-symmetric, **C**-hermitian (symmetric), and **H**-hermitian symmetric.

In these three cases, another numerical invariant (in addition to the dimension), called the signature, is required in order to determine when two inner product spaces are isometric.

Definition 2.24. *Suppose V, ε is an inner product space of one of the three special types mentioned above. A positive subspace W of V is said to be maximal positive if*

$$\dim Z \leq \dim W \quad \text{for all positive } Z \subset V.$$

A negative subspace W of V is said to be maximal negative if

$$\dim Z \leq \dim W \quad \text{for all negative } Z \subset V.$$

The signature p, q of V is defined by

p equals the dimension of a maximal positive subspace,
q equals the dimension of a maximal negative subspace.

If V, ε is positive definite (on V), then the signature is $n, 0$, and if V, ε is negative definite (on V), then the signature is $0, n$. If V, ε has signature $p = q$, the inner product space is said to be *split*, or of *split signature*.

The case of signature $p = 2, q = 1$ is pictured in Figure 2.27.

Obviously, if V, ε and $\tilde{V}, \tilde{\varepsilon}$ are isometric, then they have the same signature. In fact, as we will prove below, if V, ε and $\tilde{V}, \tilde{\varepsilon}$ have the same dimension and signature, then they are isometric. First we must examine the important concept of orthogonality.

Two vectors $u, v \in V$ are said to be *orthogonal*, written $u \perp v$, if $\varepsilon(u, v) = 0$. Note that

$$(2.25) \qquad\qquad u \perp v \text{ if and only if } v \perp u$$

for all eight types of inner products. Suppose W is a subset of V. Then

$$u \perp W \text{ means } u \perp v \text{ for all } v \in W.$$

The *perp* or *orthogonal* to W is defined by

$$(2.26) \qquad\qquad W^{\perp} \equiv \{u \in V : u \perp W\}.$$

The perp of the line $[v]$ spanned by $v \in V$ will be denoted $v^{\perp}$.

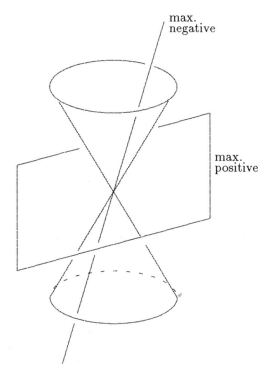

Figure 2.27

Since "$u \perp v$" just involves a pair of vectors, the key to understanding orthogonality lies in understanding two dimensions—more specifically, $\varepsilon|_W$ where $W = \text{span}\{u, v\}$. In order to sharpen our intuition, we focus attention on inner product spaces of the **R**-symmetric type.

In the euclidean plane **R**$(2, 0)$, two orthogonal vectors are pictured as being at right angles to each other.

In the Lorentzian plane **R**$(1, 1)$, first picture the *null* or *light cone*, consisting of all $a = (x, t) \in$ **R**$(1, 1)$ which satisfy $\varepsilon(a, a) = x^2 - t^2 = 0$. Such a vector $a \in$ **R**$(1, 1)$ is said to be *null* or *lightlike*. Note that $a \perp a$ for each lightlike vector a. Given a vector $b = (x, t) \in$ **R**$(1, 1)$, the vector $b' = (t, x) \in$ **R**$(1, 1)$ is orthogonal to b. The vectors $\pm b'$ are obtained pictorially by (euclidean) reflecting b through either one of the two lines making up the null cone (see Figure 2.28). The perp $b^\perp$ is just the span of $\pm b'$.

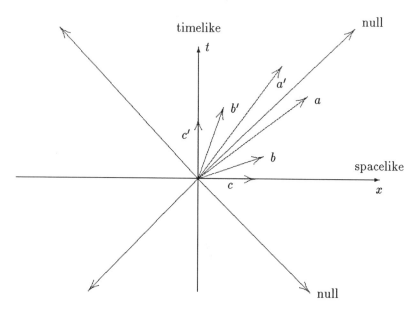

Figure 2.28

This example shows that a subspace W and its perp $W^\perp$ need not be complementary subspaces. Recall that a subspace W is null (or degenerate or lightlike) if $\varepsilon|_W$ is degenerate. If $\varepsilon|_W \equiv 0$, then W is said to be *totally null* or *isotropic*. Note that

(2.29) W is totally null if and only if $W \subset W^\perp$.

Totally null subspaces are particularly important in the split case $\mathbf{R}(p,p)$ and the symplectic cases (**R**-skew and **C**-skew).

Despite these strong counterexamples to our euclidean intuition, which says $W^\perp$ should be complementary to W, it is still true that the dimension of $W^\perp$ is complementary to the dimension of W.

The next lemma is basic for all eight types of inner products.

The Orthogonality Lemma 2.30. *Suppose V, ε is a n-dimensional inner product space, and W is a linear subspace of V. Then*

(2.31) $\dim W + \dim W^\perp = n,$

(2.32) $(W^\perp)^\perp = W,$

and the following are equivalent:

$$
(2.33) \qquad
\begin{array}{ll}
(a) & W \text{ is nondegenerate,} \\
(b) & W \cap W^\perp = \{0\}, \\
(c) & W + W^\perp = V, \\
(d) & W^\perp \text{ is nondegenerate.}
\end{array}
$$

Proof: Let F denote the field of scalars (either $\mathbf{R}$, $\mathbf{C}$, or $\mathbf{H}$). The inner product $\varepsilon : V \times V \to F$ determines a map $\flat : V \to V^*$ from V to the dual space V^* as follows. Consult Problem 1.7(b) for the definition of V^* if V is a right $\mathbf{H}$-space. For each $x \in V$ fixed, let $\flat(x) \in V^*$ be defined by

$$
(2.34) \qquad (\flat(x))(y) \equiv \varepsilon(x, y) \quad \text{for all } y \in V.
$$

Note that for all eight types of inner products $\flat(x) \in V^*$ since $\varepsilon(x, y)$ is F-linear in y. The nondegeneracy hypothesis,

$$
(2.35) \qquad \varepsilon(x, y) = 0 \text{ for all } y \in V \text{ implies } x = 0,
$$

is equivalent to

$$
(2.35') \qquad \flat : V \to V^* \text{ is one-to-one.}
$$

Since V and V^* have the same real dimension and $\flat$ is always $\mathbf{R}$-linear, $\flat$ must be a real linear isomorphism.

Given a subspace W of V, the *annihilator* of W, denoted W^0, is the subspace of V^* defined by

$$
(2.36) \qquad W^0 \equiv \{\phi \in V^* : \phi(y) = 0 \text{ for all } y \in W\}.
$$

Suppose $x \in V$ and $\phi \in V^*$ correspond under the isomorphism $\flat$. Then

$$
\phi \in W^0 \text{ if and only if } x \in W^\perp.
$$

That is,

$$
(2.37) \qquad \flat : W^\perp \to W^0 \text{ is an } \mathbf{R}\text{-linear isomorphism.}
$$

In particular,

$$
(2.38) \qquad \dim_F W^\perp = \dim_F W^0.
$$

Choose a basis $e_1, \ldots, e_m, \ldots, e_n$ for V with $e_1, \ldots, e_m$ a basis for W. Let $e_1^*, \ldots, e_n^*$ denote the dual basis for V^*. Then $e_{m+1}^*, \ldots, e_n^*$ is a basis for W^0. Therefore, $\dim_F W^0 = n - \dim_R W$, which, by (2.38), proves the first part of the Lemma, (2.31).

Obviously, $W \subset (W^\perp)^\perp$, and since the dimensions are the same, they must be equal. The proof that (a)–(d) are equivalent is left as an exercise. ■

Remark 2.39. (a) If V is the tangent space T to a (semi)Riemannian manifold at a point, the *flat* isomorphism $\flat$, defined above, between vectors and convectors is referred to as *metric equivalence* and corresponds to "lowering indices." The inverse map $\sharp = \flat^{-1}$, called *sharp*, is referred to as "raising indices."

(b) If V is the tangent space to a symplectic manifold (e.g., phase space $\equiv$ the cotangent bundle T^*M to a manifold M), then $\flat$ allows one to start with a function (*Hamiltonian function*) f and associate a vector field (the *associated Hamiltonian vector field*) by first taking df, the exterior derivative of f, and then applying sharp $\sharp = \flat^{-1}$.

The reader unfamiliar with these concepts may wish to glance at Chapter 5, "Differential Geometry."

Corollary 2.40. *If W is a nondegenerate subspace, then each vector $u \in V$ has a unique orthogonal decomposition:*

$$(2.41) \qquad u = w + z \quad \text{with } w \in W \text{ and } z \in W^\perp.$$

The linear map π that is zero on $W^\perp$ and the identity on W is called *orthogonal projection onto W*.

In the three cases **R**-symmetric, **C**-hermitian, and **H**-hermitian symmetric, where positive/negative subspaces exist, the inner product space is said to *have a signature*.

The Signature Lemma 2.42. *Suppose V, ε is an inner product space that has a signature, say p, q. Let W denote a subspace.*

(a) *W is maximal positive if and only if $W^\perp$ is maximal negative.*
(b) *If W is positive and $W^\perp$ is negative, then both are maximal.*

Corollary 2.43. *If V has signature p, q, then $p + q = n$ the dimension of V.*

The obvious guess for the signatures of the standard models can be rigorously verified using the Signature Lemma (see Problem 2).

Corollary 2.44. *The standard models* $\mathbf{R}(p,q)$, $\mathbf{C}(q,p)$, *and* $\mathbf{H}(p,q)$ *have signature* p, q.

Proof of 2.42(a): Suppose W is maximal positive. It suffices to show that $W^{\perp}$ is negative, because the maximality of $W^{\perp}$ then follows from part (b). First, we establish that $W^{\perp}$ is negative semidefinite; that is, $\varepsilon(x, x) \leq 0$ for all $x \in W^{\perp}$. Otherwise, there exists a nonzero, spacelike vector $u \in W^{\perp}$, i.e., $\varepsilon(u, u) > 0$. Since ε restricted to either of the orthogonal subspaces W or $[u] \equiv \operatorname{span} u$ is positive definite, an easy calculation shows that ε restricted to $W + [u]$ is also positive definite, contradicting the maximality of the dimension of W. Finally, we must rule out $\varepsilon(x, x) = 0$ for $x \in W^{\perp}$ unless $x = 0$. If $W^{\perp}$ contains a nonzero null vector u, then there exists a vector $y \in W^{\perp}$ with $\varepsilon(u, y) \neq 0$ because $W^{\perp}$ is nondegenerate. We may assume that $\varepsilon(u, y)$ is real and greater than zero. Now $0 \geq \varepsilon(u + ty, u + ty)/t = 2\varepsilon(u, y) + t\varepsilon(y, y)$, for all $t > 0$, which is impossible. ∎

The hypothesis that V, ε has a signature was used in this last equation to conclude that $\varepsilon(y, u) = \varepsilon(u, y)$ from the assumption that $\varepsilon(u, y)$ is real. Can you find the other use of the signature hypothesis in this proof?

Proof of 2.42(b): We shall prove that W is maximal (positive). Suppose P is another positive subspace. Since $W^{\perp}$ is negative, $\{0\} = P \cap W^{\perp} = P \cap \pi^{-1}(0)$, where π is orthogonal projection onto W. Thus, $\pi : P \to W$ is one-to-one, so that $\dim P \leq \dim W$. ∎

In a similar manner one can show:

Proposition 2.45. *If* V *has a signature (say* p, q*), then each totally null subspace* N *must have dimension* $\leq \min\{p, q\}$.

A CANONICAL FORM (THE BASIS THEOREM)

Using the Orthogonality Lemma and the Signature Lemma any inner product space may be put in canonical form.

The Basis Theorem 2.46. *Suppose* V, ε *is an inner product space of one of the eight types. Then* V *is isometric to the standard model of the same type that has the same dimension and signature.*

Corollary 2.47. *Suppose* V, ε *and* $\tilde{V}, \tilde{\varepsilon}$ *are two inner product spaces of the same type. The dimension and signature are the same if and only if* V *and* $\tilde{V}$ *are isometric.*

For each of the eight types of inner product spaces, certain bases are of particular importance and will be used to prove Theorem 2.46.

Definition 2.48. *Suppose* V, ε *is an inner product space of dimension* N *and signature* p, q *(if* V *has a signature). A basis* $\{e_1, \ldots, e_N\}$ *for* V *is said to be* orthonormal *if the linear map* f *from* V *to the standard model of the same type, dimension, and signature, defined by sending* e_j *to the* j^{th} *standard basis vector for the standard model, is an isometry.*

Thus, a basis $\{e_1, \ldots, e_N\}$ for V, ε is *orthonormal* if
R-*symmetric* ($\dim_{\mathbf{R}} V = n$):

$$
\begin{aligned}
\varepsilon(e_j, e_j) &= 1 \quad \text{for } j = 1, \ldots, p, \\
(2.49) \qquad \varepsilon(e_j, e_j) &= -1 \quad \text{for } j = p+1, \ldots, n, \\
\varepsilon(e_i, e_j) &= 0 \quad \text{for } i \neq j;
\end{aligned}
$$

R-*skew* ($\dim_{\mathbf{R}} V = 2n$):

$$
\begin{aligned}
(2.50) \qquad \varepsilon(e_{2j-1}, e_{2j}) &= 1 \quad \text{for } j = 1, \ldots, n, \\
\varepsilon(e_i, e_j) &= 0 \quad \text{for all other pairs } e_i, e_j,
\end{aligned}
$$

(orthonormal basis = *symplectic basis*);

C-*symmetric* ($\dim_{\mathbf{C}} V = n$):

$$
\begin{aligned}
(2.51) \qquad \varepsilon(e_j, e_j) &= 1 \quad \text{for } j = 1, \cdots, n, \\
\varepsilon(e_i, e_j) &= 0 \quad \text{for } i \neq j,
\end{aligned}
$$

(orthonormal basis = **C**−*orthonormal basis*);

C-*skew* ($\dim_{\mathbf{C}} V = 2n$):

$$
\begin{aligned}
(2.52) \qquad \varepsilon(e_{2j-1}, e_{2j}) &= 1 \quad \text{for } j = 1, \ldots, n, \\
\varepsilon(e_i, e_j) &= 0 \quad \text{for all other pairs } e_i, e_j,
\end{aligned}
$$

(orthonormal basis = **C**-*symplectic basis*);

C-*hermitian* (*symmetric*) ($\dim_{\mathbf{C}} V = n$):

$$
\begin{aligned}
\varepsilon(e_j, e_j) &= 1 \quad \text{for } j = 1, \ldots, p, \\
(2.53) \qquad \varepsilon(e_j, e_j) &= -1 \quad \text{for } j = p+1, \ldots, n, \\
\varepsilon(e_i, e_j) &= 0 \quad \text{for } i \neq j,
\end{aligned}
$$

(orthonormal basis = *unitary basis*);

H-*hermitian symmetric* (dim$_\mathbf{H}$ $V = n$):

(2.54)
$$\varepsilon(e_j, e_j) = 1 \quad \text{for } j = 1, \cdots, p,$$
$$\varepsilon(e_j, e_j) = -1 \quad \text{for } j = p + 1, \ldots, n,$$
$$\varepsilon(e_i, e_j) = 0 \quad \text{for } i \neq j,$$

(orthonormal basis = **H**-*unitary* or *hyperunitary basis*);

H-*hermitian skew* (dim$_\mathbf{H}$ $V = n$):

(2.55)
$$\varepsilon(e_j, e_j) = i \quad \text{for } j = 1, \ldots, n$$
$$\varepsilon(e_i, e_j) = 0 \text{ for } i \neq j,$$

(orthonormal basis = **H**-*skew basis*).

Proof of Theorem 2.46: It suffices to show that V, ε has an orthonormal basis. The proof is by induction on the dimension.

R-*symplectic*, **C**-*symplectic*: Choose e_1 any nonzero vector. Since ε is nondegenerate, there exists u with $\varepsilon(e_1, u) \neq 0$. Let e_2 denote u rescaled so that $\varepsilon(e_1, e_2) = 1$. Let $W \equiv \text{span}\{e_1, e_2\}$. Since W is nondegenerate, Lemma 2.30 implies that $V = W + W^\perp, W \cap W^\perp = \{0\}$, and that $W^\perp, \varepsilon$ is a lower dimensional (symplectic) inner product space. By the induction hypothesis $W^\perp$ has a symplectic basis $e_3, \ldots, e_{2n}$. Since $W \perp W^\perp, e_1, e_2, e_3, \ldots, e_{2n}$ is a symplectic basis for $V = W + W^\perp$.

Signature: Assume V, ε has a signature. Choose P maximal positive and N maximal negative with $N = P^\perp$ by Lemma 2.42. Since P and N are orthogonal, it suffices to show that P and N have an orthonormal basis. Thus, we may assume that V is positive if V has a signature.

The proof is completed, in the positive definite cases as well as in all remaining cases without signature, as follows. Choose a nonzero vector $u \in V$. Let e_1 denote u properly normalized so that $\varepsilon(e_1, e_1) = 1$, unless ε is **H**-hermitian skew, in which case require $\varepsilon(e_1, e_1) = i$ (see Problem 6). Let $W \equiv \text{span } e_1$. Then W, and hence $W^\perp$, is nondegenerate. Thus, by the induction hypothesis $W^\perp$ has an orthonormal basis. Since W and $W^\perp$ are orthogonal, e_1 combined with the orthonormal basis for $W^\perp$ is an orthonormal basis for $V = W + W^\perp$. ∎

THE PARTS OF AN INNER PRODUCT

Suppose ε is one of the eight types of inner products. If ε is either **C**-valued or **H**-valued, then ε has various parts.

The simplest case is when ε is a complex-valued inner product. Then ε has a real part α and an imaginary part β defined by the equation

$$(2.56) \qquad\qquad \varepsilon = \alpha + i\beta$$

(while requiring α and β to be real-valued).

Suppose ε is a quaternion valued inner product. There are several options for analyzing the parts of ε.

First, ε has a real part α and a pure imaginary part β defined by

$$(2.57) \qquad\qquad \varepsilon \equiv \alpha + \beta,$$

where $\alpha \equiv \operatorname{Re}\varepsilon$ is real-valued and $\beta \equiv \operatorname{Im}\varepsilon$ takes on values in $\operatorname{Im}\mathbf{H} \equiv \operatorname{span}\{i, j, k\}$. The imaginary part β has three components defined by

$$(2.57') \qquad\qquad \beta \equiv i\beta_i + j\beta_j + k\beta_k,$$

where the *parts* $\beta_i, \beta_j, \beta_k$ are real-valued.

Second, using the complex structure R_i (right multiplication by i) on $\mathbf{H}$, each quaternion $x \in \mathbf{H}$ has a unique decomposition $x \equiv z + jw$ where $z, w \in \mathbf{C} \subset \mathbf{H}$. Therefore,

$$(2.58) \qquad\qquad \varepsilon = \gamma + j\delta \quad \text{with } \gamma \text{ and } \delta \text{ complex-valued}$$

uniquely defines γ (*the first complex part of* ε) and δ (*the second complex part of* ε).

Of course,

$$(2.58') \qquad\qquad \gamma = \alpha + i\beta_i \quad \text{and} \quad \delta = \beta_j - i\beta_k.$$

This section is devoted to computing the types of these parts of ε. As an application, alternate definitions of the isometry group G for ε are deduced. However, this material is better motivated by geometric considerations and should be read in conjunction with Chapter 5 on geometry. In fact, readers may wish to examine Chapters 3–5 before returning to this section. Because of its importance in geometry, each type of inner product ε is decomposed into parts using notation from geometry.

C-Hermitian Symmetric

Consider the standard **C**-hermitian (symmetric) form

$$(2.59) \qquad \varepsilon(z, w) \equiv z_1\overline{w}_1 + \cdots + z_p\overline{w}_p - z_{p+1}\overline{w}_{p+1} - \cdots - z_n\overline{w}_n$$

with signature p, q on $\mathbf{C}^n$. Since ε is complex-valued, it has a real and imaginary part given by $\varepsilon = g - i\omega$. For $z \equiv x + iy$ and $w \equiv \xi + i\eta \in \mathbf{C}^n \cong \mathbf{R}^{2n}$, the real and imaginary parts $g \equiv \mathrm{Re}\,\varepsilon$ and $\omega \equiv -\mathrm{Im}\,\varepsilon$ are given by

$$(2.60) \qquad g(z, w) \equiv x_1 \xi_1 + y_1 \eta_1 + \cdots + x_p \xi_p + y_p \eta_p - \cdots - x_n \xi_n - y_n \eta_n$$

and

$$(2.61) \qquad \omega(z, w) = x_1 \eta_1 - y_1 \xi_1 + \cdots - x_n \eta_n + y_n \xi_n.$$

Thus, g is the standard $\mathbf{R}$-symmetric form on $\mathbf{R}^{2n}$ with signature $2p, 2q$. Modulo some sign changes, ω is the standard symplectic form on $\mathbf{R}^{2n}$. In this context, when $q = 0$, ω is exactly the *standard Kähler form* on $\mathbf{C}^n$ and is usually written as

$$(2.62) \qquad \omega = \frac{i}{2}\, dz_1 \wedge \overline{dz_1} + \cdots + \frac{i}{2}\, dz_n \wedge \overline{dz_n}.$$

Lemma 2.63. *Suppose ε is $\mathbf{C}$-hermitian symmetric (signature p, q) on a complex vector space V with complex structure i. Then $g \equiv \mathrm{Re}\,\varepsilon$ is $\mathbf{R}$-symmetric (signature $2p, 2q$) and $\omega \equiv -\mathrm{Im}\,\varepsilon$ is $\mathbf{R}$-skew. Moreover, each determines the other by*

$$(2.64) \qquad g(z, w) = \omega(iz, w) \quad \text{and} \quad \omega(z, w) \equiv g(iz, w).$$

Also, i is an isometry for both g and ω:

$$(2.65) \qquad g(iz, iw) = g(z, w), \quad \text{and} \quad \omega(iz, iw) = \omega(z, w).$$

Conversely, given g $\mathbf{R}$-symmetric with i an isometry, if ω is defined by (2.64), then $\varepsilon \equiv g - i\omega$ is $\mathbf{C}$-hermitian. Also, given ω $\mathbf{R}$-skew with i an isometry, if g is defined by (2.64), then $\varepsilon = g - i\omega$ is $\mathbf{C}$-hermitian.

Proof: It suffices to prove that $\mathrm{Re}\,\varepsilon$ and $\mathrm{Im}\,\varepsilon$ have the desired properties when $\mathbf{C}^n$, ε is the standard model; this has already been carried out (see (2.60) and (2.61)). Alternatively, the properties of $\mathrm{Re}\,\varepsilon$ and $\mathrm{Im}\,\varepsilon$ can be easily derived from the fact that ε is $\mathbf{C}$-hermitian symmetric with signature p, q, providing a second proof, this one without coordinates.

Reconstructing ε from g (or from ω), with i an isometry is Problem 8a. ∎

Remark 2.66. If the signature is positive definite, then Lemma 2.63 may be cryptically summarized by saying that "the confluence of any two of

(a) complex geometry,
(b) symplectic geometry,
(c) Riemannian geometry

is Kähler geometry." See Lemma 5.17 for more details.

This decomposition $\varepsilon = g - i\omega$ also provides several alternative definitions of the unitary group $U(p, q)$.

Suppose V, ε is a C-hermitian symmetric inner product space of signature p, q. Let $GL(n, \mathbf{C})$ denote $\text{End}_{\mathbf{C}}(V)$, let $O(2p, 2q)$ denote the subgroup of $GL(V, \mathbf{R})$ fixing $g \equiv \text{Re}\,\varepsilon$, and let $\text{Sp}(n, \mathbf{R})$ denote the subgroup of $GL(V, \mathbf{R})$ fixing ω.

Corollary 2.67. *The intersection of any two of the three groups*

$$GL(n, \mathbf{C}), \quad \text{Sp}(n, \mathbf{C}), \quad \text{and} \quad 0(2p, 2q)$$

is the group $U(p, q)$.

Similar results are valid in the C-symmetric and C-skew cases, but perhaps not so interesting since in either of these cases the group that fixes $\alpha \equiv \text{Re}\,\varepsilon$ and the group that fixes $\beta \equiv \text{Im}\,\varepsilon$ are just different versions of the same group under a coordinate change. These cases are left to the reader.

H-Hermitian Symmetric

Consider the standard H-hermitian symmetric form

$$(2.68) \qquad \varepsilon(x, y) \equiv \overline{x}_1 y_1 + \cdots + \overline{x}_p y_p - \cdots - \overline{x}_n y_n$$

on $\mathbf{H}^n$ with signature p, q. Note that ε is H-valued. As noted earlier, it is natural to consider $\mathbf{H}$ as two copies of $\mathbf{C}$, $\mathbf{H} \cong \mathbf{C} \oplus j\mathbf{C}$, or $x \equiv z + jw$. In particular, $\varepsilon = h + j\sigma$ with h and σ complex-valued.

Note that if $x \equiv z + jw$ and $y \equiv \xi + j\eta$ with $z, w, \xi, \eta \in \mathbf{C}$, then

$$(2.69) \qquad \begin{aligned} \overline{x}y &= (\overline{z} - jw)(\xi + j\eta) = \overline{z}\xi - jwj\eta + \overline{z}j\eta - jw\xi \\ &= \overline{z}\xi + \overline{w}\eta + j(z\eta - w\xi). \end{aligned}$$

Therefore, the first complex part of h of ε is given by

$$(2.70) \quad h(x, y) = \overline{z}_1\xi_1 + \overline{w}_1\eta_1 + \cdots + \overline{z}_p\xi_p + \overline{w}_p\eta_p - \cdots - \overline{z}_n\xi_n - \overline{w}_n\eta_n.$$

Thus h is the standard C-hermitian symmetric form on $\mathbf{C}^{2n} \cong \mathbf{H}^n$. Because of (2.69) the second complex part σ is given by

$$(2.71) \qquad \sigma(x, y) = z_1\eta_1 - w_1\xi_1 \pm \cdots \pm z_n\eta_n \pm w_n\xi_n.$$

Thus, modulo some possible sign changes, σ is the standard C-skew form on $\mathbf{C}^{2n}$.

Lemma 2.72. *Suppose* $\varepsilon = h + j\sigma$ *is* H-*hermitian symmetric on a right* H-*space* V. *Then* h, *the first complex part of* ε, *is* C-*hermitian symmetric and* σ, *the second complex part of* ε, *is* C-*skew. Moreover, each determines the other by*

$$(2.73) \qquad h(x,y) = \overline{\sigma(x,yj)} \quad \text{and} \quad \sigma(x,y) = \overline{-h(x,yj)}.$$

Also,

$$(2.74) \qquad h(xj,yj) = \overline{h(x,y)} \quad \text{and} \quad \sigma(xj,yj) = \overline{\sigma(x,y)}.$$

Conversely, given h C-*hermitian symmetric and* $h(xj,yj) = \overline{h(x,y)}$, *if* σ *is defined by (2.73), then* $\varepsilon = h + j\sigma$ *is* H-*hermitian symmetric. Also, given* σ C-*skew and* $\sigma(xj,yj) = \overline{\sigma(x,y)}$, *if* h *is defined by (2.73), then* $\varepsilon = h + j\sigma$ H-*hermitian symmetric.*

Proof: The equation $\varepsilon(x,yj) = \varepsilon(x,y)j$ can be used to prove (2.73), while $\varepsilon(xj,yj) = -j\varepsilon(x,y)j$ can be used to prove (2.74). The remainder of the proof is left to the reader (see Problem 8(b)). ∎

Suppose V, ε is an H-hermitian symmetric inner product space of signature p, q. Let $\mathrm{GL}(n, \mathbf{H})$ denote $\mathrm{End}_{\mathbf{H}}(V)$, let $U(2p, 2q)$ denote the subgroup of $\mathrm{GL}(n, \mathbf{H})$ that fixes the first complex part h of ε, and let $\mathrm{Sp}(n, \mathbf{C})$ denote the subgroup of $\mathrm{GL}(n, \mathbf{H})$ that fixes the second complex part σ of ε.

Corollary 2.75. *The intersection of any two of the three groups*

$$\mathrm{GL}(n, \mathbf{H}), \quad U(2p, 2q), \quad \text{and} \quad \mathrm{Sp}(n, \mathbf{C})$$

is the group $\mathrm{HU}(p, q)$.

It is useful to try to construct the quaternionic structure from h and σ.

Suppose V is a complex $2n$-dimensional vector space, h is a C-hermitian symmetric inner product on V, and σ is a complex symplectic inner product on V. Then h and σ define a complex antilinear map J by

$$(2.76) \qquad h(xJ,y) = \sigma(x,y).$$

Now $\sigma(xJ^2,y) = -\sigma(y,xJ^2) = -h(yJ,xJ^2) = \overline{-h(xJ^2,yJ)} = \overline{-\sigma(xJ,yJ)}$. Therefore, $J^2 = -1$ if and only if

$$(2.77) \qquad \sigma(xJ,yJ) = \overline{\sigma(x,y)}.$$

In this case, h and σ are said to be *compatible*.

Lemma 2.78. *Suppose h is a **C**-hermitian symmetric inner product and σ is a complex symplectic inner product on a complex $2n$-vector space V. If h and σ are compatible, then they determine a right **H**-structure on V and*

$$\varepsilon \equiv h + j\sigma$$

*is an **H**-hermitian symmetric inner product on V.*

Proof: It remains to verify that $h + j\sigma$ is **H**-hermitian symmetric. ∎

There is yet another description of $\mathrm{HU}(p, q)$. First, recall (1.27) that for each unit vector $u \in \mathrm{Im}\,\mathbf{H}$, right multiplication by u (denoted R_u) acting on $\mathbf{H}^n$ determines a complex structure on $\mathbf{H}^n$ and hence an isomorphism $\mathbf{H}^n \cong \mathbf{C}^{2n}$.

Second, recall that in the **C**-hermitian symmetric case the Kähler form ω is minus the imaginary part of ε and that ω is determined by the formula (2.64) in terms of $g \equiv \mathrm{Re}\,\varepsilon$. Each of the complex structures R_u on $\mathbf{H}^n \cong \mathbf{C}^{2n}$ determines a Kähler form in exactly the same manner. Let

$$(2.79) \qquad \omega_u(x, y) \equiv Re\ \varepsilon(xu, y), \quad \text{and} \quad g(x, y) \equiv Re\ \varepsilon(x, y).$$

Lemma 2.80. *For all $x, y \in \mathbf{H}^n$,*

$$\varepsilon = g + i\omega_i + j\omega_j + k\omega_k.$$

Proof: For $u \in \mathrm{Im}\,\mathbf{H}$ with $|u| = u\,\overline{u} = -u^2 = 1$,

$$\langle u, \varepsilon(x, y) \rangle = \langle 1, \overline{u}\varepsilon(x, y) \rangle = Re\ \varepsilon(xu, y) = \omega_u(x, y). \quad ∎$$

For each complex structure $u \in Im\ \mathbf{H}, |u| = 1$, the complex $\mathbf{C}(u)$ valued form

$$(2.81) \qquad h_u \equiv g + u\omega_u$$

is $\mathbf{C}(u)$-hermitian symmetric. The group that fixes h_u is a unitary group with signature $2p, 2q$ determined by the complex structure R_u.

The next corollary justifies the name "hyper-unitary" for $\mathrm{HU}(p, q)$.

Corollary 2.82. *The hyper-unitary group $\mathrm{HU}(p, q)$ is the intersection of the three unitary groups determined by the three complex structures R_i, R_j, R_k on $\mathbf{H}^n$.*

Remark 2.83. Of course, $\mathrm{HU}(p, q)$ is also the intersection, over $u \in S^2 \subset \mathrm{Im}\,\mathbf{H}$, of the unitary groups determined by all the complex structures R_u.

The simplest case states that $HU(1)$ is the intersection of all the unitary groups determined by the complex structures R_u on $\mathbf{H} \cong \mathbf{C}^2$. Recall (1.43) the isomorphism $SU(2) \cong HUU(1)$, where $SU(2)$ denotes the special unitary group determined by the complex structure R_i on $\mathbf{H} \cong \mathbf{C}^2$. Similarly, for each complex structure R_u on $\mathbf{H}$ the corresponding special unitary group is again the same $HU(1)$ and thus is *independent* of R_u.

H-Hermitian Skew

Consider the standard $\mathbf{H}$-hermitian skew form on $\mathbf{H}^n$:

$$(2.84) \qquad \varepsilon(x, y) \equiv \overline{x}_1 i y_1 + \cdots + \overline{x}_n i y_n.$$

Since i is distinguished among all the unit imaginary quaternions u, it is natural to set $\mathbf{H} \cong \mathbf{C} \oplus j\mathbf{C}$ where each copy of $\mathbf{C}$ has the complex structure R_i induced from R_i acting on $\mathbf{H}$.

Let

$$(2.85) \qquad \varepsilon = ih + j\sigma,$$

where ih and σ are the first and second complex parts of ε respectively. Set $x \equiv z + jw$ and $y \equiv \xi + j\eta$ with $z, w, \xi, \ \eta \in \mathbf{C}$. Then

$$(2.86) \qquad h(x, y) = \overline{z}_1 \xi_1 - \overline{w}_1 \eta_1 + \cdots + \overline{z}_1 \xi_n - \overline{w}_n \eta_n$$

is $\mathbf{C}$-hermitian symmetric, and

$$(2.87) \qquad \sigma(x, y) = z_1 \eta_1 + w_1 \xi_1 + \cdots + z_n \eta_n + w_n \xi_n$$

is $\mathbf{C}$-symmetric.

In summary, if $\varepsilon = ih + j\sigma$ is $\mathbf{H}$-hermitian skew then

$$(2.88) \qquad h \text{ is } \mathbf{C}\text{-hermitian symmetric and } \sigma \text{ is } \mathbf{C}\text{-symmetric.}$$

Since the form h is equivalent to the standard $\mathbf{C}$-hermitian symmetric form on $\mathbf{C}^{2n}$ with signature n, n via a complex coordinate change, let $U(n, n)$ denote the group fixing h. Since σ is equivalent to the standard $\mathbf{C}$-symmetric form on $\mathbf{C}^{2n}$ via a complex coordinate change, let $O(2n, \mathbf{C})$ denote the group that fixes σ.

Proposition 2.89. *The intersection of any two of the three groups*

$$\mathrm{GL}(n, \mathbf{H}), \quad U(n,n), \quad \text{and} \quad O(2n, \mathbf{C}),$$

is the group $\mathrm{SK}(n, \mathbf{H})$.

Note that the **H**-skew sphere of radius i is the intersection of the unit sphere $S(2p, 2p)$, given by $|z|^2 - |w|^2 = 1$, with the null cone $\sum z_i w_i = 0$, where $(z, w) \in \mathbf{C}^{2n} \cong \mathbf{H}^n$.

For each $u \in S^2 \subset \mathrm{Im}\,\mathbf{H}$,

$$(2.90) \qquad g_u(x, y) \equiv \varepsilon(xu, y) = \langle u, \varepsilon(x, y) \rangle$$

defines a real symmetric inner product with split signature, while

$$(2.91) \qquad\qquad \omega = \mathrm{Re}\,\varepsilon$$

is a real symplectic inner product.

Because of (2.90),

$$(2.92) \qquad\qquad \varepsilon = \omega + i g_i + j g_j + k g_k.$$

Finally, note that $\mathrm{SK}(n, \mathbf{H})$ is also "hyper-unitary"(but in a different sense than $\mathrm{HU}(p, q)$). The reader may wish to compute (in coordinates) ω, g_i, g_j, and g_k.

PROBLEMS

1. Give the proof of (2.33), that:
   (a) W is nondegenerate,
   (b) $W \cap W^{\perp} = \{0\}$,
   (c) $W + W^{\perp} = V$,
   (d) $W^{\perp}$ is nondegenerate are all equivalent.

2. (a) Prove that $\mathbf{R}(p, q), \mathbf{C}(p, q)$, and $\mathbf{H}(p, q)$ have signature p, q.
   (b) If V, ε has a signature, show that each positive subspace is contained in a maximal positive subspace.

3. Suppose V, ε is an inner product space with a signature. A map $f \in \mathrm{End}_F(V)$ is called an *anti-isometry* if

$$\varepsilon(f(x), f(y)) = -\varepsilon(x, y) \quad \text{for all } x, y \in V.$$

A map $f \in \text{End}_F(V)$ is said to be *anti-conformal* if, for some negative constant $\lambda < 0$,

$$\varepsilon(f(x), f(y)) = \lambda \varepsilon(w, y), \quad \text{for all } x, y \in V.$$

Show that if there exists an anti-conformal map $f \in \text{End}_F(V)$, then the signature must be split.

4. Suppose V, ε is an inner product space of **R**-symmetric type with signature p, q and $p \leq q$. Prove the following.

(a) There exists a totally null subspace of dimension p.

(b) Each totally null subspace is of dimension $\leq p$ (i.e., prove Proposition 2.45).

(c) V is split ($p = q$) if and only if $V = N_1 \oplus N_2$ with N_1, N_2 totally null.

5. Suppose V is a real vector space and $\varepsilon : V \times V \to \mathbf{R}$ and $\flat : V \to V^*$ are related by the equation

$$\varepsilon(x, y) = (\flat(x))(y).$$

(Thus $\varepsilon(x, \cdot)$ is linear since $\sharp(x) \in V^*$.) Prove the following.

(a) $\varepsilon(\cdot, y)$ is linear if and only if $\flat$ is linear.

(b) ε is symmetric if and only if $\flat^* = -\flat$. (Here $\flat^*$ is the dual map not the adjoint.)

6. Suppose V, ε is an **H**-hermitian skew inner product space. Show that there exists a vector $e_1 \in V$ with $\varepsilon(e_1, e_1) = i$.

7. A finite dimensional vector space V over F equipped with a bilinear form ε satisfying all of the properties of an inner product except (possibly) the nondegeneracy condition is called a *degenerate inner product space*. The *standard models for the degenerate inner product spaces* are all of the form $F^k \oplus F^m, \varepsilon$ where F^m, ε is one of the standard models for an inner product space. Show that each degenerate inner product space V, ε is isometric to exactly one of the degenerate standard models.

Hint: Write $V = N \oplus W$ with $N \equiv V^\perp$.

8. (a) Suppose g is an **R**-symmetric inner product on $\mathbf{R}^{2n}$ and i is a complex structure on $\mathbf{R}^{2n}$ that is orthogonal with respect to g. Show that

$$\varepsilon(z, w) \equiv g(z, w) - ig(iz, w)$$

is **C**-hermitian symmetric on $\mathbf{C}^n \cong \mathbf{R}^{2n}, i$.

(b) Suppose σ is a C-symmetric (or C-skew) inner product on $\mathbf{C}^{2n} \cong \mathbf{H}^n$, R_i that satisfies

$$\sigma(xj, yj) = \overline{\sigma(x, y)} \quad \text{for all } x, y \in \mathbf{H}^n.$$

Show that
$$\varepsilon(x, y) \equiv \overline{\sigma(x, yj)} + j\sigma(x, y)$$

is **H**-hermitian symmetric (or **H**-hermitian skew).

9. The Basis Theorem 2.46 may also be interpreted purely algebraically. Only the first case (**R**-symmetric) is discussed. Show that the two statements (a) and (b) are equivalent.

(a) (See Problem 7.) Each degenerate **R**-symmetric inner product space is isometric to one of the standard models $\mathbf{R}(p, q) \oplus \mathbf{R}^k$.

(b) (Sylvester's Theorem.) Given a symmetric matrix $A \in M_n(\mathbf{R})$, there exists an invertible matrix $B \in M_n(\mathbf{R})$ so that BAB^{-1} is diagonal with each nonzero diagonal entry either $+1$ or -1.

10. State and prove the **H**-hermitian skew analogue of Lemma 2.78.

11. Show that the intersection of any two of the unitary groups $U_i(2p, 2q)$, $U_j(2p, 2q), U_k(2p, 2q)$ (based on the three complex structures R_i, R_j, R_k respectively) is equal to the hyper-unitary group $\mathrm{HU}(p, q)$.

3. Classical Groups II

The classical groups introduced in Chapter 1 are examined more carefully in this chapter. In particular, the "Lie algebra" of each of these groups is computed explicitly.

GROUP REPRESENTATIONS AND ORBITS

Suppose G is a group and V is a real vector space.

Definition 3.1.

(a) *A representation ρ of G on V is a group homomorphism $\rho : G \to$ GL(V). Sometimes, the notation ρ is suppressed, and we say that G acts on V, when the action ρ is understood from the context.*

(b) *Two representations $\rho : G \to$ GL(V) and $\sigma : G \to$ GL(W) are said to be equivalent if there exists a **R**-linear isomorphism $L : V \to W$ with $\sigma(g) = L \circ \rho(g) \circ L^{-1}$ for all $g \in G$. The linear map L is called an intertwining operator.*

Since the groups defined so far were defined as subgroups of GL(n, F), each such group comes equipped with a *natural action* on F^n.

Suppose $\rho : G \to$ GL(V) is a action of G on V. A set of the form

$$(3.2) \qquad \{\rho(g)v : G \in G\}$$

is called an *orbit* of G, or the orbit of G through $v \in V$. The subgroup $K_v \equiv \{g \in G : \rho(g)v = v\}$ is called the *isotropy subgroup* of G at v. The group G is said to act *transitively* on a subset $X \subset V$ if X is the orbit of G through each point $v \in X$.

Two isotropy subgroups, K_u and K_v, at two points u, v in the same orbit X are isomorphic:

(3.3) $$K_u = L \, K_v \, L^{-1},$$

where $L \in G$ is chosen so that $Lv = u$. Thus, if G acts transitively on X, we write

(3.4) $$G/K \cong X$$

for the quotient of the group G by the equivalence relation $g_1 \sim g_2$ defined by $g_2 = g_1 k$ for some $k \in K$, where K represents the isotropy subgroup at a point $u \in X$.

Recall from Problem 1.3 that the connectivity of $\mathrm{GL}(n, F)$ and $\mathrm{SL}(n, F)$ (for $F \equiv \mathbf{R}, \mathbf{C}$, or $\mathbf{H}$) was derived from

(3.5) $$\mathrm{GL}(n, F)/K \cong F^n - \{0\}$$

(3.6) $$\mathrm{SL}(n, F)/K \cong F^n - \{0\} \qquad (n \geq 2).$$

Thus, $\mathrm{GL}(n, F)$ and $\mathrm{SL}(n, F)$ $(n \geq 2)$ acting on F^n have two orbits, namely $\{0\}$ and $F^n - \{0\}$.

GENERALIZED SPHERES

For each of the seven types of inner products, the proof of the Basis Theorem can be used to compute the orbits of the isometry group G (that fixes the inner product ε) acting on F^n. Of course, each orbit of G is contained in one of the level sets

(3.7) $$\{x \in F^n : \varepsilon(x, x) = c\} \quad \text{for some constant } c \in F.$$

The most important cases occur when ε is an $\mathbf{R}$-symmetric inner product.

Definition 3.8. *The positive sphere of radius $r > 0$ in $\mathbf{R}(p, q)$ $(p \geq 1)$ is:*

(3.9) $$S_r^+ \equiv \{x \in \mathbf{R}(p, q) : \langle x, x \rangle = r^2\}.$$

The negative sphere of radius $r < 0$ in $\mathbf{R}(p, q)$ $(q \geq 1)$ is:

(3.10) $$S_r^- \equiv \{x \in \mathbf{R}(p, q) : \langle x, x \rangle = -r^2\}.$$

The unit spheres $S^{\pm}$ in $\mathbf{R}(p, q)$ are obtained by setting the radius $r = \pm 1$.

The standard notation for the euclidean sphere S_r^+ in $\mathbf{R}(n,0)$ is S_r^{n-1} since this sphere has dimension $n-1$. We will use the same notation, S_r^{n-1}, for the sphere S_r^- in $\mathbf{R}(0,n)$. S^o consist of two points. Note that the positive unit sphere S^+ in $\mathbf{R}(p,q)$ is the same as the negative unit sphere S^- in $\mathbf{R}(q,p)$.

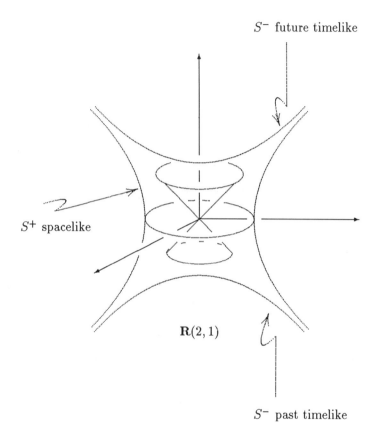

S^- future timelike

S^+ spacelike

$\mathbf{R}(2,1)$

S^- past timelike

Figure 3.11

Proposition 3.12.

(a) $(p \geq 1)$ *the unit positive sphere* S^+ *in* $\mathbf{R}(p,q)$ *is diffeomorphic to* $S^{p-1} \times \mathbf{R}^q$.

(b) $(q \geq 1)$ *the unit negative sphere* S^- *in* $\mathbf{R}(p,q)$ *is diffeomorphic to* $\mathbf{R}^p \times S^{q-1}$.

(c) *The null cone* $N \equiv \{z \in \mathbf{R}(p,q) : z \neq 0 \text{ but } \|z\| = 0\}$ *in* $\mathbf{R}(p,q)$ *is diffeomorphic to* $S^{p-1} \times S^{q-1} \times \mathbf{R}^+$.

Proof: Let

$$(3.13) \qquad \|z\|^2 \equiv |x|^2 - |y|^2, \text{ for all } z \equiv (x; y) \in \mathbf{R}(p, q) \cong \mathbf{R}^p \times \mathbf{R}^q$$

denote the square norm of $z \equiv (x, y)$ where $|x|$ and $|y|$ denote the euclidean norms of $x \in \mathbf{R}^p$ and $y \in \mathbf{R}^q$.

Proof of (a): If $z \equiv (x, y) \in S^+$ then $\|z\| = |x|^2 - |y|^2 = 1$. In particular $|x| \neq 0$ so that $(x/|x|, y) \in S^{p-1} \times \mathbf{R}^q$ is well defined. This map from S^+ to $S^{p-1} \times \mathbf{R}^q$ has inverse sending $(\sigma, y) \in S^{p-1} \times \mathbf{R}^q$ to $z \equiv (\sqrt{1 + |y|^2}\sigma, y) \in S^+$.

The proof of (b) is similar and omitted.

Proof of (c): Since

$$(3.14) \qquad |x|^2 - |y|^2 = 0, \quad \text{for all } z \equiv (x, y) \in N,$$

if $z \equiv (x, y)$ is null but non-zero then both x and y are non-zero. Thus the map from N to $S^{p-1} \times S^{q-1} \times \mathbf{R}^+$ sending $z \equiv (x, y)$ to $(\frac{x}{|x|}, \frac{y}{|y|}, r)$, where $r \equiv |x| = |y|$, is a well defined smooth map. The inverse of this map sends $(x, y, r) \in S^{p-1} \times S^{q-1} \times \mathbf{R}^+$ to $z \equiv (rx, ry) \in N$. ∎

The next result follows since S^o consists of two points, $S^n(n \geq 1)$ is connected, and $S^n(n \geq 2)$ is simply connected.

Corollary 3.15.

(a) *The unit positive sphere S^+ in n-dimensional Lorentzian space $\mathbf{R}(1, n - 1)$ has two connected components diffeomorphic to $\mathbf{R}^{n-1}$. (These components are called the future and past spheres.)*

(b) *The unit positive sphere S^+ in $\mathbf{R}(p, q)$ for $p \geq 2$ is connected.*

(c) *The unit positive sphere S^+ in $\mathbf{R}(p, q)$ for $p \geq 3$ is simply connected.*

If $\mathbf{C}(p, q), \varepsilon$ is the standard model for the $\mathbf{C}$-hermitian symmetric case of signature p, q, then $\mathbf{R}(2p, 2q) \cong \mathbf{C}(p, q)$ with $\langle \, , \rangle \equiv \mathrm{Re}\,\varepsilon$ is the standard model for the $\mathbf{R}$-symmetric case of signature $2p, 2q$. Also note that $\varepsilon(x, x)$ only takes on real values. Therefore,

$$(3.16) \qquad \{x \in \mathbf{C}(p, q) : \varepsilon(x, x) = \pm r^2\} = S_r^\pm, \text{ the sphere in } \mathbf{R}(2p, 2q).$$

Similarly,

$$(3.17) \qquad \{x \in \mathbf{H}(p, q) : \varepsilon(x, x) = \pm r^2\} = S_r^\pm, \text{ the sphere in } \mathbf{R}(4p, 4q).$$

If $\mathbf{C}^n, \varepsilon$ is the standard model for the $\mathbf{C}$-symmetric inner product space, then $S_r^{n-1}(\mathbf{C}) \equiv \{z \in \mathbf{C}^n : \varepsilon(z, z) = r\}$ is called the *complex* (or *hyperquadric*) *sphere of radius* $r \in \mathbf{C}$.

If $\mathbf{H}^n, \varepsilon$ is the standard model for the $\mathbf{H}$-hermitian skew inner product space, note that $\varepsilon(x, x)$ only takes values in $\operatorname{Im} \mathbf{H}$. Given $r \in \operatorname{Im} \mathbf{H}$, the level set

$$(3.18) \qquad S_r\left(\mathbf{H}^n\text{-skew}\right) \equiv \left\{ x \in \mathbf{H}^n : \varepsilon(x, x) = r \right\},$$

fixed by the group $\operatorname{SK}(n)$, will be called the $\mathbf{H}$-*skew sphere of radius* r. Note that if $u \in \operatorname{Im} \mathbf{H}$ is a given unit vector then the real linear isomorphism $R_u : \mathbf{H}^n \to \mathbf{H}^n$ sends the $\mathbf{H}$-skew sphere of radius $r \in \operatorname{Im} \mathbf{H}$ onto the $\mathbf{H}$-skew sphere of radius $\overline{u}ru \in \operatorname{Im} \mathbf{H}$. This fact can be used to yield a description of the general $\mathbf{H}$-skew sphere from calculating the special case $r = i$, where $S_i(\mathbf{H}^n\text{-skew}) = \{a + jb : a, b \in \mathbf{C}^n, |a|^2 - |b|^2 = 1, \text{ and } \sum a_k b_k + b_k a_k = 0\}$.

Given an inner product, the spheres are the orbits of the associated isometry group.

Theorem 3.19 (Orbit Structure). *Let $n \equiv p + q$.*

$$(3.20) \qquad \frac{O(p, q)}{O(p - 1, q)} \cong \frac{\operatorname{SO}(p, q)}{\operatorname{SO}(p - 1, q)} \cong S_r^+, \text{ the sphere in } \mathbf{R}(p, q).$$

$$(3.21) \qquad \frac{O(p, q)}{O(p, q - 1)} \cong \frac{\operatorname{SO}(p, q)}{\operatorname{SO}(p, q - 1)} \cong S_r^-, \text{ the sphere in } \mathbf{R}(p, q).$$

$$(3.22) \qquad \frac{U(p, q)}{U(p - 1, q)} \cong \frac{\operatorname{SU}(p, q)}{\operatorname{SU}(p, q - 1)} \cong S_r^+, \text{ the sphere in } \mathbf{R}(2p, 2q).$$

$$(3.23) \qquad \frac{U(p, q)}{U(p, q - 1)} \cong \frac{\operatorname{SU}(p, q)}{\operatorname{SU}(p, q - 1)} \cong S_r^-, \text{ the sphere in } \mathbf{R}(2p, 2q).$$

$$(3.24) \qquad \frac{\operatorname{HU}(p, q)}{\operatorname{HU}(p - 1, q)} \cong S_r^+, \text{ the sphere in } \mathbf{R}(4p, 4q).$$

$$(3.25) \qquad \frac{\operatorname{HU}(p, q)}{\operatorname{HU}(p, q - 1)} \cong S_r^-, \text{ the sphere in } \mathbf{R}(4p, 4q).$$

$$(3.26) \qquad \frac{O(n, \mathbf{C})}{O(n - 1, \mathbf{C})} \cong \frac{\operatorname{SO}(n, \mathbf{C})}{\operatorname{SO}(n - 1, \mathbf{C})} \cong S_r^{n-1}(\mathbf{C}) \quad \text{for } r \in \mathbf{C}^* \text{ and } n \geq 2.$$

(3.27) $$\frac{\mathrm{Sp}(n, \mathbf{R})}{K} \cong \mathbf{R}^{2n} - \{0\},$$

where K is topologically the same as $\mathbf{R}^{2n-1} \times \mathrm{Sp}(n - 1, \mathbf{R})$.

(3.28) $$\frac{\mathrm{Sp}(n, \mathbf{C})}{K} \cong \mathbf{C}^{2n} - \{0\},$$

where K is topologically the same as $\mathbf{C}^{2n-1} \times \mathrm{Sp}(n - 1, \mathbf{C})$.

(3.29) $$\frac{\mathrm{SK}(n, \mathbf{H})}{\mathrm{SK}(n - 1, \mathbf{H})} \cong S_r(\mathbf{H}^n\text{-}skew) \quad \text{for each } r \in \mathrm{Im}\,\mathbf{H}.$$

Proof: Let $G/K \cong X$ correspond to one of the cases in Theorem 3.19. First, the group G acts transitively on X. In each case, this is verified by examining the proof of the Basis Theorem 2.46. For example, $O(p, q)$ acts transitively on S_r since, given a vector $u \in S_r^+$, the vector $\tilde{e}_1 \equiv u/\sqrt{r}$ may be chosen as the first vector in an orthonormal basis for $\mathbf{R}(p, q)$. The isometry $f \in O(p, q)$ that maps the standard orthonormal basis $e_1, \ldots, e_n$ for $\mathbf{R}(p, q)$ to this orthonormal basis sends e_1 to $\tilde{e}_1$ and hence $\sqrt{r}e_1$ to u.

Second, the isotropy subgroup K of G at e_1 must be computed. In all cases, the fact that

(3.30)   if $f(e_1) = e_1$ then $f : (\mathrm{span}\ e_1)^\perp \to (\mathrm{span}\ e_1)^\perp$

is of key importance. The individual cases are left to the reader. ∎

Corollary 3.31. *Each of the groups* $\mathrm{Sp}(n, \mathbf{R}), U(p, q), \mathrm{SU}(p, q), \mathrm{SO}(n, \mathbf{C})$, $\mathrm{Sp}(n, \mathbf{C}), \mathrm{HU}(p, q)$, *and* $\mathrm{SK}(n, \mathbf{H})$ *is connected. The groups* $\mathrm{SO}(p, q)$, $p, q \geq 1$ *and* $O(n, \mathbf{C})$ *have two connected components, while the groups* $O(p, q), p, q \geq 1$, *have four connected components.*

This corollary can be deduced from the theorem by induction on the dimension.

THE BASIS THEOREM REVISITED

The Basis Theorem may be interpreted as calculating an orbit of one of the groups $\mathrm{GL}(n, F)$. Let ε denote the standard inner product on F^n. The group $\mathrm{GL}(n, F)$ acts on the space $\mathrm{BIL}(\mathbf{R}^n)$ containing ε as follows. Suppose $\eta \in \mathrm{BIL}(F^n)$. For each $A \in \mathrm{GL}(n, F)$, let $A^*\eta$ denote the action of A on η where A^* is defined by

(3.32) $(A^*\eta)(x, y) \equiv \eta(Ax, Ay)$ for all $x, y \in F^n$.

Let $I(V)$ denote the subset of $\mathrm{BIL}(V)$ consisting of all inner products of the same type (and signature if applicable) as ε. Let $G(\varepsilon)$ denote the classical group that fixes ε (i.e., the isometry group of the given type).

Theorem 3.33. $\mathrm{GL}(n, F)$ *acts transitively on* $I(V)$ *with isotropy* $G(\varepsilon)$ *at* ε. *That is*,

$$(3.34) \qquad \mathrm{GL}(n, F)/G(\varepsilon) \cong I(V).$$

Note that $\{\eta \in \mathrm{SYM}(V) : \eta$ is nondegenerate$\}$ is an open subset of $\mathrm{SYM}(V)$ if ε is symmetric; while if ε is skew, $\{\eta \in \mathrm{SKW}(V) : \eta$ is nonde$-$ generate$\}$ is an open subset of $\mathrm{SKW}(V)$.

Since this theorem is a reinterpretation of the Basis Theorem, the proof is an exercise (see Problem 10).

ADJOINTS

Recall from the proof of the Orthogonality Lemma 2.30 that the flat map $\flat : V \to V^*$ is a real linear isomorphism for all seven types of inner product spaces V, ε. In particular, if $A \in \mathrm{End}_F(V)$ is given and $y \in V$ is fixed, then $\varepsilon(y, Ax) \in V^*$ is considered as a function of x. Thus, $\varepsilon(y, Ax) = (\flat(z))(x)$ for a unique $z \in V$. By definition of the flat map $(\flat(z))(x) \equiv \varepsilon(z, x)$. Now consider $z \in V$ to be a function of the given point $y \in V$ and denote z by A^*y. For each of the eight types of inner product spaces, it is easy to check that $\varepsilon(y, Ax) = \varepsilon(z, x)$ if and only if $\varepsilon(Ax, y) = \varepsilon(x, z)$. Thus the map $A^* : V \to V$ is characterized by

$$(3.35) \quad \text{either } \varepsilon(y, Ax) = \varepsilon(A^*y, x) \text{ or } \varepsilon(Ax, y) = \varepsilon(x, A^*y) \text{ for all } x, y \in V.$$

This map A^* is called the *adjoint* (not the dual) *map* of A, and definitely depends on the given inner product ε. The proof of the next lemma is straightforward and hence omitted (i.e., should be checked once carefully enough to see that it is routine).

Lemma 3.36. *The adjoint* $A \to A^*$ *is an anti-automorphism of the algebra* $\mathrm{End}_F V$ *that squares to the identity (an involution). That is,* $A^* \in \mathrm{End}_F(V)$ *for each* $A \in \mathrm{End}_F(V)$ *and*

$$(3.37) \qquad (AB)^* = B^*A^* \quad and \quad (A^*)^* = A.$$

Note that, given $A \in \mathrm{End}_F(V)$,

$$(3.38) \quad \varepsilon(Ax, Ay) = \varepsilon(x, y) \quad \text{for all} \quad x, y \in V \quad (\text{i.e., } A \text{ fixes } \varepsilon),$$

and

$$(3.39) \qquad A^*A = \mathrm{Id} \quad (\text{i.e., } A^* = A^{-1}),$$

are equivalent. Thus, each of the groups, defined by requiring that ε be fixed, can also be described as $\{A \in \mathrm{End}_F(V) : A^* = A^{-1}\}$.

Depending on the type of the inner product space, the adjoint is referred to as the **R**-*orthogonal adjoint*, **R**-*symplectic adjoint*, **C**-*orthogonal adjoint*, **C**-*symplectic adjoint*, **C**-*hermitian adjoint*, **H**-*hermitian symmetric adjoint*, and **H**-*hermitian skew adjoint*. Given a matrix $A = (a_{ij}) \in M_n(F)$ the *matrix transpose* is denoted by $A^t = (a_{ji})$, while the *matrix conjugate* is denoted by $\overline{A} = (\overline{a}_{ij})$. Using the matrix transpose and the matrix conjugate, each of the various types of adjoints can be explicitly computed (Problem 3). For example, the **C**-hermitian (positive definite) adjoint of $A \in M_n(\mathbf{C})$ is given by $A^* = \overline{A}^t$ the conjugate transpose. Combining the fact that A fixes ε if and only if $AA^* = 1$ with the explicit computations of the adjoints for $A \in M_n(F)$, it is easy to give alternative definitions of the groups: $O(p,q)$, $\mathrm{Sp}(n, \mathbf{R})$, $O(n, \mathbf{C})$, $\mathrm{Sp}(n, \mathbf{C})$, $U(p, q)$, $\mathrm{HU}(p, q)$ and $\mathrm{SK}(n, \mathbf{H})$. For example, $U(n) = \{A \in M_n(\mathbf{C}) : A\overline{A}^t = 1\}$.

LIE ALGEBRAS

Each of the groups G considered so far is a subset of $M_n(F)$, with $F \equiv \mathbf{R}, \mathbf{C}$, or $\mathbf{H}$, defined implicitly by a certain number of polynomial equations. The objective of this section is to show that G is a submanifold of $M_n(F)$ and to compute the tangent space to G at the identity $1 \in G$ in $M_n(F)$. This tangent space is denoted $\mathfrak{g}$ and called the *Lie algebra* of G.

Given two matrices $A, B \in M_n(F)$ the commutator

$$(3.40) \qquad\qquad [A, B] \equiv AB - BA$$

is also called the *Lie bracket* of A and B. This product provides the natural algebraic structure on each of the Lie algebras $\mathfrak{g}$.

The basic identity satisfied by a Lie algebra $\mathfrak{g}$, $[\,,\,]$ is called the Jacobi identity:

$$[A, [B, C]] = [[A, B], C] + [A, [B, C]],$$

for all $A, B, C \in \mathfrak{g}$.

The reader is referred to the plethora of texts on Lie algebras for more information.

The Lie algebra of a particular group is usually denoted by using the same label used for the group—except in German lower case. For example, $\mathfrak{so}(r, s)$ denotes the Lie algebra of $\mathrm{SO}(r, s)$.

First note that

$$(3.41) \qquad\qquad \mathfrak{gl}(n, \mathbf{R}) = M_n(\mathbf{R})$$

(3.42) $$\mathfrak{gl}(n, \mathbf{C}) = M_n(\mathbf{C})$$

(3.43) $$\mathfrak{gl}(n, \mathbf{H}) = M_n(\mathbf{H}),$$

since $\mathrm{GL}(n, F)$ is just an open subset of $M_n(F)$.

Now suppose that G is one of the seven classical groups that fixes one of the standard inner products ε. Then G is given implicitly as a subset of $M_n(F)$ by

(3.44) $$G = \{X \in M_n(F) : XX^* = 1\},$$

where F is $\mathbf{R}, \mathbf{C}$, or $\mathbf{H}$ and X^* denotes the ε-adjoint.

Let $\mathrm{Sym}_\varepsilon(n, F)$ denote the real vector subspace of $M_n(F)$ consisting of the ε-symmetric matrices, i.e.,

(3.45) $$\mathrm{Sym}_\varepsilon(n, F) \equiv \{A \in M_n(F) : A = A^*\}.$$

Let $\mathrm{Skew}_\varepsilon(n, F)$ denote the real vector subspace of $M_n(F)$ consisting of the ε-skew matrices, i.e.,

(3.46) $$\mathrm{Skew}_\varepsilon(n, F) \equiv \{A \in M_n(F) : A^* = -A\}.$$

Proposition 3.47. *Suppose ε is one of the standard inner products. The ε-orthogonal group*

$$G \equiv \{A \in M_n(F) : AA^* = 1\}$$

is a real submanifold of the real vector space $M_n(F)$ with

(3.48) $$\dim_{\mathbf{R}} G = \dim_{\mathbf{R}} \mathrm{Skew}_\varepsilon(n, F).$$

The tangent space $\mathfrak{g}$ at $1 \in G$ is given by

(3.49) $$\mathfrak{g} = \mathrm{Skew}_\varepsilon(n, F).$$

Proof: Consider XX^* as a function from $M_n(F)$ to $\mathrm{Sym}_\varepsilon(n, F)$ (i.e., from $\mathbf{R}^p$ to $\mathbf{R}^q$ for some p and q). The group G is defined to be the level set of this function at level $1 \in \mathrm{Sym}_\varepsilon(n, F)$. To prove that G is a real submanifold of $M_n(F)$ we invoke the implicit function theorem. Thus we must compute the rank of the derivative of XX^* at each point $B \in G$ and show that it equals $\dim_{\mathbf{R}} \mathrm{Sym}_\varepsilon(n, F)$.

The derivative (i.e., the linearization) of the function XX^* at $B \in G$ is given by

$$(3.50) \qquad \frac{d}{dt}(B + tX)(B + tX)^* \big|_{t=0} = XB^* + BX^*.$$

The linearized map $XB^* + BX^*$ from $M_n(F)$ to $\mathrm{Sym}_\varepsilon(n, F)$ is surjective; given $A \in \mathrm{Sym}_\varepsilon(n, F)$, take $X \equiv \frac{1}{2}AB$ and note that $XB^* + BX^* = \frac{1}{2}ABB^* + \frac{1}{2}BB^*A^* = A$. Thus the rank of the linear map sending $X \in M_n(F)$ to $XB^* + BX^* \in \mathrm{Sym}_\varepsilon(n, F)$ is equal to $\dim_{\mathbf{R}} \mathrm{Sym}_\varepsilon(n, F)$, completing the proof that G is an implicitly defined submanifold of $M_n(F)$.

The tangent space to G at $B \in G$ is the kernel of the linear map $XB^* + BX^*$. Thus, the tangent space $\mathfrak{g}$ to G at $1 \in G$ is the kernel of $X + X^*$, which equals $\mathrm{Skew}_\varepsilon(n, F) \subset M_n(F)$. ∎

The submanifold G of $M_n(F)$ defined by (3.44) can also be described parametrically. Since, given $B \in G$, right multiplication by B provides a diffeomorphism of $M_n(F)$ that maps G to G and sends $1 \in G$ to $B \in G$, it suffices to give a parametric description of a neighborhood of $1 \in G$.

First, we need a lemma about power series. The *sup norm* on $M_n(F)$ is defined by

$$(3.51) \qquad \text{sup norm } A \equiv \sup_x \frac{|Ax|}{|x|} \quad \text{for } A \in M_n(F),$$

where $| \cdot |$ denotes the standard positive definite norm on $F^n (F \equiv \mathbf{R}, \mathbf{C},$ or $\mathbf{H}$). The proof is an exercise.

Lemma 3.52. *Suppose* $f(z) \equiv \sum_{n=0}^\infty a_n z^n$ *is a power series with radius of convergence* r. *The series*

$$f(A) \equiv \sum_{n=0}^\infty a_n A^n$$

converges in the sup norm on $M_n(F)$ *if the sup norm* $A < r$.

Definition 3.53. *Given* $A \in M_n(F)$, *define*

$$\exp(A) \equiv e^A \equiv 1 + A + \frac{1}{2!}A^2 + \cdots .$$

Proposition 3.54. *Let* G *denote the group that fixes the inner product* ε. *The exponential map*

$$(3.55) \qquad\qquad \exp : \mathrm{Skew}_\varepsilon(n, F) \to G$$

provides a parameterization of a neighborhood of the identity 1 for the group G.

Proof: If $A \in \mathrm{Skew}_\varepsilon(n, F)$, then $e^A \in G$, since $(e^A)(e^A)^* = e^A e^{A^*} = e^A \cdot e^{-A} = e^{A-A} = e^0 = 1$. The derivative of exp at $A = 0$ is given by

$$\frac{d}{dt} e^{tX} \bigg|_{t=0} = X.$$

This proves that exp is injective near $A = 0$. Now the proposition follows since G has dimension equal to the dimension of the parameter space $\mathrm{Skew}_\varepsilon(n, F)$. ∎

The special linear groups are defined implicitly by setting the polynomial $\det_F X$ equal to one, where $F \equiv \mathbf{R}$ or $\mathbf{C}$. The derivative, or linearization, of this function at a point A is by definition

$$(3.56) \qquad \frac{d}{dt} \det_F(A + tX) \bigg|_{t=0}.$$

At the point $A = 1$ (the identity matrix), this formula provides a convenient coordinate free definition of the *trace* of a linear map $X \in \mathrm{End}_F(V)$. That is, for $F \equiv R$ or C, we define

$$(3.57) \qquad \mathrm{trace}_F X \equiv \frac{d}{dt} \det_F(1 + tX) \bigg|_{t=0}.$$

Of course, it is important to compute that if $X \equiv (x_{ij}) \in M_n(F)$ for $F \equiv R$ or C, then

$$(3.58) \qquad \mathrm{trace}_F X = \sum_{i=1}^n x_{ii}.$$

To check this consider $\det_F(1 + tX)e_1 \wedge \cdots \wedge e_n = (e_1 + tXe_1) \wedge \cdots \wedge (e_n + tXe_n)$, where $e_1, \ldots, e_n$ is an F-basis for V.

If $X \equiv (x_{ij}) \in M_n(\mathbf{H})$, then it follows that

$$(3.59) \qquad \mathrm{trace}_\mathbf{R} X = 4 \, \mathrm{Re} \sum_{i=1}^n x_{ii}.$$

Now suppose that $A \in \mathrm{SL}(n, F)$. In particular, A is invertible. The derivative of $\det_F X$ at the point A is equal to

$$(3.60) \qquad \begin{aligned} \frac{d}{dt} \det_F(A + tX) \bigg|_{t=0} &= \frac{d}{dt} \det_F(1 + tXA^{-1})A) \bigg|_{t=0} \\ &= (\mathrm{trace}_F \ XA^{-1}) \det_F A. \end{aligned}$$

Therefore, the linearization has rank 1 for $\det_{\mathbf{R}}$ and rank 2 for $\det_{\mathbf{C}}$. Because of the implicit function theorem, this provides a proof that the special linear groups are submanifolds. The tangent space at the identity 1 is the kernel of the linear map $\mathrm{trace}_F X$, obtained by setting $A = 1$ in (3.60). This proves the next proposition.

Proposition 3.61.

$$(3.62) \qquad \mathfrak{sl}(n, \mathbf{R}) = \{A \in M_n(\mathbf{R}) : \mathrm{trace}_{\mathbf{R}} A = 0\},$$

$$(3.63) \qquad \mathfrak{sl}(n, \mathbf{C}) = \{A \in M_n(\mathbf{C}) : \mathrm{trace}_{\mathbf{C}} A = 0\},$$

$$(3.64) \qquad \mathfrak{sl}(n, \mathbf{H}) = \{A \in M_n(\mathbf{H}) : \mathrm{trace}_{\mathbf{R}} A = 0\}.$$

PROBLEMS

1. (a) Prove that $(AB)^* = B^* A^*$ and $(A^*)^* = A$ for each of the seven types of adjoints.
   (b) Show that $\varepsilon(Ax, Ay) = \varepsilon(x, y)$ if and only if $A^* = A^{-1}$.
2. (a) Show that $\det A^* = \det A$ if A^* is the adjoint with respect to ε, where ε is either of $\mathbf{R}$-symmetric type or $\mathbf{C}$-symmetric type.
   (b) Show that each $A \in O(p, q)$ has $\det_{\mathbf{R}} A = \pm 1$.
   (c) Show that each $A \in O(n, \mathbf{C})$ has $\det_{\mathbf{C}} A = \pm 1$.
   (d) Suppose $A \in \mathrm{CO}(p, q)$ with conformal factor λ. Show $(\det_{\mathbf{R}} A)^2 = \lambda^n$, where $n \equiv p + q$ is the dimension.
3. (Adjoints) Compute the adjoint of a matrix $A \in M_m(F)$ in the following cases. Note that a decomposition of the vector space F^m into the direct sum of two subspaces induces a 2×2 blocking of the matrices:

$$A = \begin{pmatrix} a & b \\ c & d \end{pmatrix}.$$

Case (a) $\mathbf{R}$-positive definite or $\mathbf{C}$-pure symmetric. Show $A^* = A^t$.

Case (b) Positive (or negative) definite, $F \equiv \mathbf{R}, \mathbf{C}$, or $\mathbf{H}$.   Show $A^* = \overline{A}^t$ (conjugate transpose). This case is a subcase of the next case, F-hermitian symmetric.

Case (c) $F(p, q) = F^p \times F^q$, with $F \equiv \mathbf{R}, \mathbf{C}$, or $\mathbf{H}$. Show

$$\begin{pmatrix} a & b \\ c & d \end{pmatrix}^* = \begin{pmatrix} \overline{a}^t & -\overline{c}^t \\ -\overline{b}^t & \overline{d}^t \end{pmatrix}.$$

Case (d) **H**-hermitian skew: $A^* = i\overline{A}^{\,t}\,\overline{i}$, where i denotes i times the identity matrix.

Case (e) **R**- or **C**-symplectic: $\mathbf{R}^{2n} = \mathbf{R}^n \oplus \mathbf{R}^n$ or $\mathbf{C}^{2n} = \mathbf{C}^n \oplus \mathbf{C}^n$, written as the direct sum of the even coordinates and the odd coordinates. Show

$$\begin{pmatrix} a & b \\ c & d \end{pmatrix}^* = \begin{pmatrix} d^t & -b^t \\ -c^t & a^t \end{pmatrix}.$$

Hint: Reduce all cases to either Case (a) or Case (b) by finding a linear map L so that $\varepsilon(x, Ly) = \langle x, y \rangle$, with $\langle\, , \rangle$ an inner product of the type in either case (a) or case (b).

4. Suppose F^n, ε is one of the standard models for an inner product space. Let G denote the isometry group of F^n, ε, and $\mathfrak{g} \equiv \mathrm{Skew}_\varepsilon(n, \mathbf{F})$ the Lie algebra of G.

    (a) Suppose ε is (either pure or hermitian) symmetric. Exhibit an isomorphism $\mathfrak{g} \cong \mathrm{SKW}(F^n)$ with the space $\mathrm{SKW}(F^n)$ of all skew bilinear forms on $\mathbf{F}^n$ of the same type as ε.

    (b) Formulate the analogue of part (a) for ε skew.

    (c) The common notation for the space of all skew bilinear forms on $\mathbf{R}^n$ is $\Lambda^2(\mathbf{R}^n)^*$, while $S^2(\mathbf{R}^n)^*$ denotes the space of all symmetric bilinear forms on $\mathbf{R}^n$. Thus, by (a) $\mathfrak{so}(p, q) \cong \Lambda^2(\mathbf{R}^n)^*$ ($n \equiv p + q$), and by (b) $\mathfrak{sp}(n, \mathbf{R}) \cong S^2(\mathbf{R}^{2n})^*$. Compute that $\dim_R \mathrm{SO}(p, q) = \frac{1}{2}n(n + 1)$ and that $\dim_R \mathrm{Sp}(n, \mathbf{R}) = 2n^2 + n$.

5. Using the definition (3.57) for $\mathrm{trace}_F\, X$ show that the formula (3.58) is valid for $X \in M_n(\mathbf{R})$ or $X \in M_n(\mathbf{C})$. If $X \in M_n(\mathbf{H})$, show that (3.59) can be used to compute $\mathrm{trace}_R\, X$.

6. (a) Show that $\mathfrak{hu}(1) \cong \mathrm{Im}\,\mathbf{H}$ and that $\exp : \mathrm{Im}\,\mathbf{H} \to S^3 \cong \mathrm{HU}(1) \subset \mathbf{H}$ is surjective.

    (b) Show that $\exp : \mathfrak{sl}(2, \mathbf{R}) \to \mathrm{SL}(2, \mathbf{R})$ is not surjective.

7. Explicitly compute $\mathfrak{o}(n)$, $\mathfrak{o}(r, s)$; $\mathfrak{u}(n)$, $\mathfrak{u}(r, s)$; $\mathfrak{hu}(n)$, $\mathfrak{hu}(r, s)$; $\mathfrak{sp}(n, \mathbf{R})$, $\mathfrak{sp}(n, \mathbf{C})$; $\mathfrak{o}(n, \mathbf{C})$; and $\mathfrak{sl}(n, \mathbf{H})$.

8. The real manifold dimension of a Lie group G is the same number as the real vector space dimension of the Lie algebra $\mathfrak{g}$. Show that

    (a)

$$\dim_R \mathrm{GL}(n, \mathbf{R}) = n^2, \quad \dim_R \mathrm{SL}(n, \mathbf{R}) = n^2 - 1,$$
$$\dim_R \mathrm{GL}(n, \mathbf{C}) = 2n^2, \quad \dim_R \mathrm{SL}(n, \mathbf{C}) = 2n^2 - 2,$$
$$\dim_R \mathrm{GL}(n, \mathbf{H}) = 4n^2, \quad \dim_R \mathrm{SL}(n, \mathbf{H}) = 4n^2 - 1.$$

(b) Using Problem 12, compute that $(n \equiv r + s)$

$$\dim_R \text{ SO}(r, s) = \frac{1}{2}n(n - 1),$$
$$\dim_R \ U(r, s) = n^2,$$
$$\dim_R \text{ SU}(r, s) = n^2 - 1,$$
$$\dim_R \text{ HU}(r, s) = 2n^2 + n,$$
$$\dim_R \ O(n, \mathbf{C}) = n(n - 1),$$
$$\dim_R \text{ SK}(n, \mathbf{H}) = 2n^2 - n,$$
$$\dim_R \text{Sp}(n, \mathbf{R}) = 2n^2 + n,$$
$$\dim_R \text{ Sp}(n, \mathbf{C}) = 4n^2 + 2n.$$

9. Give the proof of Lemma (3.52) on power series.

10. Give the proof of the seven results contained in Theorem 3.33. For example, show that

$$\text{GL}(n, \mathbf{R})/O(r, s) \cong I(r, s; \mathbf{R}),$$

the set of $\mathbf{R}$-symmetric inner products on $\mathbf{R}^n$ with signature r, s.

11. Prove that $\text{SK}(n, \mathbf{H})$ is the intersection of three distinct split unitary groups on $\mathbf{H}^n$, corresponding to the three complex structures R_i, R_j, and R_k on $\mathbf{H}^n$.

12. Show that

(a) $\mathfrak{su}(p, q) \otimes_{\mathbf{R}} \mathbf{C} \cong \mathfrak{sl}(n, \mathbf{C})$,

(b) $\mathfrak{hu}(p, q) \otimes_{\mathbf{R}} \mathbf{C} \cong \mathfrak{sp}(n, \mathbf{C})$,

(c) $\mathfrak{sk}(n) \otimes_{\mathbf{R}} \mathbf{C} \cong \mathfrak{so}(2n, \mathbf{C})$,

(d) $\mathfrak{sl}(n, \mathbf{H}) \otimes \mathbf{C} \cong \mathfrak{sl}(2n, \mathbf{C})$.

13. Suppose G is one of the classical groups considered in this chapter. Given an element $h \in G$, the mapping from G to G defined by sending $g \in G$ to $hgh^{-1} \in G$ is a diffeomorphism of G that maps the identity element $1 \in G$ into itself. The linearization of this map, denoted Ad_h, is a linear map from $\mathfrak{g} \equiv T_1 G$ into itself.

(a) Show that $\text{Ad}_h\ A = hAh^{-1}$, and that $h \mapsto \text{Ad}_h$ is a representation of the group G on the vector space $\mathfrak{g}$. This representation is called the *adjoint representation* of G.

(b) Show that for $B \in \mathfrak{g}$ fixed, the map sending $h \in G$ to $\text{Ad}_h\ B$ has the derivative at $h = 1 \in G$ given by the linear map sending $A \in \mathfrak{g} \equiv T_1 G$ to $[A, B] \equiv AB - BA$.

14. Suppose G is a *Lie group*, that is, a group equipped with a compatible differentiable manifold structure. Let $\mathfrak{X}_{\mathrm{inv}}(G)$ denote the vector space of *right invariant vector fields*, that is, vector fields on G which are fixed by the diffeomorphisms R_g (right multiplication by g) for each $g \in G$.

(a) Show that $\mathfrak{X}_{\mathrm{inv}}(G)$ is closed under the binary operator $[X, Y] \equiv XY - YX$.

(b) Under the natural map sending $X \in \mathfrak{X}_{\mathrm{inv}}(G)$ to $X_1 \in T_1(G) \cong \mathfrak{g}$, show that $\mathfrak{X}_{\mathrm{inv}}(G), [\ ,\ ]$ and $\mathfrak{g}, [\ ,\ ]$ are isomorphic.

Hint: Consult a text on Lie groups.

15. (The quaternionic unitary group) The space of imaginary quaterions $\mathrm{Im}\,\mathbf{H}$ has several realizations relating to $\mathbf{H}^n, \varepsilon$. First, $\mathrm{Im}\,\mathbf{H}$ can be identified with the *coefficient space* $\{R_u : u \in \mathrm{Im}\,\mathbf{H}\}$ of $\mathbf{R}$-linear maps of $\mathbf{H}^n$. Second, using ε, $\mathrm{Im}\,\mathbf{H}$ can be identified with the space $\{\omega_u : u \in \mathrm{Im}\,\mathbf{H}\}$ of Kähler forms on $\mathbf{H}^n$, where each 2-form $\omega_u(x, y)$ is defined to be $\langle u, \varepsilon(x, y)\rangle$ (cf. (2.79)).

(a) Show that the quaternionic unitary group $\mathrm{HU}(p, q) \cdot \mathrm{HU}(1)$ can be characterized as the subgroup of $O(4p, 4q)$ that fixes $R_i \wedge R_j \wedge R_k$, where each element $g \in \mathrm{SO}(4p, 4q)$ acts on the coefficient space $\{R_u : u \in \mathrm{Im}\,\mathbf{H}\}$ by sending R_u to $g \circ R_u \circ g^{-1}$.

Hint: Recall (1.35) that $\mathrm{HU}(1)/\mathbf{Z}_2 \cong \mathrm{SO}(\mathrm{Im}\,\mathbf{H})$ to reduce to the case where $g \in O(4p, 4q)$ leaves the coefficient space $\{R_u : u \in \mathrm{Im}\,\mathbf{H}\}$ pointwise fixed.

(b) Prove that the two representations of $\mathrm{HU}(1)$ given by

$$\chi_a(x) \equiv ax\overline{a} \quad \text{for all } x \in \mathrm{Im}\,\mathbf{H}$$

and

$$\rho_a(\omega_u) \equiv R_a^* \omega_u \quad \text{for all } u \in \mathrm{Im}\,\mathbf{H}$$

are equivalent.

(c) Prove that the following two representations of the group

$$\frac{\mathrm{HU}(1)}{\mathbf{Z}_2} \cong \mathrm{SO}(\mathrm{Im}\,\mathbf{H})$$

(see 1.35) are equivalent. First, $\rho_a(\omega_u \wedge \omega_v) \equiv R_a^*(\omega_u \wedge \omega_v)$ defines a representation on

$$\mathrm{span}\{\omega_u \wedge \omega_v : u, v \in \mathrm{Im}\,\mathbf{H}\} \subset \Lambda^4(\mathbf{H}^n)^*.$$

Second, $\sigma_a(A) \equiv \chi_a \circ A \circ \chi_{\overline{a}}$ defines a representation of $\mathrm{Sym}(\mathrm{Im}\,\mathbf{H})$, the space of symmetric 3×3 matrices on $\mathbf{R}^3 \equiv \mathrm{Im}\,\mathbf{H}$.

(d) Prove that the quaternionic unitary group $HU(p,q) \cdot HU(1)$ fixes the *quaternionic Kähler form*

$$\Phi \equiv \frac{1}{6} \left(\omega_i \wedge \omega_i + \omega_j \wedge \omega_j + \omega_k \wedge \omega_k \right) \in \Lambda^4 \left(\mathbf{H}^n \right)^* .$$

(e)* $(n \geq 2)$ Prove that if $g \in GL_{\mathbf{R}}(\mathbf{H}^n)$ and $g^* \Phi = \Phi$, then $g \in HU(p,q) \cdot HU(1)$. Thus, the quaternionic unitary group $HU(p,q) \cdot HU(1)$ is characterized as the subgroup of $GL_{\mathbf{R}}(\mathbf{H}^n)$ that fixes the quaternionic Kähler form Φ with no reference to an inner product on $\mathbf{H}^n$.

16. A representation $\rho = \rho_1 \oplus \rho_2$ that can be expressed as the direct sum of two representations ρ_1 and ρ_2 is said to be *reducible*. Otherwise, a representation is *irreducible*.

(a) Suppose $\rho = \rho_1 \oplus \rho_2 : SO(n) \to O(V_1) \oplus O(V_2)$ is a reducible representation of $SO(n)$ on a positive definite inner product space $V = V_1 \oplus V_2$, with inner product $\langle , \rangle$. Show that there exists an inner product on V different from $c\langle , \rangle$, where c is a constant, which is fixed by $SO(n)$.

(b) Using part (a), show that the representation of $SO(n)$ on $\Lambda^k \mathbf{R}^n$ is irreducible unless $k = n/2$ and $n = 0 \mod 4$. (In this case, the representations of $SO(n)$ on $\Lambda^k_+ \mathbf{R}^{2k}$ and $\Lambda^k_- \mathbf{R}^{2k}$ are both irreducible, where $\Lambda^k_\pm \mathbf{R}^{2k} \equiv \{ u \in \Lambda^k \mathbf{R}^{2k} : \star u = \pm u \}$ denotes the space of self/anti-self dual k-vectors, and $\star$ is the Hodge star operator.)

4. Euclidean/Lorentzian Vector Spaces

In this chapter, we will study **R**-symmetric inner product spaces. If the inner product is positive definite, then the inner product space is better known as a *euclidean vector space*. If the signature is $n - 1, 1$, then the inner product space is better known as a *Lorentzian vector space*. Rather than refer to a Lorentzian vector space as "pseudo" or "semi" euclidean or to a euclidean vector space as "pseudo" or "semi" Lorentzian, we shall refer to an **R**-symmetric inner product space with signature p, q simply as a *euclidean vector space with signature p, q*.

Remark. For the reader familiar with manifolds, a *Riemannian manifold* is a smooth manifold whose tangent space at each point is equipped with a positive definite euclidean inner product, and the inner product varies smoothly. A *Lorentzian manifold* is a smooth manifold whose tangent space at each point is equipped with a Lorentzian inner product, varying smoothly. To be consistent with our inner product space terminology, we shall refer to a Lorentzian manifold as a Riemannian manifold with signature $n - 1, 1$ rather than as a "pseudo" or "semi" Riemannian manifold (cf. Chapter 5).

The first topic in this chapter is very elementary: the Cauchy–Schwarz equality. The second topic is a brief introduction to special relativity, and the third topic is the Cartan–Dieudonné theorem. This last result is crucial to our discussion of the spin groups.

THE CAUCHY–SCHWARZ EQUALITY

Suppose $V, \varepsilon = \langle \, , \, \rangle$ is a euclidean vector space with signature p, q and dimension $n = p + q$.

Definition 4.1.

(a) *The norm of a vector $v \in V$ is defined to be*

$$|v| \equiv \sqrt{|\langle v, v \rangle|}.$$

(b) *The square norm or quadratic form associated with the bilinear form $\varepsilon = \langle \, , \, \rangle$ is defined by;*

$$\|v\| \equiv \langle v, v \rangle.$$

The inner product $\varepsilon = \langle \, , \, \rangle$ is completely determined by the quadratic form $\| \ \|$ by the process of *polarization*—namely, replace v by $x + y$ in $\|v\| = \langle v, v \rangle$ and use bilinearity to obtain

(4.2) $\langle x, y \rangle = \dfrac{1}{2} \left(\|x + y\| - \|x\| - \|y\| \right).$

Recall the musical isomorphism (metric equivalence or lowering indices) $\flat : V \to V^*$. This linear map naturally induces a linear map, also denoted $\flat$, on tensors:

$$\flat : \otimes^k V \to \otimes^k V^* \cong \left(\otimes^k V \right)^* \quad (\text{e.g.}, \flat(u \otimes v) = \flat(u) \otimes \flat(v)),$$

$$\flat : \Lambda^k V \to \Lambda^k V^* \cong \left(\Lambda^k V \right)^* \quad (\text{e.g.}, \flat(u \wedge v) = \flat(u) \wedge \flat(v)).$$

Similarly, $\sharp \equiv \flat^{-1} : V^* \to V$ naturally induces a linear map $\sharp$ on tensors and $\sharp\flat = \flat\sharp = \text{Id}$. Thus, $\otimes^k V$ and $\Lambda^k V$ naturally inherit a euclidean structure V with the inner product ε defined by $\varepsilon(x, y) = (\flat(x))(y)$. This inner product on tensors will also be denoted by $\varepsilon \equiv \langle \, , \, \rangle$. Suppose $e_1, \ldots, e_n$ is an orthogonal basis for V with the sign $\sigma_j \equiv \langle e_j, e_j \rangle$ either ± 1. Then, for example,

(4.3) $\{ e_i \wedge e_j : i < j \}$ is an orthonormal basis for $\Lambda^2 V$ with

$$\langle e_i \wedge e_j, e_i \wedge e_j \rangle = \sigma_i \sigma_j,$$

and more generally,

$$\left\{ e_{i_1} \wedge \cdots \wedge e_{i_p} : i_1 < \cdots < i_p \right\}$$

is an orthonormal basis for $\Lambda^p V$.

Note: If $\dim V = 2$, then the one-dimensional space $\Lambda^2 V$ is positive in both the case V positive and the case V negative, while $\Lambda^2 V$ is negative if V has signature $1, 1$.

Theorem 4.4 (Cauchy–Schwarz Equality). *Suppose V is a euclidean vector space of signature p, q. For all $x, y \in V$,*

(4.5) $$\langle x, y \rangle^2 + \langle x \wedge y, x \wedge y \rangle = \|x\| \, \|y\|.$$

First Proof: The idea of this proof is that since the theorem only involves two vectors it is really a two-dimensional result. Let W denote the span of x and y. If x and y are colinear, the proof is immediate. Thus, we may assume that W is a plane. Because of the Basis Theorem and its extension to the degenerate cases (Problem 2.7), the (possibly degenerate) inner product space W is isometric to one of the standard models:

(4.6) $\mathbf{R}(2, 0)$, $\mathbf{R}(1, 1)$, $\mathbf{R}(0, 2)$ if W is nondegenerate,

or

(4.7) $\mathbf{R}(1, 0) \times \mathbf{R}$, $\mathbf{R}(0, 1) \times \mathbf{R}$, $\mathbf{R} \times \mathbf{R}$ if W is degenerate.

Finally, it is a simple matter to verify directly (4.5) in each of these six cases.

For example, if $x, y \in \mathbf{R}(1, 1)$, then

$$\langle x, y \rangle^2 = (x_1 y_1 - x_2 y_2)^2 = x_1^2 y_1^2 + x_2^2 y_2^2 - 2x_1 y_1 x_2 y_2,$$
$$\langle x \wedge y, x \wedge y \rangle = -(x_1 y_2 - x_2 y_1)^2 = -x_1^2 y_2^2 - x_2^2 y_1^2 + 2x_1 y_1 x_2 y_2,$$

and

$$\|x\| \, \|y\| = \left(x_1^2 - x_2^2 \right) \left(y_1^2 - y_2^2 \right) = x_1^2 y_1^2 + x_2^2 y_2^2 - x_1^2 y_2^2 - x_2^2 y_1^2. \quad \blacksquare$$

Second Proof: $\langle x \wedge y, x \wedge y \rangle = (\flat(x \wedge y))(x \wedge y) = (\flat(x) \wedge \flat(y))(x \wedge y) = (\flat(x)(x))(\flat(y)(y)) - (\flat(x)(y))(\flat(y)(x)) = \langle x, x \rangle \langle y, y \rangle - \langle x, y \rangle^2$, where each of these equalities is an application of a definition. $\quad \blacksquare$

Third Proof: First, we verify the algebraic identity (Lagrange):

(4.8) $$\left(\sum_{i=1}^{n} z_i w_i \right)^2 + \sum_{i<j} (z_i w_j - z_j w_i)^2 = \left(\sum_{i=1}^{n} z_i^2 \right) \left(\sum_{j=1}^{n} w_j^2 \right)$$

for all $z, w \in \mathbf{C}^n$, by noting that

$$\sum_{i<j} (z_i w_j - z_j w_i)^2 = \frac{1}{2} \sum_{i,j=1}^{n} (z_i w_j - w_i z_j)^2$$

$$= \frac{1}{2} \sum_{i,j=1}^{n} (z_i^2 w_j^2 - 2z_i w_j z_j w_i + z_j^2 w_i^2)$$

$$= \sum_{i=1}^{n} z_i^2 \sum_{j=1}^{n} w_j^2 - \left(\sum_{i=1}^{n} z_i w_i \right)^2 .$$

Now, substituting

$$x_j \text{ for } z_j \text{ and } y_j \text{ for } w_j \quad \text{if } j = 1, \dots, p,$$

and

$$ix_j \text{ for } z_j \text{ and } iy_j \text{ for } w_j \quad \text{if } j = p+1, \dots, n,$$

with x_i, y_i real, yields the C.–S. equality for $\mathbf{R}(p, q)$. Since $V, \langle \, , \, \rangle$ is isometric to $\mathbf{R}(p, q)$, this completes the proof. ∎

Proposition 4.9. *Suppose* $x, y \in V$ *and* $W \equiv \mathrm{span}\{x, y\}$ *is a two-dimensional plane. The plane* W *is degenerate if and only if* $\langle x \wedge y, x \wedge y \rangle = 0$ *but* $x \wedge y \neq 0$. *If* W *is nondegenerate, then* W *is either positive, negative, or Lorentzian.*

Positive/Negative: The following are equivalent:

(4.10)   (i) W is a positive or a negative plane,

            (ii) $\Lambda^2 W$ is positive,

            (iii) $\langle x \wedge y, x \wedge y \rangle > 0$,

            (iv) $\langle x, y \rangle^2 < \|x\| \, \|y\|$.

Lorentzian: The following are equivalent:

(4.11)   (i) W is a Lorentzian plane,

            (ii) $\Lambda^2 W$ is negative,

            (iii) $\langle x \wedge y, x \wedge y \rangle < 0$,

            (iv) $\langle x, y \rangle^2 > \|x\| \, \|y\|$.

Proof: The fact that (i)–(iv) are all equivalent, in either (4.10) (positive/negative) or (4.11) (Lorentzian), follows from the note before Theorem 4.4 and Theorem 4.4 (C.–S. Equality).

If W is a degenerate plane, then there exists a nonzero vector $u \in W$ with $\langle u, z \rangle = 0$ for all $z \in W$. Pick $v \in W$ with $u \wedge v = x \wedge y$. Then, by the C.–S. Equality, $\langle x \wedge y, x \wedge y \rangle = \langle u \wedge v, u \wedge v \rangle = \langle u, v \rangle^2 - \|u\|\,\|v\| = 0$. Conversely, if $x \wedge y \neq 0$ but $\langle x \wedge y, x \wedge y \rangle = 0$, then W must be a degenerate plane, since W cannot have dimension ≤ 2 ($x \wedge y \neq 0$) and W cannot be a positive, negative, or Lorentzian plane. ∎

Corollary 4.12 (Cauchy–Schwarz Inequality). *If V is positive euclidean, then*

$$\langle u, v \rangle \leq |u|\,|v| \quad \text{for all } u, v \in V,$$

with equality if and only if one of the vectors is a positive multiple of the other.

Remark. Since $-|u|\,|v| \leq \langle u, v \rangle \leq |u|\,|v|$,

$$(4.13) \qquad \cos\theta \equiv \frac{\langle u, v \rangle}{|u|\,|v|}$$

uniquely defines an *angle* θ *between* any two non-zero vectors u *and* v with $0 \leq \theta \leq \pi$. Note that $\sin\theta = |u \wedge v|/|u|\,|v|$.

A *curve* M in a euclidean vector space V is, by definition, a one-dimensional oriented submanifold with boundary. Since (by definition) a curve is oriented, each point on a curve possesses a well-defined tangent ray (or half-line) and, if this tangent line is non-null, a well-defined unit tangent vector. The *length* of a curve M is defined to be

$$(4.14) \qquad l(M) \equiv \int_\alpha^\beta |x'(t)|\, dt,$$

where $x : [\alpha, \beta] \to M \subset V$ parameterizes M. The *initial point* $a \equiv x(\alpha)$ and the *terminal point* $b \equiv x(\beta)$ form the boundary of M. If the tangent line to M is never null, then the parameterization can be chosen so that $|x'(s)| = 1$ and is called *parameterization by arclength*. The length of the line segment from a to b is $|b - a|$.

Proposition 4.15. *Suppose V is positive definite.*

$$|b - a| \leq l(N)$$

for any curve N with initial point a and terminal point b. Equality occurs if and only if N = the line segment from a to b.

The proof of this result is fundamental in the theory of "calibrations" (see Chapter 7).

Proof: Let M denote the line segment from a to b. Let $\vec{M}$ denote the unit tangent vector to M (i.e., $\vec{M} = b - a/|b - a|$). Let $\phi \equiv \flat(\vec{M})$ denote the corresponding 1-form tangent to M. Then $|b - a| = l(M) = \int_M \phi$, which equals $\int_N \phi$ by the fundamental theorem of calculus for paths since $d\phi = 0$. Choose the parameterization $x(s)$ of N by arclength. Then

$$\int_N \phi = \int_0^l \langle \vec{M}, x'(s) \rangle \, ds.$$

By the C.–S. Inequality, this is $\leq \int_0^l ds = l(N)$. ∎

SPECIAL RELATIVITY

The terminology and intuition provided by special relativity is very helpful in understanding Lorentzian vector spaces, and, in particular, in understanding dimension four (Minkowski space).

A Lorentzian vector space V (signature $n - 1, 1$) is said to be *time oriented* if one of the two components of the *timecone* $\{v \in V : \langle v, v \rangle < 0\}$ is designated as the *future timecone*. Vectors in the future timecone are said to be *future timelike* vectors.

Corollary 4.16 (Backwards Cauchy–Schwarz Inequality). *Suppose V is a time-oriented Lorentzian vector space. If u and v are future timelike vectors, then*

$$-\langle u, v \rangle \geq |u|\,|v|,$$

with equality if and only if u and v are positive multiples of each other.

Remark. For u, v future timelike

$$(4.17) \qquad\qquad \cosh \theta = \frac{-\langle u, v \rangle}{|u|\,|v|}$$

uniquely determines an *angle* $0 \leq \theta < +\infty$ *between u and v.* Note that $\sinh \theta = |u \wedge v|/|u|\,|v|$.

Suppose V is a time-oriented Lorentzian vector space. A curve M in V is called a *worldline* of a *particle* if its tangent is future timelike at each point. The arclength parameter is called *proper time* and usually denoted τ. The length of M is called the *proper time* of the worldline or particle. If the worldline M is a line segment, then the particle is said to be in *free fall*. Of course, the proper time of a particle in free fall from a to b is $|b - a|$.

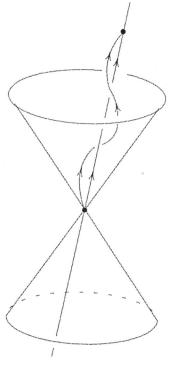

Figure 4.18

Proposition 4.19 (Twin Paradox). *The proper time of a particle is maximized by free fall.*

Proof: See Problem 3. ∎

Corollary 4.20 (Backwards Triangle Inequality). *If u and v are future timelike vectors, then*

$$|u + v| \geq |u| + |v|,$$

with equality if and only if u and v are positive multiples of one another.

The null lines in a Lorentzian vector space V are called *light rays* (or the *worldline of a photon*). Thus, the union of all the light rays is the *lightcone* (or nullcone)

$$\{x \in V : \|x\| = 0\}.$$

In Minkowski space $\mathbf{R}(3, 1)$, the lightcone is given by

$$\{(x, t) \in \mathbf{R}(3, 1) : \|x\| = t^2\}.$$

The speed of light has been normalized to be one. One of Einstein's axioms of special relativity, referred to as the *Constancy of Light*, states that

If a linear coordinate change $A \in \mathrm{GL}(\mathbf{R}(3, 1))$ on Minkowski space is *physically admissible* (or provides a new *inertial coordinate system*), then A must preserve the set of light rays.

Any invertible linear map A on $V, \| \cdot \|$ determines a new square norm $\| \ \|'$ by

$$\|z\|' \equiv \|Az\| \qquad \text{for all } z \in V.$$

If A preserves the nullcone for $\| \cdot \|$, then $\| \cdot \|'$ has the same nullcone as $\| \cdot \|$. Under this hypothesis, the next theorem states that $\| \cdot \|'$ and $\| \cdot \|$ differ by a constant c. That is, A must preserve $\| \cdot \|$ up to a constant. In summary, the next theorem has as a consequence:

Einstein's axiom, the Constancy of Light, implies that each inertial coordinate system on $\mathbf{R}(3, 1)$ is obtained from the standard coordinate system by a coordinate change A that preserves $\| \cdot \|$ up to a constant.

If additionally A preserves futuretime, then it is easy to show that the constant is positive, i.e., A must be conformally Lorentzian, $A \in \mathrm{CO}(3, 1)$. (An additional axiom, called the *Axiom of Relativity*, implies that the constant is 1; that is, $A \in \mathrm{SO}^\uparrow(3, 1)$ the *Lorentz group*.)

Lemma 4.21. *Suppose $V, \| \cdot \|$ has signature p, q with $p, q \geq 1$ and $\| \cdot \|'$ is another quadratic form on V. If each null vector for $\| \cdot \|$ is also null for $\| \ \|'$, then for some constant c,*

$$\|x\|' = c\|x\|.$$

Proof: First assume that V has dimension 2. Then $V, \| \cdot \|$ is isometric to the Lorentz plane $\mathbf{R}(1,1)$. Let $\mathbf{R} \oplus \mathbf{R}$ denote $\mathbf{R}^2$ with the square norm $\|x\| \equiv x_1 x_2$. Then $\mathbf{R}(1,1)$ and $\mathbf{R} \oplus \mathbf{R}$ are isometric. Using the model $\mathbf{R} \oplus \mathbf{R}$, the hypothesis that $\| \cdot \|'$ vanishes on the null cone for $\| \cdot \|$, says that $\|x\|' = Ax_1^2 + Bx_1 x_2 + Cx_2^2$ must vanish when $x_1 = 0$ or $x_2 = 0$. Thus $\|x\|' = B\|x\|$.

Now assume V has dimension > 2. It suffices to prove that for any pair of vectors u, v with one spacelike and one timelike the ratios $\|u\|'/\|u\|$ and $\|v\|'/\|v\|$ are the same. Repeated use of this will yield $\|x\|' = c\|x\|$ for all x. Since span$\{u, v\}$ is a Lorentz plane, there exists a constant c (depending on u and v) such that

$$\|x\|' = c\|x\| \quad \text{for all } x \in \text{span}\{u, v\}. \quad \blacksquare$$

A future timelike vector might also be called the *instantaneous* worldline of a particle. This particle can also be thought of as an "observer." It is useful to define an *observer* to be a curve whose unit tangent is always future timelike. That is, "observer" and "worldline of a particle" are mathematically identical concepts. Finally, an *instantaneous observer* is another name for a (unit) future timelike vector. To say that an instantaneous observer u is provided or given should be thought of as the same as to say that an orthogonal decomposition of Minkowski space is provded or given. Namely,

$$\mathbf{R}(3,1) = S \oplus \text{span } u,$$

where $S \equiv (\text{span } u)^\perp$ is a spacelike 3-plane referred to as *rest space*. Armed with this expanded definition of an instantaneous observer, it is usually possible to guess the classical concepts that are "observed" or "deduced" from a relativistic concept. Two examples are discussed. For both, suppose an instantaneous observer u is given at a point z in Minkowski space with associated rest space S.

In the first example, suppose M is the worldline of a particle. Let $\vec{M}$ denote the unit (future timelike) tangent vector to M at the point z. Then $\vec{M}$ has the orthogonal decomposition $\vec{M} = tu + w$, and the observed velocity of the particle is $v = w/t \in S$.

As a second example, consider the relativistic concept of an *electromagnetic field* on Minkowski space. By definition, this is just a two-form F on $\mathbf{R}(3,1)$; that is, for each point $z \in \mathbf{R}(3,1), F(z) \in \Lambda^2 \mathbf{R}(3,1)$. Now, given an instantaneous observer u at $z \in \mathbf{R}(3,1)$, what can be "observed"? The vector field E that is metrically equivalent to the one-form $u \llcorner F$ is called the (*observed*) *electric field*.

Recall that if a euclidean space $V, \langle \, , \, \rangle$ is oriented, then there exists a unique unit volume element $\lambda \in \Lambda^n V$ defined by $\lambda \equiv e_1 \wedge \ldots \wedge e_n$ where $e_1, \ldots, e_n$ is an oriented orthonormal basis for V. Moreover, the equation

$$\alpha \wedge (\star \beta) = \langle \alpha, \beta \rangle \lambda$$

defines a linear isomorphism

$$\star : \Lambda^k V \to \Lambda^{n-k} V$$

called the Hodge $\star$ operator.

Now consider the unit volume element $\lambda \equiv dx^1 \wedge dx^2 \wedge dx^3 \wedge dt$ along with the standard inner product $\langle \, , \, \rangle$ on $\mathbf{R}(3,1)$. Note that $\star$ maps $\Lambda^2 \mathbf{R}(3,1)$ isomorphically on $\Lambda^2 \mathbf{R}(3,1)$. Let $\flat(u)$ denote the one-form metrically equivalent to u. The vector field B that is metrically equivalent to $\star(F \wedge \flat(u))$ is called the (*observed*) *magnetic field*. If the instantaneous observer $u = e_4$ is taken to be the standard unit timelike basis vector in $\mathbf{R}(3,1)$, then

$$F = B^3 dx^1 \wedge dx^2 + B^1 dx^2 \wedge dx^3 + B^2 dx^3 \wedge dx^1$$
$$+ E^1 dx^1 \wedge dt + E^2 dx^2 \wedge dt + E^3 dx^3 \wedge dt.$$

The classical Maxwell equations are most succinctly written

$$dF = 0 \qquad \text{and} \qquad d(\star F) = 0.$$

We conclude this section with a sketch of the proof of the isomorphism between the *Lorentz group* $\mathrm{SO}^{\uparrow}(3,1)$ and the special complex orthogonal group $\mathrm{SO}(3, \mathbf{C})$. First note that $\mathrm{SO}^{\uparrow}(3,1)$ can be defined as the subgroup of $\mathrm{SO}(3,1)$, which preserves the future timecone. In particular, $-1 \notin \mathrm{SO}^{\uparrow}(3,1)$. Later, the alternate description of $\mathrm{SO}^{\uparrow}(3,1)$, as the connected component of the identity in $\mathrm{SO}(3,1)$, is used.

The standard action of $\mathrm{SO}^{\uparrow}(3,1)$ on $\mathbf{R}(3,1)$ induces an action of $\mathrm{SO}^{\uparrow}(3,1)$ on $\Lambda^2 \mathbf{R}(3,1)$ by the usual "pull back" map on forms. The following facts are straightforward to verify. The square, $\star^2$, of the star operator $\star$ on $\Lambda^2 \mathbf{R}(3,1)$ is minus the identity. Consequently, $\star$ provides $\Lambda^2 \mathbf{R}(3,1)$ with a complex structure $J \equiv \star$. The real valued symmetric bilinear form $\star(\alpha \wedge \beta)$ on $\Lambda^2 \mathbf{R}(3,1)$ is nondegenerate, and $J \equiv \star$ is an anti-isometry with respect to this bilinear form. Consequently,

$$\varepsilon(\alpha, \beta) \equiv \star(\alpha \wedge \beta) + i \star ((\star \alpha) \wedge \beta)$$

is a $\mathbf{C}$-symmetric inner product on the complex vector space $\Lambda^2 \mathbf{R}(3,1)$, $J \equiv \star$. Since $\mathrm{SO}^{\uparrow}(3,1)$ commutes with $J \equiv \star$, we obtain a group homomorphism from $\mathrm{SO}^{\dagger}(3,1)$ to $\mathrm{SO}(3,\mathbf{C})$, the subgroup of $\mathrm{GL}_{\mathbf{C}}(\Lambda^2\mathbf{R}(3,1))$ (where $J \equiv \star$) that fixes ε.

THE CARTAN–DIEUDONNÉ THEOREM

Given $u \in V$ nonnull, then $u^{\perp}$ is a nondegenerate hyperplane and by Corollary 2.40, each vector $x \in V$ has a unique orthogonal decomposition $x = w + \lambda u$ with $w \in u^{\perp}$ and $\lambda \in \mathbf{R}$.

Definition 4.22. *The map R_u defined by $R_u x \equiv w - \lambda u$ is called the reflection through (the hyperplane) $u^{\perp}$ or reflection along u. That is, R_u is the identity on $u^{\perp}$ and minus the identity on the line through u. Therefore,*

$$R_u x = x - \frac{2\langle x, u \rangle}{\|u\|} u.$$

Note that $R_u \in O(V)$, $\det R_u = -1$, and $R_u^2 = I$.

By definition, the identity $I \in O(V)$ is said to be the product of $r = 0$ reflections along lines.

Theorem 4.23 (Cartan–Dieudonné).

(a) (Weak Form) *Each orthogonal transformation $A \in O(V)$ can be expressed as the product of a finite number r of reflections*

(4.24) $$A = R_{u_1} \circ \cdots \circ R_{u_r}$$

along nonnull lines, so that $\|u_j\| \neq 0$ for all $j = 1, \ldots, r$.

(b) (Strong Form) *The number of reflections required is at most n; that is, $r \leq n$.*

(c) (Sherk Version) *First note that the fixed set (or axis of revolution) $F_A \equiv \{x \in V : Ax = x\} = \ker(A - I)$ is always orthogonal to $\mathrm{image}(A - I)$. Now suppose that (4.24) is any representation of A (as the product of r reflections) with r minimal. Then*

   (1) *If $A - I$ is not skew, then $r = n - \dim F_A$. In fact, $u_1, \ldots, u_r$ is a basis for $\mathrm{image}(A - I)$.*

   (2) *If $A - I$ is skew, then $r = n - \dim F_A + 2$, but F_A must be totally null so $\dim F_A \leq n/2$. This case cannot occur if V is positive or negative definite, or positive or negative Lorentzian.*

Remark 4.25. The part of this theorem that will be used later to study the Spin and Pin groups is simply the weak form: each $A \in O(V)$ can be expressed as a (finite) product of reflections. Consequently, some readers may wish to skip the proof of parts (b) and (c).

The proof of the theorem begins with the case of dimension $n = 2$. As a prelude to Chapter 6 on normed algebras, this case provides us with an excuse to discuss the complex numbers $\mathbf{C}$ and the less familiar Lorentz numbers $\mathbf{L}$. In order to emphasize the analogy between $\mathbf{C}$ and $\mathbf{L}$, a brief discussion of the well-known complex numbers is included (for later comparison with $\mathbf{L}$).

The Complex Numbers C

The complex numbers $\mathbf{C}$ are defined to be $\mathbf{R}(2, 0)$ with the extra structure of multiplication, given by

$$(a, b)(c, d) \equiv (ac - bd, ad + bc).$$

Let $1 \equiv (1, 0)$ and $i \equiv (0, 1)$, so that $(a, b) = a + bi$ and $i^2 = -1$. *Conjugation* is defined by $\overline{z} = a - ib$ for $z = a + ib$. Note that $\overline{zw} = \overline{z}\,\overline{w}$, $z\overline{z} = \|z\|$, and hence $\|zw\| = \|z\|\,\|w\|$. If $z \neq 0$, then $z^{-1} = \overline{z}/\|z\|$, so that $\mathbf{C}$ is a (commutative) field. Also, note that

$$\langle z, w \rangle = \operatorname{Re} z\overline{w} \equiv \frac{1}{2}\left(z\overline{w} + \overline{z}w\right).$$

Let $e^{i\theta} = \cos\theta + i\sin\theta$ denote a point on the unit circle and note that $M_{e^{i\theta}}$, multiplication by $e^{i\theta}$, is an orthogonal transformation since $\|e^{i\theta}\| = 1$. As a 2×2 real matrix,

$$M_{e^{i\theta}} = \begin{pmatrix} \cos\theta & -\sin\theta \\ \sin\theta & \cos\theta \end{pmatrix},$$

so that $M_{e^{i\theta}} \in \mathrm{SO}(2)$. Since $M_{e^{i\theta}e^{i\psi}} = M_{e^{i(\theta+\psi)}}$, the map $\theta \mapsto M_{e^{i\theta}}$ induces the group isomorphism,

$$\frac{\mathbf{R}}{2\pi\mathbf{Z}} \cong \mathrm{SO}(2).$$

Lemma 4.26. *Suppose $A \in O(2)$ and define θ by $A1 = e^{i\theta}$. If $\det A = 1$, then*

$$(4.27) \qquad\qquad R_{ie^{i\theta/2}} \circ A = R_i,$$

where R_i (reflection along i) is just conjugation $Cz = \overline{z}$. If $\det A = -1$, then

$$(4.28) \qquad\qquad R_{ie^{i\theta/2}} \circ A = \mathrm{Id}.$$

Proof: It suffices to show that $R_{ie^{i\theta}/2} \circ A$ fixes 1, because R_i and Id are the only two orthogonal transformations fixing 1. The reflection of $e^{i\theta} = A1$ along $ie^{i\theta/2}$ (or through span $e^{i\theta/2}$) is 1 since both unit vectors $e^{i\theta}$ and 1 have the same inner product with $e^{i\theta/2}$, namely, cos $\theta/2$. ∎

Corollary 4.29. *Suppose $A \in SO(2)$. Then*

$$(4.30) \qquad\qquad A = R_{ie^{i\theta/2}} \circ R_i$$

is the product of two reflections. Moreover, $A = M_{e^{i\theta}}$.

Suppose $A \in O(2)$ and det $A = -1$. Then

$$(4.31) \qquad\qquad A = R_{ie^{i\theta/2}}$$

is a reflection. Also, $A = C \circ M_{e^{-i\theta}}$.

Proof: Solving (4.27) and (4.28) for A, we obtain A as the product of reflections proving the first equalities in (4.30) and (4.31). To prove $A = M_{e^{i\theta}}$, note that both satisfy det $= 1$ and map 1 to $e^{i\theta}$ so that both satisfy (4.27). The proof of $A = C \circ M_{e^{-i\theta}}$ is similar. ∎

The Lorentz Numbers L

The Lorentz numbers **L** are defined to be the inner product space $\mathbf{R}(1,1), \langle\,,\,\rangle$ with the extra structure of multiplication given by

$$(a,b)(c,d) \equiv (ac + bd, ad + bc).$$

Let $1 \equiv (1,0)$ and $\tau \equiv (0,1)$, so that $(a,b) = a + b\tau$ and $\tau^2 = 1$. Conjugation is defined by $\overline{z} = a - b\tau$ for $z = a + b\tau$. Note that $\overline{zw} = \overline{z}\,\overline{w}$ and $z\overline{z} = \|z\|$, so that $\|zw\| = \|z\|\,\|w\|$. Thus, if $\|z\| \neq 0$ (z nonnull), then $z^{-1} = \overline{z}/\|z\|$ exists, while for $\|z\| = 0$ (z null) z cannot have an inverse. Also note that $\langle z, w \rangle = \operatorname{Re} z\overline{w} \equiv \frac{1}{2}(z\overline{w} + \overline{z}w)$.

Let $e^{\tau\theta} = \cosh\theta + \tau\sinh\theta$ (calculate the formula power series for $e^{\tau\theta}$ to see that this definition is appropriate). Note that $M_{e^{\tau\theta}}$, multiplication by $e^{\tau\theta}$, is an orthogonal transformation since $\|e^{\tau\theta}\| = 1$. As a 2×2 real matrix,

$$M_{e^{\tau\theta}} = \begin{pmatrix} \cosh\theta & \sinh\theta \\ \sinh\theta & \cosh\theta \end{pmatrix},$$

so that det $M_{e^{\tau\theta}} = 1$. Define a timelike vector $z = a + b\tau$ to be *future timelike* if $b > 0$. Since $M_{e^{\tau\theta}}(\tau) = \sinh\theta + \tau\cosh\theta$, multiplication by

$e^{\tau\theta}$ preserves the futuretime cone. Thus, $M_{e^{\tau\theta}} \in \mathrm{SO}^\uparrow(1,1)$. In fact, since $e^{\tau\theta}e^{\tau\psi} = e^{\tau(\theta+\psi)}$, the map sending $\theta \mapsto M_{e^{\tau\theta}}$ determines the group isomorphism

(4.32) $$\mathbf{R} \cong \mathrm{SO}^\uparrow(1,1).$$

Note that the set of unit spacelike vectors $\{z = a + b\tau : \|z\| = a^2 - b^2 = 1\}$ consists of two disjoint curves $\{e^{\tau\theta} : \theta \in \mathbf{R}\}$ and $\{-e^{\tau\theta} : \theta \in \mathbf{R}\}$. Therefore, given $A \in O(1,1)$, the unit spacelike vector $A1$ uniquely determines an angle θ by $A1 = \pm e^{\tau\theta}$.

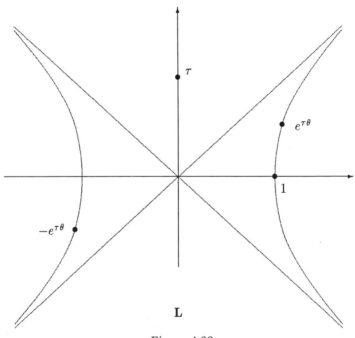

Figure 4.33

Lemma 4.34. *Suppose* $A \in O(1,1)$ *and* $A1 = \pm e^{\tau\theta}$ *defines* $\theta \in \mathbf{R}$.

(a) *If* $A1 = e^{\tau\theta}$ *and* $\det A = 1$, *then*

$$R_{\tau e^{\tau\theta/2}} \circ A = R_\tau.$$

(b) *If* $A1 = -e^{\tau\theta}$ *and* $\det A = 1$, *then*

$$R_{\tau e^{\tau\theta/2}} \circ A = R_1.$$

(c) If $A1 = e^{\tau\theta}$ and $\det A = -1$, then

$$R_{\tau e^{\tau\theta/2}} \circ A = \mathrm{Id}.$$

(d) If $A1 = -e^{\tau\theta}$ and $\det A = -1$, then

$$R_{\tau e^{\tau\theta/2}} \circ A = -\mathrm{Id}.$$

Proof: The reflection of $e^{\tau\theta}$ along the line $\tau e^{\tau\theta/2}$ (or through the line $e^{\tau\theta/2}$) is 1 since both $e^{\tau\theta}$ and 1 have the same inner product with $e^{\tau\theta/2}$ (since $\overline{e^{\tau\theta}} = e^{-\tau\theta}$ and $\langle e^{\tau\theta}, e^{\tau\theta/2}\rangle = \mathrm{Re}\, e^{\tau\theta}e^{-\tau\theta/2} = \langle e^{\tau\theta/2}, 1\rangle$). Thus, $(R_{\tau e^{\tau\theta/2}} \circ A)(1) = \pm 1$, where $A1 = \pm e^{\tau\theta}$ has the same $\pm$ sign. Since $R_{\tau e^{\tau\theta/2}} \circ A$ is orthogonal, it must also map the τ-axis into itself and hence be $\pm$ Id on the τ-axis. The four cases now follow. ■

Corollary 4.35. *Suppose $A \in O(1,1)$. If $\det A = 1$, then A is the product of two reflections. Either*

$$A = R_{\tau e^{\tau\theta/2}} \circ R_\tau \quad \text{or} \quad A = R_{\tau e^{\tau\theta/2}} \circ R_1.$$

If $\det A = -1$, then A is a reflection. Either

$$A = R_{\tau e^{\tau\theta/2}} \quad \text{or} \quad A = R_{e^{\tau\theta/2}}.$$

This completes the proof of Theorem 4.23 in the special case of dimension two using the complex numbers **C** and the Lorentz numbers **L**.

General Proof of Theorem 4.23: The general proof proceeds by induction on the dimension n of V. Given $A \in O(V)$, there are cases to consider.

Case 1: A is said to have a *nonnull axis of rotation* if there exists a nonnull vector x with $Ax = x$.

Proof of Case 1: In this case, the fixed set F_A can be written as $F_A = N \oplus H$, where N is totally null and H is nondegenerate. Since A is the identity on H, A maps $H^\perp$ to $H^\perp$. By the induction hypothesis, the theorem is true for $A|_{H^\perp} \in O(H^\perp)$. In particular, if (4.24) is valid for $A|_{H^\perp} \in O(H^\perp)$, then the same representation is valid for $A \in O(V)$. Note that R_u, reflection along a line u in $H^\perp$, naturally extends to reflection along the same line u, but now considered a line in V.

Case 2: Suppose $u \equiv Av - v$ and v are nonnull vectors. Then $R_u A$ fixes v.

Proof of Case 2: Note that $u \equiv Av - v$ and $w \equiv Av + v$ are orthogonal since $A \in O(V)$. Therefore, Av has the orthogonal decomposition

$$(4.36) \qquad\qquad Av = \frac{1}{2}u + \frac{1}{2}w$$

with respect to $V \equiv \operatorname{span} u \oplus (\operatorname{span} u)^{\perp}$. Thus, $R_u Av = -\frac{1}{2}u + \frac{1}{2}w = v$ fixes v. By the induction hypothesis, $R_u A|_{\operatorname{span} u^{\perp}}$ can be expressed as the product of $\leq n-1$ reflections. Therefore A can be expressed as the product of $\leq n$ reflections.

Now the proof of the weak form (i.e., each $A \in O(V)$ is the product of a finite number of reflections) can be completed. Pick any nonnull vector $v \in V$. If $u \equiv Av - v$ is nonnull, then $R_u A|_{\operatorname{span} v^{\perp}}$, and hence $R_u A$, can be expressed as a product of a finite number of reflections. If $w \equiv Av + v = -((-A)v - v)$ is nonnull, then $R_w(-A)|_{\operatorname{span} v^{\perp}}$, and hence $R_w(-A)$, can be expressed as a product of a finite number of reflections. Finally, note that both u and w cannot be null, since they are orthogonal and $v = -\frac{1}{2}u + \frac{1}{2}w$.

Before preceeding with Case 3, we need to make several observations. Recall that for any linear map $L \in \operatorname{End}(V)$,

$$(4.37) \qquad\qquad \operatorname{image} L = (\ker L^*)^{\perp}$$

and

$$(4.38) \qquad\qquad \dim \operatorname{image} L + \dim \ker L = n.$$

Since $A^* = A^{-1}$, the equations $Ax - x$ and $A^*x = x$ are equivalent. Thus, $\ker(A - I) = \ker(A^* - I)$. Combined with (4.35) this proves that

$$(4.39) \qquad\qquad \operatorname{image}(A - I) = F_A \quad \text{for } A \in O(V).$$

Note that if $A \in O(V)$ does not belong to Case 2, then $u \equiv Av - v$ is null for each nonnull vector v. By continuity, this implies that each $u \equiv Av - v$ is null. Therefore, if $A \in O(V)$ does not belong to Case 2, then $\operatorname{image}(A - I)$ is totally null. If Case 1 is not applicable, then $F_A = \ker(A - I)$ is totally null.

Case 3: Suppose both $\operatorname{image}(A - I)$ and $\ker(A - I)$ are totally null.

Proof of Case 3: Since by (4.38) each is the orthogonal compliment of the other, they must be equal, i.e.,

$$N \equiv \operatorname{image}(A - I) = \ker(A - I).$$

By (4.38) this implies that $\dim V = 2p$, where $p = \dim N$.

Since N is totally null, Proposition 2.45 implies that V has split signature p, p.

Note that A is the identity on N. Pick M complimentary to N. Given $y \in M, Ay = (Ay - y) + y$ is the direct sum decomposition for $Ay \in V = N \oplus M$ since $Ay - y \in N$. Let $b : M \to N$ denote $A - I$ restricted to M. Then $V = N \oplus M$ determines the 2×2 blocking of A as

$$(4.40) \qquad A = \begin{pmatrix} I & b \\ 0 & I \end{pmatrix}.$$

Therefore $\det A = 1$. This proves that if A belongs to Case 3, then $\det A = 1$.

Choose any reflection R through a nondegenerate hyperplane. Then $\det RA = -1$. Thus, RA must be in Case 1 or Case 2. Therefore $RA = R_1 \cdots R_k k \leq n$. Now $\det RA = -1$ implies k is odd. Since $n = 2p$ is even, we must have $k \leq n - 1$. Therefore $A = RR_1 \cdots R_k$ with $k \leq n - 1$. This completes the proof of what is usually called the Cartan–Dieudonné Theorem, i.e., part (b).

Remark. By choosing a *null basis* for $V = N \oplus M$, i.e., $n_1, \ldots, n_p$ a basis for N and $M = \operatorname{span}\{m_1, \ldots, m_p\}$ totally null with the planes $\operatorname{span}\{n_i, m_i\}$ nondegenerate and orthogonal, one can show that $A \in O(V)$ belongs to Case 3 if and only if A is of the form (4.40) with b a skew matrix. See Problem 10 for an example of an orthogonal transformation $A \in O(V)$ that belongs to Case 3 (dimension of V equals four).

Now we complete the proof of the theorem. Since A satisfies part (c) of the theorem if and only if $A|_{H^\perp}$ satisfies part (c), we may assume that the fixed point set F_A is totally null. Several statements equivalent to "$L \equiv A - I$ is skew" are provided in Problem 11. In particular, $A - I$ is skew if and only if $\operatorname{image}(A - I) \subset \ker(A - K) \equiv F_A$. Therefore, because of the assumption that F_A is totally null, the cases where $A - I$ is skew fall under Case 3. This case is discussed in Problem 11.

Now assume that $A - I$ is not skew, F_A is totally null, and hence Case 2 is applicable. Suppose $j \equiv Lv$ is nonnull with $L \equiv A - I$ and $\|v\| \neq 0$. Then $R_u A$ fixes v.

We must show that, for some choice of u, $R_u A - I$ is not skew:

$$(4.41) \qquad \begin{array}{l} \text{If } R_u A - I \text{ is skew then, for all } v \in V, \\ \|Lv\|(L + L^*)(y) = -4\langle Lv, y \rangle \, \langle L^* v, y \rangle \text{ for all } y \in V. \end{array}$$

Proof of (4.41): Assume $u \equiv Lv$ is nonnull. Note

$$R_u = I - 2\frac{\langle u, \cdot \rangle u}{\|u\|\|}.$$

Therefore,

$$R_u A - I = L - 2\frac{\langle u, A(\cdot)\rangle u}{\|u\|} = L - 2\frac{\langle Lv, A(\cdot)\rangle Lv}{\|u\|}.$$

This gives

(4.42) $$R_u A - I = L + 2\frac{\langle v, L(\cdot)\rangle Lv}{\|u\|}$$

since $L^* A = (A^{-1} - I)A = I - A = -L$. Thus, $R_u A - I$ is skew if and only if

(4.43) $$(L + L^*)(y) + 4\|u\|^{-1}\langle L^* v, y\rangle \langle Lv, y\rangle = 0.$$

Since (4.41) is valid for all $v \in$ with $\|Lv\| \neq 0$, by continuity it is valid for all $v \in V$.

Note that $\{v \in V : Lv = v\} = \{v \in V : L^* v = V\} = F_A$. Consequently, if $R_u A$ is skew, then

(4.44) $$Lv \text{ null implies } Lv = 0,$$

or each nonzero $u \in$ image L is also nonnull. Consequently, the subspace image $L \equiv \text{image}(A - I)$ is either positive or negative definite. In this case, the theorem is valid for A. Therefore, we may choose $u \equiv Av - v$ with u and v nonnull so that $R_u A - I$ is not skew.

It remains to show that

(4.45) $$\text{rank}(R_u A - I) < \text{rank}(A - I).$$

Now $R_u A$ still fixes F_A since $u \in \text{Im}(A - I) = F_A\perp$. In addition, $R_u A$ fixes v and $v \notin F_A$. Thus

$$\dim F_{R_u A} > \dim F_A,$$

which proves (4.45). This completes the proof of Theorem 4.23. ∎

Definition 4.46. *Given a nondegenerate subspace W of V, reflection along W (or through $W^\perp$), denoted R_W, is defined to be -1 on W and $+1$ on $W^\perp$. If $u_1, \ldots, u_k$ is a basis for W, then reflection along W will also be denoted by $R_{u_1 \wedge \cdots \wedge u_k}$.*

Just as $O(V)$ is generated by reflections along lines, $SO(V)$ is generated by reflections along planes.

Theorem 4.47. *If* $n \equiv \dim_{\mathbb{R}} V \geq 3$, *then each* $f \in SO(V)$ *can be expressed as the product of an even number* $k \leq n$ *of reflections along non-degenerate planes:*

$$f = R_{u_1 \wedge v_1} \circ \cdots \circ R_{u_k \wedge v_k}, \quad k \text{ even and } \leq n.$$

Proof: By the Cartan–Dieudonné Theorem, $f = R_{u_1} \circ \cdots \circ R_{u_k}$ with each $u_k \in V$ nonnull and with k even. If $n = 3$, then either $k = 0$ and $f = 1$, or, $k = 2$ and $f = R_{u_1} \circ R_{u_2}$. Now $-R_u = R_{v \wedge w}$, where $\mathrm{span}(v \wedge w) = (\mathrm{span}\ u)^\perp$. Thus

$$f = R_{u_1} \circ R_{u_2} = (-R_{u_1}) \circ (-R_{u_2}) = R_{v_1 \wedge w_1} \circ R_{v_2 \wedge w_2}.$$

If $n \geq 4$, then given a pair of nonnull vectors u_1, u_2 there exists a nonnull vector v orthogonal to both u_1 and u_2 (see Problem 4.10). Therefore,

$$f = R_{u_1} \circ R_{u_2} = R_{u_1} \circ R_v \circ R_v \circ R_{u_2} = R_{u_1 \wedge v} \circ R_{u_2 \wedge v}$$

as desired. ∎

GRASSMANIANS AND $SO^\dagger(p, q)$, THE REDUCED SPECIAL ORTHOGONAL GROUPS

Suppose $V, \langle\, , \,\rangle$ is an $\mathbb{R}$-symmetric inner product space with signature p, q that is not definite, i.e., $p, q \geq 1$. Let $G_+(p, V)$ denote the *grassmanian* of all oriented p-planes through the origin that are maximally positive. Given $\xi \in G_+(p, V)$, represent $\xi \equiv e_1 \wedge \cdots \wedge e_p \in \Lambda^p V$ by choosing an oriented orthonormal basis for V. Let $P \equiv \mathrm{span}\ \xi$ denote the p-plane without orientation.

If $\xi \in G_+(p, V)$ is given, then each $\eta \in G_+(P, V)$ can be graphed over $P \equiv \mathrm{span}\ \xi$ because π, the orthogonal projection onto P along $P^\perp$, is injective when restricted to $Q \equiv \mathrm{span}\ \eta$ (ker $\pi \cap Q = \{0\}$). Consequently, there is a well-defined notion of when ξ, η are *compatibly oriented*. If $e_1, \ldots, e_p$ is an oriented orthonormal basis for ξ and $Q \equiv \mathrm{span}\ \eta$ is the graph of a linear map $A : P \to P^\perp$, then either $\{e_1 + Ae + 1, \ldots, e_p + Ae_p\}$ or $\{-e_1 - Ae_1, e_2 + Ae_2, \ldots, e_p + Ae_p\}$ is an oriented basis for η. In the first case, we say that ξ and η have the same (*spacelike*) *orientation*. Note that in terms of the natural inner product $\langle\, , \,\rangle$ on $\Lambda^p V$, ξ and η have the same orientation if and only if $\langle \xi, \eta \rangle$ is positive and the opposite orientation if and only if $\langle \xi, \eta \rangle$ is negative. Thus, $G_+(p, V)$ is naturally divided into two sets.

It is also easy to see that $G_+(p, V)$ has two connected components. Shrink the graphing map A to zero. That is, if ξ and η are compatibly oriented with $Q \equiv \text{span } \eta$ the graph of A over $P \equiv \text{span } \xi$, then $\eta_t \equiv$ graph tA defines a path in $G_+(p, V)$ connecting η to ξ.

Similarly, the grassmannian, $G_-(q, V)$, of oriented maximal negative subspaces of V has two connected components. Selecting $G_+^\uparrow(p, V)$ as one of the two components of $G_+(p, V)$ is called the choice of *spacelike orientation*, while selecting $G_-^\uparrow(q, V)$ as one of the two components of $G_-(q, V)$ is called the choice of *timelike orientation* (or *future orientation*).

An orthogonal transformation $A \in O(V)$ is said to preserve the spacelike orientation if A preserves the two components of the grassmannian $G_+(p, V)$ of oriented maximal positive subspaces. An orthogonal transformation $A \in O(V)$ is said to preserve the timelike orientation if A preserves the two components of the grassmannian $G_-(q, V)$ of oriented maximal negative subspaces. In this context, $A \in O(V)$ with $\det A = 1$ is said to preserve the *total orientation* of V.

Definition 4.48. *The orthogonal group $O(V)$ in the cases that are not definite $(p, q \geq 1)$ has subgroups defined by*

(4.49) $SO(V) \equiv \{A \in O(V) : A$ *preserves the total orientation*$\}$,

(4.50) $O^+(V) \equiv \{A \in O(V) : A$ *preserves the spacelike orientation*$\}$,

(4.51) $O^-(V) \equiv \{A \in O(V) : A$ *preserves the timelike orientation*$\}$.

(The notation $O^\uparrow(V)$ is also used for $O^-(V)$. The arrow indicates that the direction of time is preserved.)

The intersection of any two of these three subgroups is the same group and is denoted by

(4.52) $SO^\uparrow(V) \equiv \{A \in O(V) : A$ preserves the total, spacelike, and timelike orientation$\}$.

$O^+(V)$ is called the *spacelike reduced orthogonal group*, $O^-(V)$ is called the *timelike reduced orthogonal group*, and $SO^\uparrow(V)$ is called the *reduced special orthogonal group*.

Proposition 4.53.

(a) $O^+(p, q)/SO(p) \times O(q) \cong G_+^\uparrow(p, V)$.

(b) $O^-(p, q)/O(p) \times SO(q) \cong G_-^\uparrow(q, V)$

The straightforward proof is omitted.

In summary, the group $O(p, q)$ (with signature that is not definite) has four connected components given by the four possibilities for $A \in O(V) : A$ preserves (or reverses) space orientation and A preserves (or reverses) time orientation.

Lemma 4.54. *Suppose $A \in O(V)$ is expressed as the product of reflections* $A = R_{u_1} \circ \cdots \circ R_{u_r}$.

(a) *The number of spacelike vectors in $\{u_1, \ldots, u_r\}$ is even if and only if $A \in O^+(V)$.*

(b) *The number of timelike vectors in $\{u_1, \ldots, u_r\}$ is even if and only if $A \in O^-(V)$.*

In particular, the parity of the number of reflections along spacelike lines (or along timelike lines) is independent of the representation of A as the product of reflections.

Definition 4.55. *Given $A \in O(V)$, let $\|A\|_S$ denote the parity of the number of reflections along timelike vectors when A is expressed as the product of reflections.*

Note that

$$(4.56) \qquad \| \cdot \|_S : O(V) \to \mathbf{Z}_2$$

is a group homomorphism. The number $\|A\|_S$ is called the *spinorial norm* of A.

The two group homomorphisms $\det : O(V) \to \mathbf{Z}_2$ and $\| \cdot \|_S : O(V) \to \mathbf{Z}_2$ distinguish the four components of $O(V)$. In addition, each of the three groups

$$SO(V), \quad O^\dagger(V), \quad \text{and} \quad O^+(V) \quad \blacksquare$$

consist of the identity component $SO^\dagger(V)$ plus one additional component given by

$$(4.57) \qquad \det = 1 \quad \text{and} \quad \| \cdot \|_S = -1 \quad \text{for } SO(V),$$

$$(4.58) \qquad \det = -1 \quad \text{and} \quad \| \cdot \|_S = 1 \quad \text{for } O^-(V) \equiv O^\dagger(V),$$

$$(4.59) \qquad \det = -1 \quad \text{and} \quad \| \cdot \|_S = -1 \quad \text{for } O^+(V).$$

PROBLEMS

1. (a) Verify (4.3) directly from the definition of the inner product on $\Lambda^2 V$.

   (b) Show that $\Lambda^2 V$ has signature $\binom{p}{2} + \binom{q}{2}, pq$. In particular, $\Lambda^2 V$ has signature $3, 3$ if V has signature $3, 1$ or $1, 3$.

2. (A canonical form for $O(n)$ acting on $\mathfrak{o}(n)$) Given $\alpha \in \Lambda^2 \mathbf{R}^*(n)$, find an orthonormal basis $\omega^1, \ldots, \omega^n$ for $\mathbf{R}^*(n)$ and $\lambda_1 \geq \lambda_2 \geq \cdots \geq \lambda_r > 0$, so that

$$\alpha = \lambda_1 \omega^1 \wedge \omega^2 + \cdots + \lambda_r \omega^{2r-1} \wedge \omega^{2r}.$$

Hint: If the maximum value λ_1 of $\alpha(x \wedge y)$ taken over all $|x \wedge y| = 1$ is attained at $e_1 \wedge e_2$, then show that $\alpha(e_1 \wedge u) = 0$ for all $u \in \mathbf{R}(n)$ with $u \perp e_2$ by considering the function $\alpha(e_1 \wedge (\cos\theta \, e_2 + \sin\theta \, u))$ of θ.

3. Give the proof of the twin paradox, Proposition 4.19.

4. Suppose V is a time-oriented Lorentzian vector space. The *backwards triangle inequality* says that $|u + v| \geq |u| + |v|$ for all future timelike vectors, and equality holds if and only if u and v are positive multiples of one another. Deduce this result

    (a) algebraically from the backwards C.–S. Inequality;

    (b) as a special case of the twin paradox (for piecewise smooth curves).

    Given a, b with $b - a$ future timelike, show that there exist (piecewise smooth) particles from a to b with arbitrarily small proper time.

5. Suppose $V, \| \ \|$ is a Lorentzian vector space. If $\| \cdot \|'$ is a square norm on V with $\|x\|' \leq M$ for all unit timelike vectors, then show that $\| \cdot \|' = c \| \cdot \|$ for some constant c.

6. (a) Show that $\overline{zw} = \overline{z}\,\overline{w}$ and that $z\overline{z} = \overline{z}\,z = \|z\|$ for all $z, w \in \mathbf{L}$.

    (b) Show that
    $$\mathrm{SO}(1,1) = \left\{ \pm M_{e^{\tau\theta}} : \theta \in \mathbf{R} \right\}.$$

    (c) Let $Cz \equiv \overline{z}$ denote conjugation on $\mathbf{L}$. Show that $C \in O(1,1)$ and that
    $$O(1,1) = \left\{ \pm M_{e^{\tau\theta}} : \theta \in \mathbf{R} \right\} \cup \left\{ \pm C M_{e^{\tau\theta}} : \theta \in \mathbf{R} \right\}.$$

7. (*Double Numbers*) Consider the algebra $\mathbf{R} \oplus \mathbf{R}$ of diagonal 2×2 matrices
    $$\begin{pmatrix} u_1 & 0 \\ 0 & u_2 \end{pmatrix},$$
    with matrix multiplication, and with the inner product determined by the quadratic form $u_1 u_2$. Define conjugation by
    $$C \begin{pmatrix} u & 0 \\ 0 & v \end{pmatrix} = \begin{pmatrix} v & 0 \\ 0 & u \end{pmatrix}.$$

    Exhibit an isomorphism between $\mathbf{L}$ and $\mathbf{R} \oplus \mathbf{R}$ preserving the multiplications, the inner products, and the conjugations. Exhibit $e^{\tau\theta}$ as a 2×2 matrix.

8. Suppose V is a euclidean vector space of signature p, q. Recall that the *unit sphere* is defined to be $\{u \in V : |u| = 1\}$. Show that if x is a point on the unit sphere, then x is normal to the unit sphere at x.

9. Consider $\mathbf{R}(4,0) \cong \mathbf{H}$.

(a) Show that reflection along a line spanned by $a \in \mathbf{H}$ with $|a| = 1$ is given by $R_a x = -a\bar{x}a$.

(b) Use part (a) to give a proof that $\chi : S^3 \times S^3 \to SO(4)$ defined by $\chi_{a,b}(x) = ax\bar{b}$ for all $x \in \mathbf{H}$, as in Remark 1.33, is a surjection.

10. (a) Exhibit a pair of nonnull vectors $u_1, u_2 \in \mathbf{R}(2,1)$ that do not have a nonnull vector $v \in \mathbf{R}(2,1)$ orthogonal to both of them.

(b) If $n \equiv \dim V \geq 4$, show that each pair u_1, u_2 of nonnull vectors has a nonnull vector $v \in V$ orthogonal to both of them.

(c) Find $A \in SO(2,1)$ with the fixed axis $F_A \equiv \{x \in \mathbf{R}(2,1) : Ax = x\}$ a null line.

11. Let $V \equiv M_2(\mathbf{R})$ with $\|v\| \equiv \det v$. Let $A \in \mathrm{End}(V)$ denote left multiplication by the matrix $\begin{pmatrix} 1 & 1 \\ 0 & 1 \end{pmatrix}$.

(a) Show that $A \in SO(V)$.

(b) Show that image $(A - \mathrm{Id})$ is totally null of dimension 2.

(c) Show that $F_A \equiv \{x \in V : Ax = x\}$ is totally null of dimension 2.

(d) Show that A cannot be written as the product of less than four reflections along (nonnull) lines.

12. Given $A \in O(V)$, show that the following are equivalent:

(a) $A - I$ is skew,

(b) $\frac{1}{2}(A + A^*) = I$,

(c) A^{-1} is the identity on image$(A - I)$,

(d) image$(A - I) \subset F_A \equiv \ker(A - I)$.

(e) $F_A^{\perp} \subset F_A$.

13. Suppose $A, B \in O(V)$ both leave the subspace $F \subset V$ fixed and agree on $F^{\perp}$.

(a) Show that $A = B$ if V is positive definite or Lorentzian.

(b) Find a counterexample to $A = B$ if $V \equiv \mathbf{R}(2,2)$.

14. Suppose $V, \langle \, , \, \rangle$ is positive definite euclidean space of dimension n. Given a subspace W of V, let R_W denote reflection along W through $W^{\perp}$ and let P_W denote orthogonal projection from V onto W. Show that $P_W = \frac{1}{2}(I - R_W)$ or $R_W = I - 2P_W$.

15. Let $G(k, V)$ denote the *grassmanian of* (not necessarily oriented) k-*planes in* V through the origin.

(a) Show that $G(k, \mathbf{R}^n)$, the grassmannian of non-oriented k-planes in $\mathbf{R}^n$, can be identified with a subset (actually a submanifold) of the vector space of self-adjoint linear maps $\mathrm{Sym}(\mathbf{R}^n) \equiv \{A \in \mathrm{End}_\mathbf{R}(\mathbf{R}^n) : A^* = A\}$. More precisely, identify $G(k, \mathbf{R}^n)$ with

$$\left\{ A \in \mathrm{Sym}(\mathbf{R}^n) : A^2 = A \text{ and } \mathrm{trace}_\mathbf{R}\ A = k \right\}.$$

(b) Similarly, identify $G(k, \mathbf{C}^n)$, the *grassmannian of complex k-planes in $\mathbf{C}^n$*, with

$$\left\{ A \in \mathrm{Herm}(\mathbf{C}^n) : A^2 = A \text{ and } \mathrm{trace}_\mathbf{C}\ A = k \right\}.$$

(c) Identify the *grassmannian of quaternionic k-planes in $\mathbf{H}^n$*, denoted $G(k, \mathbf{H}^n)$ with the set of projection operators

$$\left\{ A \in \mathrm{Herm}(\mathbf{H}^n) : A^2 = A \text{ and } \mathrm{trace}_\mathbf{R}\ A = 4k \right\}.$$

In the special case $k = 1$, $G(k, F^n)$ is called *projective space* and denoted $\mathbf{P}^n(F) \equiv G(1, F^n)$—see Chapter 5 and Problem 6.16 for more information.

16. Suppose $V, \|\cdot\|$ has split signature p, p. Given N, totally null of dimension p, show that there exist M complimentary to N, also totally null of dimension p, and an isometry from V to $\mathbf{R}^p \times \mathbf{R}^p$, with square norm $\|(x, y)\| = x_1 y_1 + \cdots + x_p y_p$, which sends N to $\mathbf{R}^p \times \{0\}$ and M to $\{0\} \times \mathbf{R}^p$.

17. Show that each $A \in O(n, \mathbf{C})$ can be expressed as the product

$$A = R_{u_1} \circ \cdots \circ R_{u_r} \quad \text{with } r \leq n,$$

where each u_j is nonnull. The complex reflection R_u along a nonnull line u is defined by exactly the same formula as in the real case.

5. Differential Geometry

This chapter provides a brief description of several types of geometries. Its purpose is to add motivation for the other chapters. As such, it may be either browsed over quickly or studied carefully in conjunction with some of the standard texts in geometry. The chapter should only be considered as introductory and motivational, despite the fact that the reader is assumed to be familiar with the notion of a manifold.

While the novice in geometry may wish to use this chapter as a reference point for further study, the more experienced student may be disappointed in its introductory nature, especially after reading the following table of geometries.

Each of the geometries discussed is labeled with a particular group (see Table 5.1). However, a particular group may be associated with one, two, or even three different but closely related geometries. The connection between the group and the geometry is meant only to be suggestive. A deeper relationship is provided by the notions of principle G-bundles, reduction of the structure group, and integrability. The reader is encouraged to pursue these topics.

REAL n-MANIFOLDS: THE GROUP $GL(n, R)$

Briefly, for the sake of completeness, the general notion of a (*smooth*) real *n-dimensional manifold M* is defined as follows.

Definition 5.2. *M must be a topological space equipped with a countable atlas that is smoothly compatible. An atlas is a collection of charts, where in turn each chart consists of two things: an open subset of M and a homeomorphism of this open set with an open subset of $\mathbf{R}^n$. Of course, each point of M is required to belong to at least one chart. Two overlapping charts induce a homeomorphism from one open set in $\mathbf{R}^n$ onto another, called a transition function. The atlas is said to be smoothly compatible if each transition function is a diffeomorphism.*

This completes the definition of a *real n-manifold*. The group $GL(n, \mathbf{R})$ is associated with this geometry, since the linearization, at a point, of each transition function belongs to $GL(n, \mathbf{R})$. In later discussions in this chapter, the reader is assumed to be familiar with the notion of the tangent space $T_x M$ to M at a point $x \in M$.

Table 5.1

Manifold	Group
Real n-manifolds	$GL(n, \mathbf{R})$
Oriented real n-manifolds	$GL^+(n, \mathbf{R})$
Complex (and almost-complex) n-manifolds	$GL(n, \mathbf{C})$
Quaternionic (and almost-quaternionic) n-manifolds	$GL(n, \mathbf{H}) \cdot \mathbf{H}^*$
Manifolds with volume	$SL(n, \mathbf{R})$ and $SL(n, \mathbf{C})$
Riemnnian manifolds (of signature p, q)	$O(p, q)$
Conformal manifolds (of signature p, q)	$CO(p, q)$
Real symplectic manifolds	$Sp(n, \mathbf{R})$
Complex Riemannian n-manifolds	$O(n, \mathbf{C})$
Complex symplectic manifolds	$Sp(n, \mathbf{C})$
Kähler manifolds (of signature p, q)	$U(p, q)$
Special Kähler manifolds (signature p, q)	$SU(p, q)$
HyperKähler manifolds (of signature p, q)	$HU(p, q)$
Quaternionic Kähler manifolds (of signature p, q)	$HU(p, q) \cdot HU(1)$
Quaternionic skew hermitian manifolds	$Sk(\mathbf{H})$

ORIENTED REAL n-MANIFOLDS: THE GROUP $GL^+(n, \mathbf{R})$

Suppose M is a real n-manifold equipped with an atlas so that each transition function preserves orientation; that is, the linearization of each transition function, at each point, belongs to the subgroup $GL^+(n, \mathbf{R})$, consisting of all invertible matrices with positive determinant. Then M is called an *oriented real n-manifold.* Note that each tangent space $T_x M$ carries a natural orientation as follows. Choose any chart with open set containing x. Then the associated homeomorphism between this open set and an open subset of $\mathbf{R}^n$ induces an orientation on $T_x M$, and this orientation on $T_x M$ is independent of the choice of chart since the transition functions have derivatives in $\text{GL}^+(n, \mathbf{R})$.

An oriented real n-manifold may also be (equivalently) defined as follows. Assume M is a real n-manifold. Suppose there exists a never-vanishing n-form Ω on M (but Ω is not necessarily distinguished), then M is said to be *orientable*. In this case, any other never-vanishing n-form Ω' is a multiple f of Ω where f is a never-vanishing function. If f is positive, then we say Ω' and Ω are equivalent. Thus, on an orientable manifold there exist precisely two equivalence classes of never-vanishing n-forms. If a choice of one of these two equivalence classes is made, then M is called an *oriented manifold*.

COMPLEX (AND ALMOST COMPLEX) n-MANIFOLDS: THE GROUP $\text{GL}(n, \mathbf{C})$

The general notion of a complex n-manifold is defined by mimicking the definition of a real n-manifold, except $\mathbf{R}^n$ is replaced by $\mathbf{C}^n$ and the transition functions are required to be biholomorphisms (from an open subset of $\mathbf{C}^n$ to an open subset of $\mathbf{C}^n$). There are several useful, and of course equivalent, definitions of a holomorphic map. The simplest, from our point of view, requires that the linearization or derivative of the map, at each point, should be complex linear. Thus, a *complex n-manifold M* is, first of all, a real $2n$-dimensional manifold, but with the additional property that each transition function (from an open subset of $\mathbf{C}^n \cong \mathbf{R}^{2n}$ to another open subset of $\mathbf{C}^n \cong \mathbf{R}^{2n}$) has its derivative ($\equiv$ differential) in the subgroup $\text{GL}(n, \mathbf{C})$ of $\text{GL}(2n, \mathbf{R})$. The obvious choice is to label the geometry of all complex n-manifolds with the group $\text{GL}(n, \mathbf{C})$.

On a complex vector space T, complex scalar multiplication by i is a real linear map (denoted I) that squares to minus the identity. Conversely, given any real linear map I (on a real vector space) that squares to minus the identity, it is easy to convert the real vector space to a complex vector space by setting the given linear map I equal to multiplication by i. Such a map I is called a *complex structure*. Now, on a complex manifold M each tangent space $T_x M$ (considering M as a real $2n$-manifold) is naturally equipped with a complex structure due to the fact that the linearization of a transition function commutes with the complex structure on $\mathbf{C}^n$.

An *almost-complex n-manifold* is a real $2n$-manifold that is equipped with a complex structure I on each tangent space $T_x M$ (that varies smoothly with the point $x \in M$).

The differential forms (with complex coefficients) on an almost complex manifold are particularly interesting. The space of degree one forms (at a point) is naturally the direct sum of two vector spaces: the space $T^{*1,0}$ of forms of type $1,0$ and $T^{*0,1}$ the space of forms of type $0,1$ (see Problem 3). More generally the space of degree k forms, denoted $\Lambda^k T^*$, is the direct sum $\sum_{p+q=k} \oplus \Lambda^{p,q} T^*$, where $\Lambda^{p,q} T^*$ is the space of forms of bidegree p, q.

There are convenient nontrivial ways of testing whether a given almost complex n-manifold is a complex n-manifold, which are provided by the "Newlander–Nirenberg Theorem."

There is a further weakening of the concept of an almost-complex n-manifold that is not of much interest itself but does provide a basis for a useful quaternion analogue. A *"weak" almost-complex n-manifold* is a real $2n$-manifold that is equipped with two complex structures $\pm I$, one the negative of the other, on each tangent space $T_x M$ (and they should locally vary smoothly with x). The point is (see Problem 2) that there may not be any way globally to make a choice of I versus $-I$.

Remark 5.3. For the reader comfortable with the notion of a vector bundle on a manifold, the distinction between almost-complex and the "weaker form" of almost-complex can be described as follows: Let $\text{End}(TM)$ denote the endomorphism vector bundle on M, i.e., the vector bundle with fiber $\text{End}(T_x M)$ at each point $x \in M$. A two-dimensional subbundle C of $\text{End}(TM)$ is said to be a bundle of $\mathbf{C}$-algebras on M if, for each point x, the fiber C_x is real algebra isomorphic to $\mathbf{C}$. Now a *"weak" almost-complex n-manifold* is a real $2n$-manifold equipped with a subbundle $C \subset \text{End}(TM)$ of $\mathbf{C}$-algebras. Since $z \to \overline{z}$ is a real algebra automorphism of $\mathbf{C}$ sending i to $-i$ and the isomorphism $C_x \cong \mathbf{C}$ is not canonical, the set of operators $\{I, -I\} \subset C_x$ corresponding to $\{i, -i\} \subset \mathbf{C}$ is distinguished, but not the individual operators I and $-I$.

If, in addition, a global choice of I is possible, then by identifying $I \in C_x$ with $i \in \mathbf{C}$ (and by identifying the identity in each C_x with $1 \in \mathbf{C}$) we have a global trivialization of the bundle C. Thus, we have the following equivalent definition of an almost complex manifold.

Definition 5.4. *An almost-complex n-manifold is a real $2n$-manifold equipped with a trivial subbundle $M \times \mathbf{C} \subset \text{End}(TM)$ of $\mathbf{C}$-algebras.*

QUATERNIONIC (AND ALMOST-QUATERNIONIC) n-MANIFOLDS: THE GROUP $GL(n, \mathbf{H}) \cdot \mathbf{H}^*$

Now the situation is more delicate, and the concept that is analogous in a straightforward manner to the concept of a complex n-manifold is barren, i.e., there are essentially no examples. The example of *quaternionic projective space* $\mathbf{P}^n(\mathbf{H})$ provides some guidance.

Let $[x]$ denote the quaternionic line $[x] \equiv \{x\lambda : \lambda \in \mathbf{H}\}$ through the origin and the point $x \equiv (x_0, \ldots, x_n) \in \mathbf{H}^{n+1} - \{0\}$. By definition, $\mathbf{P}^n(\mathbf{H}) \equiv \{[x] : x \in \mathbf{H}^{n+1} - \{0\}\}$.

The standard atlas for $\mathbf{P}^n(\mathbf{H})$ consists of $n + 1$ charts U_k, ϕ_k, $k = 0, \ldots, n$, defined as follows. Let $U_k \equiv \{[x] \in \mathbf{P}^n(\mathbf{H}) : x_k \neq 0\}$, and define maps $\phi_k : U_k \to \mathbf{H}^n$ by sending $[x] \in U_k$ to $\phi_k([x]) \equiv (x_0 x_k^{-1}, \ldots, \widehat{1}, \ldots, x_n x_k^{-1}) \in \mathbf{H}^n$. Now the transition functions are all of the same form. For example, the transition function $\psi : \phi_0(U_0) \to \phi_1(U_1)$ is given by $\psi(y_1, \ldots, y_n) = (y_1^{-1}, y_2 y_1^{-1}, \ldots, y_n y_1^{-1})$. The linearization $\psi'(y)$ (or derivative) of ψ at y sends $u \in \mathbf{H}^n$ to

$$-(y_1^{-1} u_1 y_1^{-1}, y_2 y_1^{-1} u_1 y_1^{-1} - u_2 y_1^{-1}, \ldots, y_n y_1^{-1} u_1 y_1^{-1} - u_n y_1^{-1}).$$

In particular, $\psi'(y) \equiv A_y \circ R_{y_1^{-1}}$, where $A_y u \equiv -\psi(y) u_1 + (0, u_2, \ldots, u_n)$ is $\mathbf{H}$-linear and $R_{y_1^{-1}} u = u y_1^{-1}$ is simply right multiplication by y_1^{-1}.

In summary, the linearizations $\psi'(y)$ belong to $GL(n, \mathbf{H}) \cdot \mathbf{H}^*$ but *not* $GL(n, \mathbf{H})$. Thus, $\mathbf{P}^n(\mathbf{H})$ is a "quaternionic n-manifold" in the sense that the transition functions for the standard atlas have linearizations in the enhanced quaternionic linear group $GL(n, \mathbf{H}) \cdot \mathbf{H}^*$. Unfortunately, even adopting this more general notion of "quaternionic n-manifold," it follows that each such manifold is locally equivalent to $\mathbf{P}^n(\mathbf{H})$ (see Besse [3], p. 411). Consequently, it is most useful to reserve the terminology "quaternionic n-manifold" for yet another notion due to Salamon. Since the precise definition involves the concept of a torsion-free connection, we only state the definition and refer the reader to the literature (see Salamon [16] or Besse [3]).

A *quaternionic n-manifold* M is an almost-quaternionic n-manifold whose $GL(n, \mathbf{H}) \cdot \mathbf{H}^*$ reduced frame bundle admits a torsion-free connection. If $n = 1$, this condition is automatic.

The useful notion of "almost-quaternionic" is simpler to define. First consider $\mathbf{P}^n(\mathbf{H})$ again. Suppose a point $[x] \in \mathbf{P}^n(\mathbf{H})$ is fixed. Right multiplication on the tangent space $T_{[x]} \mathbf{P}^n(\mathbf{H})$ by a quaternion λ can be defined in several ways using different charts. Using U_k, ϕ_k, it is natural to define $u\lambda$ by representing the tangent vector $u \in T_{[x]} \mathbf{P}^n(\mathbf{H})$ as a vector in $\phi_k(U_k) \subset \mathbf{H}^n$ and multiplying by λ on the right. Let this multiplication on $T_{[x]}\mathbf{P}^n(\mathbf{H})$, for $x \in U_k$, be denoted by R_λ^k. Then

(5.5) $$R_\lambda^j \equiv R_{x_j x_k^{-1} \lambda x_k x_j^{-1}}^k$$

(see Problem 5). Therefore, right multiplication by a scalar $\lambda \in \mathbf{H}$ is *not* well-defined on $T_{[x]}\mathbf{P}^n(\mathbf{H})$. However, the vector space $\mathbf{H} \subset \mathrm{End}(T_{[x]}M)$ defined by $H \equiv \{R_\lambda^k \in \mathrm{End}(T_{[x]}M) : \lambda \in \mathbf{H}\}$ for each k with $x_k \neq 0$ is independent of k and (non-canonically) isomorphic as a real algebra to the algebra $\mathbf{H}$, since $\lambda \mapsto a\lambda a^{-1}$, with $a \equiv x_j x_k^{-1}$ fixed, is an algebra automorphism of $\mathbf{H}$.

This example makes it clear, since a productive definition of "almost-quaternionic n-manifold" should include $\mathbf{P}^n(\mathbf{H})$, that one should elect to utilize the quaternionic analogue of the notion of "weak" almost-complex.

Definition 5.6. *An almost-quaternionic n-manifold M is a real $4n$-manifold equipped with a subbundle C of $\mathrm{End}(TM)$ with $C_x \cong \mathbf{H}$ (as real algebras) for each point $x \in M$. The subbundle C is called the coefficient bundle or the almost-quaternionic structure bundle. Suppose this subbundle C is trivial and M is equipped with a particular trivilization (a global bundle isomorphism)*

$$C \cong M \times \mathbf{H}$$

so that operators I, J, K equal to right multiplication by $i, j, k \in \mathbf{H}$ are globally defined. Then M will be referred to as an almost-quaternionic manifold with trivialized coefficient bundle. This stronger motion of an almost-quaternionic manifold with trivialized coefficient bundle is the exact analogue of almost-complex.

MANIFOLDS WITH VOLUME: THE GROUPS SL(n, **R**) AND SL(n, **C**)

A real n-manifold M equipped with a never-vanishing n-form Ω is called a (*real*) *manifold with volume*. Of course, such a manifold is automatically oriented. A interesting result of Moser says that if a compact n-manifold M is a manifold with volume in two different ways (i.e., two different never-vanishing n-forms Ω and Ω' are prescribed), then M, Ω and M, Ω' are equivalent (i.e., there exists a diffeomorphism f of M with $f^*(\Omega') = \Omega$) if and only if the Ω volume of M is the same as the Ω' volume of M' (see Problem 10).

Similarly, a complex n-manifold M equipped with a never-vanishing $n, 0$-form Ω is called a *complex manifold with (complex) volume*. If, in addition, Ω is closed under exterior differentiation d, then M is said to be a *complex manifold with trivialized canonical bundle*.

Remark 5.7. If an almost-complex manifold M of complex dimension n is equipped with a never-vanishing d-closed form Ω of bidegree $n, 0$, then the hypothesis of the Newlander–Nirenberg Theorem can be verified so that M must, in fact, be a complex manifold. (And hence Ω must be a holomorphic n-form.)

RIEMANNIAN MANIFOLDS (OF SIGNATURE p, q): THE GROUP $O(p, q)$

A *Riemannian manifold M of signature p, q* is a real n-manifold (with $n \equiv p+q$) equipped with a real symmetric inner product of signature p, q on each tangent space $T_x M$ (and varying smoothly with the point $x \in M$). In particular, this includes the usual concept of Riemannian manifold ($q = 0$) and the concept of Lorentzian manifold ($q = 1$) as particular cases. Refer to Chapter 4 for additional information.

CONFORMAL MANIFOLDS (OF SIGNATURE p, q): THE GROUP $\mathrm{CO}(p, q)$

A *conformal manifold of signature p, q* is a real n-manifold with $n \equiv p + q$ equipped with a (smoothly varying) ray $[g]$ of $\mathbf{R}$-symmetric inner products of signature p, q, where g is a Riemannian inner product and $[g]$ denotes the ray $\{\lambda g : \lambda \in \mathbf{R}^+\}$. The group associated with an oriented conformal manifold is $\mathrm{CSO}(p, q)$.

REAL SYMPLECTIC MANIFOLDS: THE GROUP $\mathrm{Sp}(n, \mathbf{R})$

We shall give two definitions of a symplectic manifold. A theorem of Darboux states that the two definitions are equivalent. First, a *symplectic manifold* is a real $2n$-manifold whose atlas has the property that each transition function is a symplectomorphism from an open subset of $\mathbf{R}^{2n}, \omega$ (the standard symplectic vector space) to another such open subset of $\mathbf{R}^{2n}, \omega$. A *symplectic map* is a map whose linearization at each point belongs to the subgroup $\mathrm{Sp}(n, \mathbf{R})$ of $\mathrm{GL}(2n, \mathbf{R})$, or equivalently a map that fixes the degree 2 form ω under "pull back." Thus, the global degree 2-form Ω obtained by pulling back the standard symplectic form ω on $\mathbf{R}^{2n}, \omega$ (for each chart) is well-defined. Moreover, the 2-form Ω on M has the properties that

$$(5.8) \qquad \begin{array}{l} \text{At each point } x \in M, \Omega \text{ defines a real skew inner product} \\ \text{on } T_x M, \text{ the tangent space to } M. \end{array}$$

(5.9) Ω is closed under exterior differentiation, i.e., $d\Omega \equiv 0$ on M.

The second definition of a *symplectic manifold* M requires that M be a real $2n$-manifold equipped with a (smoothly varying) real skew inner product Ω on $T_x M$ at each point $x \in M$ (i.e., a nondegenerate 2-form on M) with $d\Omega = 0$ on M.

Symplectic geometry is important for two reasons. First, it provides a framework for classical mechanics. As noted in Remark 2.39 (b), a function h (the Hamiltonian) can be converted to a vector field by first taking the exterior derivative dh and then (using the symplectic form) converting dh to a vector field V_h. Many important flows are the solutions to such Hamiltonian vector fields. The main point is this fact: Another function f is constant on the flow lines of V_h (i.e., a *first integral* or *conserved quantity*) if and only if the original Hamiltonian function h is constant in the directions V_f, i.e., $V_f(h) = 0$. There are far-reaching global consequences. See the beautiful book by Arnold entitled *Classical Mechanics* [1] for more information. The second reason symplectic geometry is important is hinted at in Problem 8. In particular, symplectic manifolds provide a geometric framework for solving first order nonlinear partial differential equations. As one might suspect, these two reasons for the importance of symplectic geometry are closely related.

COMPLEX RIEMANNIAN MANIFOLDS: THE GROUP $O(n, \mathbf{C})$

A *complex Riemannian n-manifold* is a complex manifold equipped with a complex symmetric inner product on each tangent space $T_x M$ (and varying smoothly with the point $x \in M$). These manifolds arise naturally by "complexifying" a (real analytic) Riemannian manifold.

COMPLEX SYMPLECTIC MANIFOLDS: THE GROUP $\mathrm{Sp}(n, \mathbf{C})$

A *complex symplectic manifold* is a complex $2n$-manifold equipped with a complex symplectic inner product σ on each tangent space (and varying smoothly), where the bidegree $2, 0$ form σ is required to be closed under exterior differentiation. These manifolds arise naturally by "complexifying" a (real analytic) symplectic manifold. Compact examples occur in algebraic geometry.

Remark 5.10.

(a) On a complex symplectic manifold M, σ the $2, 0$-form σ is automatically $\bar{\partial}$-closed (i.e., holomorphic) so that M, σ is also called a *holomorphic symplectic manifold*.

(b) An almost-complex symplectic manifold M, σ is automatically a complex manifold because of Remark 5.7: since σ^n is d-closed, never-vanishing and of bidegree $n, 0$.

KÄHLER MANIFOLDS (OF SIGNATURE p, q): THE GROUP $U(p, q)$

A *Kähler manifold* M is a complex n-manifold equipped with a complex hermitian inner product h on each tangent space $T_x M$ (and varying smoothly), with the property that the 2-form

$$(5.11) \qquad \omega \equiv -\operatorname{Im} h \quad \text{(the Kähler form)} \quad \text{is } d\text{-closed.}$$

Note that

$$(5.12) \qquad g \equiv \operatorname{Re} h$$

is a Riemannian structure with signature $2p, 2q$ on M. The hermitian structure is given by

$$(5.13) \qquad h = g - i\omega.$$

Given a Riemannian structure g, a real symplectic structure ω, and an almost-complex structure J on a manifold M:

$$(5.14) \qquad \omega \text{ and } J \text{ are } \textit{compatible} \text{ if } J \text{ is an } \omega \text{ isometry,}$$

$$(5.15) \qquad g \text{ and } J \text{ are } \textit{compatible} \text{ if } J \text{ is a } g \text{ isometry,}$$

$$(5.16) \qquad \begin{array}{l} \omega \text{ and } g \text{ are } \textit{compatible} \text{ if the linear operator } J \text{ defined by} \\ \omega(x, y) = g(Jx, y) \text{ is an isometry for either} \\ \text{(or both) } \omega \text{ and } g. \end{array}$$

Lemma 5.17. *Given any two of the following three structures—(1) a Riemannian structure, (2) a real symplectic structure, and (3) an (almost)* complex structure, that are compatible—they determine the third structure. In fact, they determine an (almost) Kähler structure.

Proof: Apply Lemma 2.63. ∎

Kähler manifolds have been studied extensively.

SPECIAL KÄHLER MANIFOLDS (OF SIGNATURE p, q): THE GROUP SU(p, q)

Suppose M is a Käher manifold with signature p, q and complex dimension n. If, in addition, in a neighborhood of each point, there exists a form σ of bidegree $n, 0$ that is both

(5.18)
$$\text{(a) } d-\text{closed, and}$$
$$\text{(b) of constant size } |\sigma| = c,$$

then M is called a *special Kähler manifold of signature* p, q. Note that $d\sigma = 0$ and $\bar{\partial}\sigma = 0$ (i.e., σ holomorphic) are equivalent since there are no forms of bidegree $n + 1, 0$.

If M is a simply connected Kähler manifold, then the definition of special Kähler can be simplified. In this case, M is a special Kähler manifold if and only if there exists a global holomorphic $n, 0$ form σ of constant size $|\sigma| \equiv 1$. In particular, note that in this case the canonical bundle $\Lambda^{n,0}$ is holomorphically trivialized (by σ).

Remark 5.19. Suppose M is a Kähler manifold of complex dimension n and that σ is an $n, 0$ form on M. One can show that either of the two conditions:

(a) σ is parallel,
(b) the first chern form of $\Lambda^{n,0}$ vanishes (i.e., Ricci flat),

are equivalent to the condition (5.18) required in the definition of special Kähler manifolds.

Remark 5.20 (Calabi–Yau). A beautiful theorem of Yau provides a means of constructing special Kähler manifolds in the important positive definite case. Suppose M is a (positive definite) compact Kähler manifold with Kähler form ω. Assume that the canonical bundle $\Lambda^{n,0}$ is trivial. Let σ denote the global, never-vanishing holomorphic n-form (unique up to a constant). Yau proved that there exists a new Kähler form $\tilde{\omega}$ that is homologous to ω (i.e., $\tilde{\omega} - \omega = d\alpha$ for some global 1-form α) such that $M, \tilde{\omega}$ is a special Kähler manifold, i.e., σ has constant size with respect to the new Kähler metric based on $\tilde{\omega}$. (Yau's theorem can be restated in a stronger form if M is not simply connected.)

HYPERKÄHLER MANIFOLDS (OF SIGNATURE p, q): THE GROUP HU(p, q)

There are quite a few useful ways of defining a hyperKähler manifold. The next definition contains in a certain sense the maximal amount of information.

Definition 5.21. *A hyperKähler manifold M is a real $4n$-manifold equipped with the following extra structure. First, assume that the quaternions* **H** *act (smoothly) on the tangent bundle on the right, giving each tangent space T_x the structure of a right quaternionic vector space (i.e., M is equipped with an almost quaternionic structure with trivial coefficient bundle, cf. Problem 7). Second, suppose ε is an* **H**-*hermitian inner product of signature p, q on each tangent space (which varies smoothly with the point $x \in M$). Thus, $g \equiv \operatorname{Re} \varepsilon$ provides M with the structure of a Riemannian manifold of signature $4p, 4q$; and each operator R_u (right multiplication by u) with $u \in S^2 \subset \operatorname{Im} \mathbf{H}$ a unit imaginary quaternion, provides M with the structure of an almost-complex manifold. Third, require that each associated (Kähler) form ω_u, defined by*

$$(5.22) \qquad \omega_u(w, z) \equiv \operatorname{Re} \varepsilon(wu, z) \quad \text{for all } w, z \in T_x M,$$

be closed under exterior differentiation. Fourth, require that the almost-complex manifold M, R_u, with $u \in S^2 \subset \operatorname{Im} \mathbf{H}$, be, in fact, a complex manifold. This completes the definition of a hyperKähler manifold.

Remark 5.23. The fourth and final requirement can be proved to be a consequence of the first three requirements, by utilizing Remark 5.10b—see the definition of σ_u given by (5.27) below.

Recall from the section in Chapter 2 on "The Parts of an Inner Product" that an **H**-hermitian symmetric inner product ε has first and second complex parts h and σ defined by

$$(5.24) \qquad \varepsilon = h + j\sigma.$$

Further, with respect to the complex structure R_i, h is a **C**-hermitian symmetric inner product and σ is a **C**-skew (complex symplectic) inner product. In particular,

$$(5.25) \qquad h = g + i\omega_I$$

provides a natural Kähler structure on M, and

$$(5.26) \qquad \sigma = \omega_J - i\omega_K$$

provides a complex symplectic structure (with almost-complex structure I). More generally, for each $u \in S^2 \subset \mathrm{Im}\,\mathbf{H}$, $h_u \equiv g + i\omega_u$ provides a Kähler structure on M (thus the name hyperKähler), while

$$(5.27) \qquad \varepsilon = h_u + u\sigma_u$$

defines a complex symplectic structure σ_u (with complex structure R_u). Also note that

$$(5.28) \qquad \varepsilon = g + i\omega_i + j\omega_j + k\omega_k.$$

On a hyperKähler manifold, with the complex structure I chosen, i.e., distinguished, the complex hermitian structure h and the complex symplectic structure σ defined by (5.26) are *compatible* in the sense that

$$(5.29) \qquad h(xJ, y) = \sigma(x, y)$$

and

$$(5.30) \qquad \sigma(xJ, yJ) = \overline{\sigma(x, y)}.$$

Conversely, suppose a complex hermitian structure h and a complex symplectic structure σ are given. Define a linear map J by requiring (5.29) to be valid. In Chapter 2, it was shown that (5.30) is valid if and only if $J^2 = -1$ (see Equation (2.77)). Now once (5.29) and (5.30) are satisfied one can define ε to be $h + j\sigma$ and show that ε is $\mathbf{H}$-hermitian symmetric. This proves that the next definition of a hyperKähler manifold is equivalent to the earlier definition.

A *hyperKähler manifold* M is a Kähler manifold equipped with a complex symplectic structure σ that is compatible with the hermitian structure $h \equiv g + i\omega$.

QUATERNIONIC KÄHLER MANIFOLDS (OF SIGNATURE p, q): THE GROUP $\mathrm{HU}(p, q) \cdot \mathrm{HU}(1)$

A *quaternionic Kähler manifold of signature p, q* is, by definition,

(a) an almost-quaternionic manifold (with coefficient bundle C),
(b) equipped with a Riemannian structure of signature $4p, 4q$, with each coefficient $u \in C_x$ of unit length an orthogonal map,
(c) such that the coefficient bundle C is parallel with respect to the riemannian structure.

Recall by Problem 3.15 (a) that the quaternionic unitary group, $\mathrm{HU}(p,q) \cdot \mathrm{HU}(1)$ can be defined to be the subgroup of $\mathrm{SO}(4p,4q)$ that fixes $\mathbf{H} \cong C_x$.

By choosing a particular algebra isomorphism $C_x \cong \mathbf{H}$ between the fiber of the coefficient bundle C and the quaternions $\mathbf{H}$, we may define the *quaternionic 4-form* Φ by

$$\Phi \equiv \frac{1}{6}\left(\omega_i^2 + \omega_j^2 + \omega_k^2\right),$$

where the two forms ω_u with $u \in \operatorname{Im}\mathbf{H}$ are defined by

$$\omega_u(x,y) \equiv \langle xu, y\rangle \quad \text{for all tangent vectors } x, y,$$

using the Riemannian metric $\langle\,,\,\rangle$.

Since any other isomorphism $C_x \cong \mathbf{H}$ differs by an element $a \in \mathrm{HU}(1)$ acting on $\mathbf{H}$ by $\chi_a(x) = ax\bar{a}$, Problem 3.15 (c) can be used to show that Φ does not depend on the choice of isomorphism $C_x \cong \mathbf{H}$.

Thus, the quaternionic 4-form Φ is a well-defined global 4-form on any quaternionic Kähler manifold. Also note that Φ^n is a nonzero volume form. Moreover, Φ is parallel since the coefficient bundle is parallel.

Based on Problem 3.15, one can show that $(n \geq 2)$ an almost-quaternionic n-manifold equipped with a 4-form Φ, that (at each point) can be expressed as $\frac{1}{6}(\omega_i^2 + \omega_j^2 + \omega_k^2)$ for some choice of metric $\langle\,,\,\rangle$, in fact, uniquely determines this metric. If Φ is parallel, then the manifold is quaternionic Kähler. In fact, for $n \geq 3$, the weaker condition $d\Phi = 0$ implies that Φ is parallel (see Salamon [16]). The case $n = 2$ is unknown.

Note: Since $\mathrm{HU}(p,q) \cdot \mathrm{HU}(1)$ is the quaternionic analogue of the unitary group $U(p,q)$, quaternionic Kähler geometry is the analogue of Kähler geometry. (Similarly, since $\mathrm{HU}(p,q)$ is the quaternionic analogue of $\mathrm{SU}(p,q)$, hyper-Kähler geometry is the quaternionic analogue of special Kähler geometry.) However, a quaternionic Kähler manifold need not be Kähler manifold.

QUATERNIONIC SKEW HERMITIAN MANIFOLDS: THE GROUP $\mathrm{SK}(n,\mathbf{H})$

An **H**-*hermitian skew manifold* M is a real $4n$-manifold equipped with the following extra structure. First, assume that the quaternions $\mathbf{H}$ act (smoothly) on the tangent bundle on the right giving each tangent space $T_x M$ the structure of a right quaternion vector space (cf. Problem 7). Second, suppose ε is an **H**-hermitian skew inner product on each tangent space (that varies smoothly). Third, require that the 2-form defined by

$$(5.31) \qquad \omega \equiv \operatorname{Re} \varepsilon$$

be closed under exterior differentiation. Fourth, and last, require that each of the almost-complex structures R_u, right multiplication by a unit imaginary quaternion, be, in fact, a complex structure on M.

For each $u \in S^2 \subset \operatorname{Im} \mathbf{H}$,

$$(5.32) \qquad g_u(z, w) \equiv \operatorname{Re} \varepsilon(zu, w) \qquad \text{for all } z, w \in T_x M$$

defines a split Riemannian structure on M. In fact, $g_u - u\omega$ is a C-hermitian symmetric inner product on the complex manifold M, R_u with Kähler form $\omega \equiv \operatorname{Re} \varepsilon$ independent of u.

Note that

$$(5.33) \qquad \varepsilon = \omega + ig_i + jg_j + kg_k.$$

Consult the **H**-hermitian skew case in the section *The Parts of an Inner Product* in Chapter 2.

COINCIDENCES OF GEOMETRIES IN LOW DIMENSIONS

In this section, we discuss the coincidences of certain geometries in low dimensions. These coincidences can be labeled with group coincidences (cf. Proposition 1.40, see Problem 9).

$\operatorname{Sp}(1, \mathbf{R}) \cong \operatorname{SL}(2, \mathbf{R})$ and $\operatorname{Sp}(1, \mathbf{C}) \cong \operatorname{SL}(2, \mathbf{C})$

Suppose M is a real (respectively complex) manifold of real (complex) dimension 2. Then the notion of a real (complex) symplectic manifold M is exactly the same as the notion of a real (complex) manifold with volume. That is, M must be equipped with a particular never-vanishing 2-form.

Suppose M is a complex manifold of complex dimension 2. Then the notion of an almost-complex symplectic manifold (or equivalently a holomorphic symplectic manifold—see Remark 5.10b) is exactly the same as the notion of a complex surface with trivialized canonical bundle. That is, M must be a complex surface equipped with a never-vanishing holomorphic $2, 0$-form.

$SO(2) \cong U(1)$ **and** $CSO(2) \cong GL(1, \mathbf{C})$

Theorem 5.34.

(a) *The notion of an oriented conformal manifold of real dimension 2 is the same as the notion of a complex manifold of complex dimension 1.*

(b) *The notion of an oriented Riemannian manifold of real dimension 2 is the same as the notion of a Kähler manifold of complex dimension 1.*

Proof: First, assume that M is a one-dimensional complex manifold with complex structure J. Then M is oriented (apply (1.26) to the linearization of a holomorphic transition function). Let ω denote a choice of orientation preserving volume form. All other choices must be of the form $\phi\omega$ where ϕ is a smooth positive function on M. Each skew 2-form ω determines a Riemannian structure g by

(5.35) $g(x, y) \equiv \omega(x, Jy)$ for all tangent vectors x, y.

Since $\phi\omega$ determines ϕg, the orientation class $[\omega] \equiv \{\phi\omega : \phi > 0\}$ determines a conformal structure $[g] \equiv \{\phi g : \phi > 0\}$ on M.

If M is Kähler, then M is equipped with a particular ω that, by (5.35), determines a particular Riemannian structure g on M.

Conversely, assume that M is equipped with an orientation class $[\omega]$ and a conformal class $[g]$. Given a choice of Riemannian structure $g \in [g]$, renormalize ω so that ω is a unit volume form. Define a linear map J by

(5.36) $\omega(x, Jy) = g(x, y)$ for all tangent vectors x, y.

If ϕg is any other representative of the conformal class $[g]$, then $\phi\omega$ is the unit volume form (because the dimension is 2) in the metric ϕg. Thus, ϕg and g determine the same map J. To prove that J is an almost-complex structure, i.e., $J^2 = -1$, fix g and select an oriented orthonormal basis e_1, e_2. Then $\omega(e_1, e_2) = 1$ so that (5.36) implies $Je_1 = e_2, Je_2 = -e_1$. (Thus, J is counterclockwise rotation by 90°.) Therefore $J^2 = -1$. It is a standard classical result that, for complex dimension one, each almost-complex manifold is a complex manifold. This result is a special case of the Newlander–Nirenberg Theorem. The reader is referred to the literature.

If M is an oriented Riemannian 2-manifold, then the metric g yields the almost-complex structure J by (5.36). Note that in addition to the property $J^2 = -1$, J is an isometry with respect to both g and ω. Therefore,

$$h \equiv g - i\omega$$

is a complex hermitian inner product. Of course, $d\omega = 0$ since $d\omega$ is a 3-form on a 2-manifold. This proves that M is a Kähler manifold. ∎

SO(4) $\cong$ HU(1) $\cdot$ HU(1) and CSO(4) $\cong$ GL(1, **H**) $\cdot$ **H***

Theorem 5.37.

(a) *The notion of an oriented conformal manifold of real dimension 4 is the same as the notion of an (almost) quaternionic manifold of quaternionic dimension 1.*

(b) *The notion of an oriented Riemannian manifold of real dimension 4 is the same as the notion of a quaternionic Kähler manifold of quaternionic dimension one.*

Proof: Suppose M is an almost quaternionic manifold of quaternionic dimension one. Let $C \subset \mathrm{End}(TM)$ denote the coefficient bundle (i.e., the almost-quaternionic structure) so that, at each point $x \in M$, $C_x \cong \mathbf{H}$ as real algebras. This isomorphism is not canonical, but if $f_1 : C_x \to \mathbf{H}$ and $f_2 : C_x \to \mathbf{H}$ are two such isomorphisms, then $f_2 \circ f_1^{-1} \in \mathrm{Aut}(\mathbf{H})$ is an automorphism of the algebra $\mathbf{H}$. Later, in Chapter 6, in the context of normed algebras, it is easy to compute that $\mathrm{Aut}(\mathbf{H}) \equiv \mathrm{SO}(\mathrm{Im}\,\mathbf{H})$. Assuming this fact, it follows that the standard real inner product $\langle\,,\,\rangle$ and the standard orientation (on $\mathbf{H}$) induce a real inner product and orientation on C_x.

Now pick any nonzero tangent vector u. Then any other nonzero tangent vector v is of the form $R_a u = v$ for a unique choice of $R_a \in C_x$. Thus, we may define

(5.38) $$g(v_1, v_2) \equiv \langle R_{a_1}, R_{a_2} \rangle = \langle a_1, a_2 \rangle,$$

where $v_j = R_{a_j} u$. The inner product depends on u, but if u is replaced by $\tilde{u} = u\lambda$, then $v = ua = u\lambda\lambda^{-1}a = \tilde{u}\lambda^{-1}a$, so that a is replaced by $\tilde{a} \equiv \lambda^{-1}a$. Thus, the new inner product $\tilde{g}$ equals $|\lambda|^{-2}$ times the old inner product g, so that we have determined a conformal structure $[g]$ on the tangent space. Similarly, the orientation $\{1, i, j, k\}$ on $\mathbf{H}$ determines an orientation on the tangent space.

If, in addition, M is quaternionic Kähler, then M is automatically equipped with a volume form $\Phi \equiv \frac{1}{3}(\omega_i^2 + \omega_j^2 + \omega_k^2)$. Now, there exists a unique choice of Riemannian structure g in the conformal class determined by the quaternionic structure, so that the given 4-form Φ is of unit norm.

Next suppose M is an oriented conformal manifold of real dimension 4 with conformal structure $[g]$. Given a choice of $g \in [g]$, let Φ denote the *unit* oriented volume form. The inner product g and the volume Φ determine a linear map $\star$ on forms by

(5.39) $$\alpha \wedge \star\beta = g(\alpha, \beta)\Phi,$$

where g also denotes the inner product induced by g on forms. If g on tangent vectors is rescaled by λ so that $\tilde{g} = \lambda g$, then g on forms of degree k rescales by $g = \lambda^{-2k}g$, so that the new unit volume form $\tilde{\Phi} = \lambda^4\Phi$. Therefore, if $\alpha, \beta \in \Lambda^2$ are of degree 2, $g(\alpha,\beta)\Phi = g(\alpha,\beta)\tilde{\Phi}$ is independent of the rescaling Λ. This proves that the $\star$ operator on Λ^2 is uniquely determined by the oriented conformal structure on M. Using an oriented orthonormal basis $e_1, \ldots, e_4$, it is easy to prove that $\star : \Lambda^2 \to \Lambda^2$ squares to 1, and that the ± 1 eigenspaces are both 3-dimensional. Thus, $\Lambda^2 = \Lambda^+ \oplus \Lambda^-$ decomposes into the space of *self-dual forms* and the *anti-self-dual forms*. Each *self-dual* form $\alpha \in \Lambda^+$ determines a linear map J_α on the tangent space by

$$(5.40) \qquad \alpha(u,v) = g(J_\alpha u, v) \quad \text{for all tangent vectors } u, v.$$

The coefficient bundle $C \subset \operatorname{End}(TM)$ is defined by $C_x \equiv \operatorname{span} 1 \oplus \{J_\alpha : \alpha \in \Lambda^+\} \subset \operatorname{End} T_x M$. Note that if $\tilde{g} = \lambda g$ is a rescaling of g then $\tilde{J}_\alpha = \lambda^{-1}J_\alpha$. Therefore, the bundle C only depends on the conformal structure $[g]$ and not the choice of $g \in [g]$. In terms of an oriented orthonormal basis $\alpha^1, \alpha^2, \alpha^3, \alpha^4$ for the cotangent space (based on a choice $g \in [g]$), it is easy to compute that $\beta_1 \equiv \alpha^1 \wedge \alpha^2 + \alpha^3 \wedge \alpha^4$, $\beta_2 \equiv \alpha^1 \wedge \alpha^3 - \alpha^2 \wedge \alpha^4$, and $\beta_3 \equiv \alpha^1 \wedge \alpha^4 + \alpha^2 \wedge \alpha^3$ provide a basis for Λ^+. The operators $J_{\beta_1}, J_{\beta_2}, J_{\beta_3}$ satisfy $J_{\beta_1}^2 = J_{\beta_2}^2 = J_{\beta_3}^2 = -1$ and $J_{\beta_1}J_{\beta_2} = J_{\beta_3}$, etc., so that $C_x \cong \mathbf{H}$ as real algebras. This proves that M is naturally an almost quaternionic manifold. In this dimension $n = 1$, each almost-quaternionic manifold is a quaternionic manifold.

Now, if M is an oriented Riemannian manifold with metric g, the previous discussion applies yielding an almost-quaternionic structure on M. In addition, the 4-form

$$\Phi \equiv \frac{1}{3}\left(\beta_1^2 + \beta_2^2 + \beta_3^2\right)$$

is just the unit volume element and hence is independent of the choice of oriented orthonormal basis $\alpha^1, \alpha^2, \alpha^3, \alpha^4$. Combining the coefficient bundle C, the form Φ, and the metric g provides the quaternionic Kähler structure on M. ∎

Remark 5.41. The right **H**-structure determined by $C \subset \operatorname{End}(TM)$ determines $\operatorname{End}_{\mathbf{H}}(T_x M)$, the space of quaternionic linear maps. This subbundle $\operatorname{End}_{\mathbf{H}}(TM)$ of $\operatorname{End}(TM)$ is exactly the bundle span $\{1\} \oplus \{J_\alpha : \alpha \in \Lambda^-\}$ of operators obtained from the space of anti-self-dual 2-forms, i.e., Λ^-.

$SU(2) \cong HU(1)$

Theorem 5.42.

(a) *Suppose M is a hyperKähler manifold of quaternion dimension one. Then M is naturally a special Kähler (positive definite) manifold for a 2-sphere of different complex structures on M.*

(b) *Suppose M is a (positive definite) special Kähler surface. Then M is naturally a hyperKähler manifold.*

Proof:

(a) Suppose M, ε is a hyperKähler manifold of quaternionic dimension one. For each complex structure R_u on M (with $u \in S^2 \subset \mathrm{Im}\,\mathbf{H}$ a unit imaginary quaternion), M, h_u is a Kähler manifold with complex hermitian inner product $h_u \equiv g - u\omega_u$ and M, σ_u is a complex holomorphic symplectic manifold with complex symplectic form σ_u defined by $\varepsilon = h_u + u\sigma_u$. (See Lemma 2.72.) Moreover, $|\sigma_u|$ is globally constant since it equals the norm of $dz^1 dz^2$ on the standard model when $u \equiv i$.

(b) Suppose M is a special Kähler surface with complex structure I, hermitian structure $h = g + i\omega_I$, and unit holomorphic volume form σ. Define a linear map J on the tangent space at each point p by

$$(5.43) \qquad h(xJ, y) = \sigma(x, y) \quad \text{for all tangent vectors } x, y.$$

Choose any isomorphism $T_p M \cong \mathbf{C}^2$ and note that $\sigma \equiv e^{i\theta} dz^1 \wedge dz^2$. If σ is replaced by $\lambda \equiv dz^1 \wedge dz^2$ in (5.43), then

$$(5.44) \qquad h(x\tilde{J}, y) = \lambda(x, y)$$

obviously defines a standard quaternionic structure $I, \tilde{J}, K \equiv I\tilde{J}$ on $\mathbf{C}^2$.

Since $\sigma = e^{i\theta}\lambda$, J must equal $e^{i\theta}\tilde{J}$. Therefore, J satisfies the same identities, namely $IJ = -JI$ and $J^2 = -1$, as $\tilde{J}$. This proves that I, J provides a quaternionic structure on the tangent space $T_p M$.

Now one can show that $\varepsilon \equiv h + j\alpha$ defines an $\mathbf{H}$-hermitian symmetric inner product on each tangent space. ∎

PROBLEMS

1. (a) Prove that the two definitions of oriented (real) manifold given in the text are equivalent.

(b) Prove that each complex n-manifold is automatically oriented as a real $2n$-manifold.

2. Prove that each non-orientable Riemannian 2-manifold (of signature $2, 0$) is a "weak" almost-complex manifold as described in the text.

3. Suppose a real $2n$-dimensional vector space V with complex structure J is given. Let $V_{\mathbf{C}} = V \otimes_{\mathbf{R}} \mathbf{C} = V \oplus iV$ denote the complexification of V. That is, $V_{\mathbf{C}} \equiv \{u + iv : u, v \in V\}$. $V_{\mathbf{C}}$ is naturally equipped with a conjugation that fixes $V \subset V_{\mathbf{C}}$, i.e., $\overline{u + iv} = u - iv$. Define $\operatorname{Re} z = \frac{1}{2}(z + \bar{z})$ and $\operatorname{Im} z = (z - \bar{z})/2i$ for all $z \in V_{\mathbf{C}}$. Extend J to be complex linear on $V_{\mathbf{C}}$. Let $V^{1,0}$ denote the $+i$ eigenspace of J and $V^{0,1}$, denote the $-i$ eigenspace of J on $V_{\mathbf{C}}$, so that $V_{\mathbf{C}} = V^{1,0} \oplus V^{0,1}$. Let π denote the projection of $V_{\mathbf{C}}$ onto $V^{1,0}$ along $V^{0,1}$. Show that V with complex structure J and $V^{1,0}$ with complex structure i are complex isomorphic via the map $\pi|_V$ and that $2\operatorname{Re}$ provides the inverse.

4. When V, J is taken as $T_p M$, the tangent space to a complex manifold, then utilizing the coordinate functions $z^1, \ldots, z^n$ with $z = x + iy$, a real basis for V is given by: $\frac{\partial}{\partial x^1}, \ldots, \frac{\partial}{\partial x^n}, \frac{\partial}{\partial y^1}, \ldots, \frac{\partial}{\partial y^n}$. Show that

(a)

$$\frac{\partial}{\partial z^1}, \ldots, \frac{\partial}{\partial z^n}, \quad \text{where} \quad \frac{\partial}{\partial z} = \frac{1}{2}\left(\frac{\partial}{\partial x} - i\frac{\partial}{\partial y}\right),$$

is a complex basis for $V^{1,0}$.

(b) The isomorphism $V, J \cong V^{1,0}$, i of complex vector spaces exhibited in Problem 3 does not preserve the bracket $[,]$ of vector fields.

5. Verify (5.5), i.e., $R^j_\lambda = R^k{}_{x_j x_k^{-1} \lambda x_k x_j^{-1}}$.

6. Prove that $\Omega \equiv d\alpha$, where α is a 1-form defined by $\alpha(V) \equiv \phi(\pi_* V)$ at each point $\phi_x \in T^* M$, provides a symplectic structure on the cotangent bundle $T^* M$ to a manifold M.

7. Suppose M is an almost-quaternionic manifold equipped with an H-hermitian inner product ε that is either symmetric or skew. Prove that the coefficient bundle $C \subset \operatorname{End}(TM)$ is trivial, i.e., exhibit I, J, K globally on M.

8. (*Lagrangian submanifolds*) Consider the standard model $\mathbf{R}^{2n}$, ω of a symplectic vector space as a symplectic manifold, where

$$\omega \equiv dx^1 \wedge dy^1 + \cdots + dx^n \wedge dy^n.$$

Suppose M is the graph of a smooth function $y = f(x)$ over an open simply connected subset U of $\mathbf{R}^n$, i.e., $f : U \to \mathbf{R}^n$ and $M \equiv \{(x, f(x)) \in \mathbf{R}^{2n} : x \in U\}$. Prove that the following are equivalent.

(a) $f = \nabla\phi$ for some scalar-valued function ϕ defined on U.

(b) The 1-form $\sum f_i(x)dx^i$ is d-closed on U.

(c) The 1-form $\alpha \equiv \sum y^i dx^i$ on $\mathbf{R}^{2n}$ restricts to a d-closed form on M.

(d) The 2-form ω vanishes when restricted to M (i.e., M is an isotropic or totally null submanifold of $\mathbf{R}^{2n}, \omega$).

In particular, note that ϕ satisfies a nonlinear partial differential equation of the form $P(x, \nabla\phi(x)) = 0$ if and only if both of the following geometric conditions on the graph M are satisfied:

(1) M is contained in $\{(x, y) \in U \times \mathbf{R}^n : P(x, y) = 0\}$, and

(2) M is isotropic.

9. (a) Extract a sublist of the list of group isomorphisms provided by Proposition 1.40 with the stronger property that the corresponding lowest dimensional representations are isomorphic. For example, SO(2) acting on $\mathbf{R}(2)$ and $U(1)$ acting on $\mathbf{C}(1)$ are isomorphic representations, but the representation Sk(1) on $\mathbf{H}$ is different, so that SO(2) $\cong U(1)$ should be included on the new list but SO(2) $\cong$ Sk(1) should be deleted.

(b) Compare this new list with the list of low dimensional coincidences of geometries presented in this chapter.

10. Prove Moser's Theorem on compact manifolds with volume.
Hint:

(a) Show $\Omega' = \Omega + d\alpha$ for some global $(n-1)$ form α.

(b) Define a vector field V by $\alpha = V \lrcorner \Omega$. Make use of the corresponding flow ϕ_t and the beautiful formula

$$L_V(\alpha) = d(V \lrcorner \alpha) + V \lrcorner d\alpha$$

(valid for any form α), where L_V denotes the Lie derivative with respect to the vector field V.

6. Normed Algebras

Suppose $V, \langle \, , \, \rangle$ is a euclidean vector space of signature p, q. The *norm* of a vector $|v| \equiv \sqrt{|\langle v, v \rangle|}$ is always positive. In this section, we shall deal with the *square norm* or *quadratic form*

$$(6.1) \qquad \qquad \|x\| \equiv \langle x, x \rangle.$$

By a *normed* or *euclidean algebra*, we mean a (not necessarily associative) finite dimensional algebra over $\mathbf{R}$ with multiplicative unit 1, and equipped with an inner product $\langle \, , \, \rangle$ of general signature whose associated square norm $\| \ \|$ satisfies the multiplicative property

$$(6.2) \qquad \qquad \|xy\| = \|x\| \, \|y\| \quad \text{for all } x, y.$$

The inner product $\langle x, y \rangle$ can be determined from the square norm $\|z\|$ by *polarization*. That is, replace z by $x + y$ in the formula $\|z\| = \langle z, z \rangle$ and obtain

$$\|x + y\| = \langle x + y, x + y \rangle = \|x\| + 2\langle x, y \rangle + \|y\|$$

or

$$2\langle x, y \rangle = \|x + y\| - \|x\| - \|y\|.$$

Similarly, the identity $\|zw\| = \|z\| \, \|w\|$ can be polarized.

101

Lemma 6.3. *The following identities are all equivalent:*

(6.2) $$\|xy\| = \|x\|\,\|y\|$$

(6.4) $$\langle xw, yw\rangle = \langle x, y\rangle\|w\|$$

(6.5) $$\langle wx, wy\rangle = \|w\|\langle x, y\rangle$$

(6.6) $$\langle xz, yw\rangle + \langle yz, xw\rangle = 2\langle x, y\rangle\,\langle z, w\rangle$$

Note that the final identity is linear in all of the variables x, y, z, and w.

Proof: Setting $z = w$ in the last identity yields (6.4), and one obtains (6.5) similarly. Setting $x = y$ in (6.4) yields the multiplicative identity (6.2). Thus, it remains to deduce all the identities from the multiplicative property $\|xy\| = \|x\|\,\|y\|$ via polarization.

Replace x by $x + y$ and y by w in (6.2). This proves that

$$\|(x + y)w\| = \langle(x + y)w, (x + y)w\rangle = \langle xw, xw\rangle + 2\langle xw, yw\rangle + \langle yw, yw\rangle$$
$$= \|xw\| + \|yw\| + 2\langle xw, yw\rangle = \|x\|\,\|w\| + \|y\|\,\|w\|$$
$$+ 2\langle xw, yw\rangle$$

must equal

$$\|x + y\|\,\|w\| = (\|x\| + 2\langle x, y\rangle + \|y\|)\|w\|,$$

yielding (6.4). The proof of (6.5) is similar.

The identity (6.4) is quadratic in w and hence may be further polarized by replacing w by $z + w$. Thus,

$$\langle x(z + w), y(z + w)\rangle = \langle xz, yz\rangle + \langle xz, yw\rangle + \langle xw, yz\rangle + \langle xw, yw\rangle$$
$$= \|z\|\langle x, y\rangle + \|w\|\langle x, y\rangle + \langle xz, yw\rangle + \langle xw, yz\rangle$$

must equal

$$\|z + w\|\langle x, y\rangle = \|z\|\langle x, y\rangle + 2\langle z, w\rangle\langle x, y\rangle + \|w\|\langle x, y\rangle,$$

yielding (6.6).

In summary, the defining property $\|xy\| = \|x\|\,\|y\|$, which is quadratic in x and y, may be re-expressed as an identity, (6.6), linear in x, y, z, and w. ∎

Given an algebra, let R_w denote the linear operator right multiplication by w. Similarly, let L_w denote left multiplication by w. The identity (6.4) may be rewritten

(6.4$'$) $$\langle R_w x, R_w y\rangle = \|w\|\langle x, y\rangle.$$

That is, the key axiom for a normed algebra may be stated as:

Right multiplication by w is conformal with conformal factor $\|w\|$.

Similarly,

$$(6.5') \qquad\qquad \langle L_w x, L_w y \rangle = \|w\| \langle x, y \rangle.$$

In terms of the operators R_z and R_w the fully polarized identity (6.6) becomes

$$(6.6') \qquad\qquad \langle R_z x, R_w y \rangle + \langle R_w x, R_z y \rangle = 2 \langle x, y \rangle \langle z, w \rangle.$$

Note: If $\|w\| = -1$, then $R_w \notin O(A)$.

Interchanging x with z and y with w in (6.6) yields

$$(6.6'') \qquad\qquad \langle L_z x, L_w y \rangle + \langle L_w x, L_z y \rangle = 2 \langle x, y \rangle \langle z, w \rangle.$$

We will adopt the following notational conventions. Let $\operatorname{Re} V$ denote the span of $1 \in V$. Since $\|x\| = \|1 \cdot x\| = \|1\| \, \|x\|$, the multiplicative identity 1 cannot be null. Let $\operatorname{Im} V$ denote the orthogonal compliment of $\operatorname{Re} V$. Then, by Lemma 2.30, $\operatorname{Im} V$ is a nondegenerate hyperplane, and each $x \in V$ has a unique orthogonal decomposition:

$$x = x_1 + x', \quad \text{with } x_1 \in \operatorname{Re} V, \ x' \in \operatorname{Im} V.$$

Occasionally, we let $\operatorname{Re} x$ denote x_1 and $\operatorname{Im} x$ denote x'. *Conjugation* is defined by

$$(6.7) \qquad\qquad \overline{x} = x_1 - x'.$$

Thus

$$(6.8) \qquad \begin{aligned} x_1 &\equiv \operatorname{Re} x = \frac{1}{2}\left(x + \overline{x}\right), \\ x' &\equiv \operatorname{Im} x = \frac{1}{2}\left(x - \overline{x}\right). \end{aligned}$$

The adjoints of R_w and L_w are

$$(6.9) \qquad\qquad R_w^* = R_{\overline{w}}, \quad L_w^* = L_{\overline{w}}.$$

To prove (6.9), first note that these identities are linear in w, and obviously true when $w \in \operatorname{Re} V$. Thus, we may assume $w \perp \operatorname{Re} V$ or $w \in \operatorname{Im} V$. Now setting $z = 1$ in (6.6') yields

$$\langle x, R_w y \rangle + \langle R_w x, y \rangle = 0$$

or $R_w^* = -R_w = R_{\overline{w}}$. Similarly, $L_w^* = L_{\overline{w}}$.

The elementary facts concerning conjugation are contained in the next lemma.

Lemma 6.10.

(a) $\overline{\overline{x}} = x$ and $\langle \overline{x}, \overline{y} \rangle = \langle x, y \rangle$.
(b) $\langle x, y \rangle = \operatorname{Re} x\overline{y} = \operatorname{Re} \overline{x}y$.
(c) $\overline{xy} = \overline{y}\ \overline{x}$.
(d) $x\overline{x} = \overline{x}x = \|x\|$.

Proof: (a) is true because conjugation is reflection through a hyperplane. Part (b) is a special case of $R_y^* = R_{\overline{y}}$. Part (c) is true because

$$\langle \overline{xy}, z \rangle = \langle xy, \overline{z} \rangle = \langle y, \overline{x}\ \overline{z} \rangle = \langle yz, \overline{x} \rangle = \langle z, \overline{y}\ \overline{x} \rangle.$$

Part (d) is now an easy direct calculation. ∎

The *associator*

$$[x, y, z] \equiv (xy)z - x(yz)$$

measures the lack of associativity in an algebra. For a normed algebra, there is always a weak form of associativity, namely

Lemma 6.11. *The associator* $[x, y, z]$ *vanishes if any two of the arguments are set equal, or equivalently, the trilinear form* $[x, y, z]$ *is alternating.*

An algebra on which the associator is alternating is called *alternative*. Lemma 6.11 states that any normed algebra is alternative.

Proof: Note that the associator vanishes if one of its variables is real. Hence, it suffices to show that the associator vanishes when two of its variables are set equal to $w \in \operatorname{Im} V$ pure imaginary. Note, we show that $[x, w, \overline{w}] = 0$. Since $w\overline{w} = \|w\|$, $[x, w, \overline{w}] = (xw)\overline{w} - x\|w\|$. Since $R_w^* = R_{\overline{w}}$ and R_w is conformal,

$$\langle (xw)\overline{w}, z \rangle = \langle xw, zw \rangle = \|w\|\langle x, z \rangle,$$

which proves $[x, w, \overline{w}] = 0$. Similarly, $[w, \overline{w}, z] = 0$. Since $[x, w, w] = 0$, polarization yields $[x, y, z] = -[x, z, y]$. Therefore $[w, y, w] = -[w, w, y] = 0$. ∎

THE CAYLEY–DICKSON PROCESS

Using the facts that $L_w^* = L_{\overline{w}}$ and $R_w^* = R_{\overline{w}}$ (as well as Lemma 6.10a), the identities (6.6') and (6.6'') can be written as

Lemma 6.12.
$$x(\overline{y}w) + y(\overline{x}w) = 2\langle x, y\rangle w,$$
$$(w\overline{y})x + (w\overline{x})y = 2\langle x, y\rangle w.$$

In particular, we obtain the key identities that motivate the Cayley–Dickson process:

Corollary 6.13. *If $x \perp y$, then $x\overline{y} = -y\overline{x}$ and*

$$(6.14) \qquad x(\overline{y}w) = -y(\overline{x}w) \text{ and } (w\overline{y})x = -(w\overline{x})y, \text{ for all } w.$$

These crucial equations enable us to interchange a pair of orthogonal vectors x and y.

Lemma 6.15. *Suppose A is a normed subalgebra (with $1 \in A$) of the normed algebra B and that $\varepsilon \in A^\perp$ is a unit vector orthogonal to A with $\|\varepsilon\| = \pm 1$. Then $A\varepsilon$ is orthogonal to A and*

$$(6.16) \quad (a + b\varepsilon)(c + d\varepsilon) = (ac \mp \overline{d}b) + (da + b\overline{c})\varepsilon \quad \text{for all } a, b, c, d \in A.$$

Note that if $\|\varepsilon\| = 1$, then $-\overline{d}b$ occurs, while if $\|\varepsilon\| = -1$, then $+\overline{d}b$ occurs.

Proof: First, we prove $A \perp A\varepsilon$. Since $1 \in A, x \in A$ if and only if $\overline{x} \in A$. Thus, if $a, b \in A$, then $\langle a, b\varepsilon\rangle = \langle \overline{b}a, \varepsilon\rangle = 0$ because $\overline{b}a \in A$.

Note that $\varepsilon^2 = \mp 1$ if and only if $\|\varepsilon\| = \pm 1$ since $\varepsilon \perp 1$ is pure imaginary (i.e., $\overline{\varepsilon} = -\varepsilon$ and $\varepsilon\overline{\varepsilon} = \|\varepsilon\|$ implies $\varepsilon^2 = -\varepsilon\overline{\varepsilon} = -\|\varepsilon\|$).

Finally, to prove (6.16) we use the key identities (6.14) several times, applied to the terms in

$$(a + b\varepsilon)(c + d\varepsilon) = ac + (b\varepsilon)(d\varepsilon) + a(d\varepsilon) + (b\varepsilon)c.$$

That is,

$$(b\varepsilon)(d\varepsilon) = -\overline{d}((\overline{b\varepsilon})\varepsilon) = \overline{d}((\varepsilon\overline{b})\varepsilon) = -\overline{d}((\varepsilon\overline{\varepsilon})b) = -\|\varepsilon\|\overline{d}b$$
$$a(d\varepsilon) = a(\varepsilon\overline{d}) = \varepsilon(\overline{a}\overline{d}) = \overline{(\overline{a}\overline{d})}\varepsilon = (da)\varepsilon$$
$$(b\varepsilon)c = (b\overline{c})\varepsilon.$$

Lemma 6.15 mandates the following doubling process.

Definition 6.17 (Cayley–Dickson). *Suppose A is a normed algebra. Motivated by Lemma 6.15, we define two algebras $A(+)$ and $A(-)$. As vector spaces both*
$$A(\pm) \equiv A \oplus A.$$

Multiplication is defined by

$$(6.18) \qquad (a,b)(c,d) \equiv (ac \mp \overline{d}b, da + b\overline{c}),$$

with the coefficient of $\overline{d}b$ taken to be -1 for $A(+)$ and $+1$ for $A(-)$.

Both $A(+)$ and $(A-)$ are easily seen to be algebras with multiplicative unit $1 \equiv (1,0)$. The algebra A is naturally a subalgebra of $A(\pm)$. In fact, given $a \in A$, we also let a denote $(a,0)$, embedding A as a subalgebra of both $A(+)$ and $A(-)$. Let ε denote $(0,1)$. Then

$$(a,b) = a + b\varepsilon.$$

Note that $\varepsilon^2 = -1$ for $A(+)$ and $\varepsilon^2 = +1$ for $A(-)$.

There is a natural square norm or quadratic form to impose on $A(+)$ and on $A(-)$. If $(a,b) = a + b\varepsilon \in A(\pm)$, let

$$(6.19) \qquad \|(a,b)\| = \|a\| \pm \|b\|.$$

Note that $\|\varepsilon\| = 1$ for $A(+)$ and $\|\varepsilon\| = -1$ for $A(-)$.

The associated inner product is given by

$$(6.19') \qquad \langle x, y \rangle = \langle a, c \rangle \pm \langle b, c \rangle,$$

if $x = a + b\varepsilon$ and $y = c + d\varepsilon$. Conjugation is defined by $\overline{(a,b)} = (\overline{a}, -b)$ or $\overline{a + b\varepsilon} = \overline{a} - b\varepsilon$. Real and imaginary parts are defined in terms of conjugation or equivalently in terms of the orthogonal splitting of $A(\pm)$ into $\mathbf{R}$ and $\mathbf{R}^{\perp}$.

Next, we collect together some of the formulas valid in $A(\pm)$. The proofs are straightforward calculations using the definition (6.18) of multiplication.

Lemma 6.20. *Suppose A is a normed algebra. Let $A(\pm)$ denote the algebra defined via the Cayley–Dickson process. Suppose $x \equiv a + \alpha\varepsilon$, $y = b + \beta\varepsilon$, $z = c + \gamma\varepsilon \in A(\pm)$.*

$$(6.21) \qquad \overline{xy} = \overline{y}\ \overline{x}.$$

$$(6.22) \qquad x\overline{x} = \overline{x}x = \|x\|.$$

$$(6.23) \qquad \frac{1}{2}(x\overline{y} + y\overline{x}) = \operatorname{Re} x\overline{y} = \langle x, y \rangle.$$

$$(6.24) \qquad \frac{1}{2}[x,y] = \frac{1}{2}[a,b] \pm \operatorname{Im}\overline{\alpha}\beta + (\beta\operatorname{Im}a - \alpha\operatorname{Im}b)\varepsilon.$$

(6.25)
$$[x, y, z] = \pm [a, \overline{\gamma}\beta] \pm [b, \overline{\alpha}\gamma] \pm [c, \overline{\beta}\alpha]$$
$$+ \alpha[\overline{b}, \overline{c}]\varepsilon + \beta[a, \overline{c}]\varepsilon + \gamma[a, b]\varepsilon \pm (\alpha\overline{\beta}\gamma - \gamma\overline{\beta}\alpha)\varepsilon,$$

assuming A is associative.

(6.26)
$$[x, \overline{x}, y] = \pm[a, \overline{\beta}, \alpha] + [\alpha, \overline{b}, a]\varepsilon.$$
$$\|x\| \, \|y\| - \|xy\| = \pm 2\langle a, [\overline{\beta}, \alpha, \overline{b}]\rangle.$$

Corollary 6.27. *Suppose $A(\pm)$ is the algebra defined via the Cayley–Dickson process from a normed algebra A.*

(6.28) $A(\pm)$ is commutative if and only if $A = \mathbf{R}$.

(6.29) $A(\pm)$ is associative if and only if A is commutative and associative.

(6.30) $A(\pm)$ is alternative, $A(\pm)$ is normed, and A is associative are all equivalent.

THE HURWITZ THEOREM

Corollary 6.27 can be used to deduce the properties of the euclidean algebras obtained via the Cayley–Dickson process. First, we list (some of) these algebras.

Definition 6.31.

$\mathbf{C} = \mathbf{R}(+)$	*the complex numbers.*
$\mathbf{H} = \mathbf{C}(+)$	*the quaternions or Hamiltonians.*
$\mathbf{O} = \mathbf{H}(+)$	*the octonians or Cayley numbers.*
$\mathbf{L} = \mathbf{R}(-)$	*the Lorentz numbers.*
$M_2(\mathbf{R}) = \mathbf{C}(-)$	*real 2×2 matrices.*
$\tilde{\mathbf{O}} = \mathbf{H}(-)$	*the split octonians.*

Remark. The normed algebra $M_2(\mathbf{R})$ of real 2×2 matrices (with $\|A\| \equiv \det_{\mathbf{R}} A$ defined to be the square norm or quadratic form) occurs in several different ways via the Cayley–Dickson process. By Corollary 6.38(a) below,

$$M_2(\mathbf{R}) = \mathbf{C}(-) = \mathbf{L}(+) = \mathbf{L}(-).$$

To show $M_2(\mathbf{R}) = \mathbf{C}(-)$, identify $\mathbf{C}$ with the 2×2 reals of the form

$$\begin{pmatrix} a & -b \\ b & a \end{pmatrix} = a + ib$$

and choose

$$\varepsilon \equiv \begin{pmatrix} 1 & 0 \\ 0 & -1 \end{pmatrix}$$

to be the matrix corresponding to conjugation on $\mathbf{R}^2 \cong \mathbf{C}$. Check that $\varepsilon \perp \mathbf{C}$ and $\varepsilon^2 = 1$. Then by Lemma 6.15, $\mathbf{C}(-)$ and $M_2(\mathbf{R})$ are algebra and norm isomorphic.

Corollary 6.32.

$$(6.33) \qquad \begin{array}{c} \mathbf{C} = \mathbf{R}(+) \text{ and } \mathbf{L} = \mathbf{R}(-) \text{ are commutative, associative,} \\ \text{and normed.} \end{array}$$

$$(6.34) \qquad \begin{array}{c} \mathbf{H} = \mathbf{C}(+) \text{ and } M_2(\mathbf{R}) \text{ are not commutative but associative} \\ \text{and normed.} \end{array}$$

$$(6.35) \qquad \begin{array}{c} \mathbf{O} = \mathbf{H}(+) \text{ and } \tilde{\mathbf{O}} = \mathbf{H}(-) \text{ are neither commutative} \\ \text{nor associative but are alternative and normed.} \end{array}$$

Proof: Apply Corollary 6.27. ∎

Remark 6.36. $\mathbf{O}(+)$ and $\mathbf{O}(-)$ also fail to be alternative or normed.

Theorem 6.37 (Hurwitz). *The only normed or euclidean algebras over* $\mathbf{R}$ *are*

$$\mathbf{R}, \quad \mathbf{C} \text{ and } \mathbf{L}, \quad \mathbf{H} \text{ and } M_2(\mathbf{R}), \quad \mathbf{O} \text{ and } \tilde{\mathbf{O}}.$$

Proof: Suppose B is a normed algebra. Let $A_1 \equiv \operatorname{Re} B = \mathbf{R}$. If $A_1 = B$, we are finished. If not, choose $\varepsilon_1 \in A_1^{\perp}$ a unit vector, i.e., $-\varepsilon_1^2 = \|\varepsilon_1\| = \pm 1$. Let $A_2 = A_1 + A_1\varepsilon_1$. By Lemma 6.15, A_2 is a normed subalgebra of B isomorphic to either $\mathbf{C} = \mathbf{R}(+)$ or $\mathbf{L} = \mathbf{R}(-)$. If $A_2 = B$, we are finished. Suppose not. Choose $\varepsilon_2 \in A_2^{\perp}$ a unit vector, and let $A_3 \equiv A_2 + A_2\varepsilon_2$.

Case $\mathbf{R}(+)$: Suppose $A_2 = \mathbf{C} = \mathbf{R}(+)$. Then by Lemma 6.15, A_3 is a normed subalgebra isomorphic to $\mathbf{H} = \mathbf{C}(+)$ or $M_2(\mathbf{R}) = \mathbf{C}(-)$.

Case $\mathbf{R}(-)$: Suppose $A_2 = \mathbf{L} = \mathbf{R}(-)$. Then $\varepsilon_1^2 = 1$, $\varepsilon_2^2 = \pm 1$, and $\varepsilon_3 \equiv \varepsilon_1\varepsilon_2$ satisfies $\varepsilon_3^2 = \mp 1$ since $\varepsilon_1\varepsilon_2 + \varepsilon_2\varepsilon_1 = -2\langle \varepsilon_1, \varepsilon_2 \rangle$. In particular, either $\varepsilon_2^2 = -1$ or $\varepsilon_3^2 = -1$. Say $\varepsilon_3^2 = -1$. Then interchange ε_1 and ε_3 (note this does not change A_3) in the above constructions so we are back in Case **R**(+). Thus, either $A_3 = \mathbf{C}(+)$ or $A_3 = \mathbf{C}(-)$. If $A_3 = B$, we are finished. Suppose not. Choose $\varepsilon_3 \in A_3^\perp$ a unit vector, and let $A_4 \equiv A_3 + A_3\varepsilon_3$.

Case C(+): If $A_3 = \mathbf{C}(+) = \mathbf{H}$, then by Lemma 6.15 A_4 is a normed algebra isomorphic to either $\mathbf{O} = \mathbf{H}(+)$ or $\tilde{\mathbf{O}} = \mathbf{H}(-)$.

Case C(−): If $A_3 = \mathbf{C}(-) = M_2(\mathbf{R})$, then interchanging either ε_3 or $\varepsilon_4 \equiv \varepsilon_2\varepsilon_3$ with ε_2 reduces us to Case **C**(+). If $A_4 = B$, we are finished. If $A_4 \neq B$, then by repeating the above process one more time we obtain either $\mathbf{O}(+)$ or $\mathbf{O}(-)$ as a normed subalgebra A_5 of B. Since $\mathbf{O}(+)$ and $\mathbf{O}(-)$ are not normed algebras, this is impossible. ∎

Corollary 6.38.

(a) *As normed algebras* $M_2(\mathbf{R}) = \mathbf{R}(-, -) = \mathbf{R}(+, -) = \mathbf{R}(-, +)$.
(b) *The seven normed algebras* $\mathbf{R}(\pm, \pm, \pm)$, *excluding* $\mathbf{R}(+, +, +) = \mathbf{O}$, *are all the same, namely* $\tilde{\mathbf{O}}$.

Remark. Because of the Hurwitz Theorem, all normed algebras are subalgebras of either $\mathbf{O}$ or $\tilde{\mathbf{O}}$. Consequently, it is frequently clearer to state results that are valid for all normed algebras as results for $\mathbf{O}$ and $\tilde{\mathbf{O}}$. The next theorem is an example, as are "cross products" in the next section.

The weak form of associativity for $\mathbf{O}$ and $\tilde{\mathbf{O}}$, corresponding to the fact that the associator vanishes if any two of the three variables are set equal, can be strengthened.

Theorem 6.39 (Artin). *A subalgebra with unit, generated by any two elements of either* $\mathbf{O}$ *or* $\tilde{\mathbf{O}}$, *is associative.*

Proof: Suppose A is generated by $1, x, y$. Let $S \equiv \text{span}\{\text{Im}\,x, \text{Im}\,y\}$. If $\dim S = 0$, then $A = \mathbf{R}$ and the proof is complete. If $\dim S = 1$, choose a nonzero vector $\varepsilon_1 \in S$ and note that since $\varepsilon_1^2 = -\|\varepsilon_1\|$ is real, $A = \{a + b\varepsilon_1 : a, b \in \mathbf{R}\}$. Now it is easy to verify directly that A is associative. If $\dim S = 2$, choose an orthogonal basis $\varepsilon_1, \varepsilon_2$ for S and note that by repeated use of Corollary 6.13 as in the proof of Lemma 6.15, A is spanned by (as a vector space) $1, \varepsilon_1, \varepsilon_2$, and $\varepsilon_1\varepsilon_2$. Now using the facts that $[\varepsilon_1, \varepsilon_2, \varepsilon_1\varepsilon_2], [\varepsilon_1, \varepsilon_1, \varepsilon_2]$, etc. vanish, we see that A is associative. ∎

Actually we can prove much more.

Proposition 6.40. *If* $S \equiv \text{span}\{\text{Im}\,x, \text{Im}\,y\}$ *is nondegenerate, then the subalgebra* A *is determined as follows:*

$$(6.41) \qquad \text{If } \dim S = 1, \text{ and } S \text{ is positive, then } A \cong \mathbf{C}.$$

(6.42)   If $\dim S = 1$, *and* S *is negative, then* $A \cong \mathbf{L}$.

(6.43)   If $\dim S = 2$, *and* S *is positive, then* $A \cong \mathbf{H}$.

                If $\dim S = 2$, *and* S *is negative or Lorentz,*

(6.44)   *then* $A = M_2(\mathbf{R})$.

Further, if $\dim S = 1$ *and* S *is null, then*

(6.45) $$A = \left\{ \begin{pmatrix} a & b \\ 0 & a \end{pmatrix} : a, b \in \mathbf{R} \right\}.$$

Proof: If $\dim S = 1$ and S is nondegenerate, choose $\varepsilon_1 \in S$ to be a unit vector. Then by Lemma 6.15, $A \cong \mathbf{R} + \mathbf{R}\varepsilon_1$ is either $\mathbf{C}$ or $\mathbf{L}$ depending on $\|\varepsilon_1\|$.

Suppose S is degenerate of dimension one. Choose a nonzero null vector u in S and a pure imaginary unit vector ε with $\langle u, \varepsilon \rangle \neq 0$. Then $\text{span}\{\varepsilon, u\}$ is a Lorentz plane $\cong \mathbf{L}$ containing S as a null line. Therefore, we may choose an orthonormal basis $\varepsilon_1, \varepsilon_2$ for $\mathbf{L}$ with $\varepsilon_1 + \varepsilon_2 \in S$. The algebra generated by $\varepsilon_1, \varepsilon_2$ is isomorphic to $M_2(\mathbf{R})$ with

$$\varepsilon_1 = \begin{pmatrix} 0 & 1 \\ -1 & 0 \end{pmatrix}, \quad \varepsilon_2 = \begin{pmatrix} 0 & 1 \\ 1 & 0 \end{pmatrix}.$$

Therefore,

$$S \cong \left\{ \begin{pmatrix} 0 & b \\ 0 & 0 \end{pmatrix} : b \in \mathbf{R} \right\}$$

and

$$A \cong \left\{ \begin{pmatrix} a & b \\ 0 & a \end{pmatrix} : a, b \in \mathbf{R} \right\}.$$

Now assume $\dim S = 2$. If S is nondegenerate, then S has an orthonormal basis $\varepsilon_1, \varepsilon_2$, and by the above arguments and Lemma 6.15, either $A = \mathbf{L} + \mathbf{L}\varepsilon_2$ or $A = \mathbf{C} + \mathbf{C}\varepsilon_2$. It follows that either $A = M_2(\mathbf{R})$ or $A \cong \mathbf{H}$. ∎

CROSS PRODUCTS

The results of this section are formulated for the octonians $\mathbf{O}$. However, all of the results are valid with $\mathbf{O}$ replaced by the split octonians $\tilde{\mathbf{O}}$. First, we consider the cross product of two octonians.

Definition 6.46. *Given* $x, y \in \mathbf{O}$, *the cross product of* x *and* y *is defined by*

$$(6.47) \qquad x \times y \equiv \frac{1}{2}\left(\overline{y}x - \overline{x}y\right) \equiv \operatorname{Im}\overline{y}x.$$

The next lemma justifies the term "cross product".

Lemma 6.48.

(a) $x \times y$ *is alternating on* $\mathbf{O}$.
(b) $\|x \times y\| = \|x \wedge y\|$ *for all* $x, y \in \mathbf{O}$.

Proof:
(a) The cross product is alternating because $x \times x = 0$ for all $x \in \mathbf{O}$.
(b) Since both sides are alternating, it suffices to prove (b) when $x \perp y$. Then $0 = \langle x, y\rangle = \operatorname{Re}\overline{x}y = \frac{1}{2}(\overline{x}y + \overline{y}x)$ so that $x \times y = \overline{y}x$. Therefore, $\|x \times y\| = \|\overline{y}x\| = \|y\|\,\|x\| = \|x \wedge y\|$ because of the Cauchy–Schwarz equality. ∎

Remark. The proof that $\|x \times y\| = \|x \wedge y\|$ explains why we needed the conjugates in the definition of $x \times y$; namely, we wanted the vanishing of $\langle x, y\rangle$ to imply that $x \times y$ could be expressed as one term $\overline{y}x$.

Note that

$$(6.49) \qquad \operatorname{Re} x \times y = 0 \text{ for all } x, y \in \mathbf{O}.$$

Frequently, the cross product is restricted to $\operatorname{Im}\mathbf{O}$, in which case $x \times y$ can be expressed in a variety of interesting ways.

Lemma 6.50. *If* $x, y \in \operatorname{Im}\mathbf{O}$, *then*

$$(6.51) \qquad x \times y = \frac{1}{2}[x, y] = xy + \langle y, x\rangle,$$

where $[x, y] \equiv xy - yx$ *is the commutator of* x *and* y.

Proof: $\langle x, y\rangle = -\frac{1}{2}(xy + yx)$ for $x, y \in \operatorname{Im}\mathbf{O}$. ∎

Remark 6.52. If $x, y \in \mathbf{R}^3 \equiv \operatorname{Im}\mathbf{H} \subset \operatorname{Im}\mathbf{O}$, then the cross product $x \times y$ is just the usual *vector cross product* on $\mathbf{R}^3$ based on the "right hand rule." This explains the choice of sign in the definition of $x \times y$.

Lemma 6.53. *If $x, y \in \operatorname{Im} \mathbf{O}$, then*

(a) $x \times y \in \operatorname{Im} \mathbf{O}$ *is orthogonal to* $\operatorname{span}\{x, y\}$.

(b) $x \times (x \times y) = -\|x\| y + \langle x, y \rangle x$.

Part (b) says that (for $\|x\| = 1$) the square of left cross product by x equals minus orthogonal projection from $\operatorname{Im} \mathbf{O}$ to $(\operatorname{span} x)^{\perp}$. See Problem 13, where this propety is used to reconstruct octonian multiplication from the cross product on $\operatorname{Im} \mathbf{O}$.

Proof:

(a) $\langle x, x \times y \rangle = \frac{1}{2}\langle x, xy \rangle - \frac{1}{2}\langle x, yx \rangle = \frac{1}{2}\|x\|\langle 1, y \rangle - \frac{1}{2}\|x\|\langle 1, y \rangle = 0$.

(b) Let $C_x y \equiv x \times y$. If $y = x$, then $C_x^2 = 0$. If $y \perp x$, then $xy = -yx$, so that $C_x^2 y = -\|x\| y$.

The natural extension of the cross product to three octonians x, y, z is the alternation of $x(\overline{y}z)$. (Note that this is not equal to $x \times (y \times z)$!) However, by repeated use of Corollary 6.13, the six-term expression obtained by alternating $x(\overline{y}z)$ can be simplified to two terms. Therefore, we adopt as our definition of $x \times y \times z$ the following two-term expression.

Definition 6.54. *Given $x, y, z \in \mathbf{O}$, the triple cross product of x, y, and z is defined by*

(6.55) $$ x \times y \times z \equiv \frac{1}{2}\left[x(\overline{y}z) - z(\overline{y}x) \right]. $$

Lemma 6.56.

(a) $x \times y \times z$ *is alternating on* $\mathbf{O}$.

(b) $\|x \times y \times z\| = \|x \wedge y \wedge z\|$ *for all* $x, y, z \in \mathbf{O}$.

Proof: (a) Since the subalgebra generated by any two elements is associative,

$$ x \times x \times z = \frac{1}{2}\left[x(\overline{x}z) - z\|x\| \right] = 0, $$

and

$$ z \times x \times x = \frac{1}{2}\left[z\|x\| - x(\overline{x}z) \right] = 0. $$

Obviously, $x \times z \times x = 0$.

(b) Since both sides are alternating, we may assume that x, y, and z are pairwise orthogonal. Repeated use of the key identities in Corollary 6.13 yield $-z(\overline{y}x) = x(\overline{y}z)$. That is,

(6.57) $$ x \times y \times z = x(\overline{y}z) \quad \text{if } x, y, z \text{ are orthogonal.} $$

Therefore,

$$(6.58) \qquad \|x \times y \times z\| = \|x(\overline{y}z)\| = \|x\| \, \|y\| \, \|z\| = \|x \wedge y \wedge z\|. \quad \blacksquare$$

Now consider the real-valued trilinear form ϕ on $\operatorname{Im} \mathbf{O}$ defined by

$$(6.59) \qquad \phi(x, y, z) = \langle x, yz \rangle \quad \text{for all } x, y, z \in \operatorname{Im} \mathbf{O}.$$

This form is called the *associative* 3-form for $\mathbf{O}$. The terminology is explained in the next chapter on calibrations. Since ϕ vanishes when any two of the variables $x, y, z \in \operatorname{Im} \mathbf{O}$ are set equal, ϕ must be alternating. That is,

$$(6.60) \qquad \phi \in \Lambda^3(\operatorname{Im} \mathbf{O})^*.$$

Perhaps Gureirch [8] was the first to consider this 3-form (in coordinates as in (6.74)).

It is natural to consider the triple cross product restricted to $\operatorname{Im} \mathbf{O}$.

Lemma 6.61. *If $x, y, z \in \operatorname{Im} \mathbf{O}$, then*

$$(6.62) \qquad \operatorname{Re} x \times y \times z = \phi(x \wedge y \wedge z),$$

$$(6.63) \qquad \operatorname{Im} x \times y \times z = \frac{1}{2}[x, y, z].$$

Proof: It suffices to prove the lemma when x, y, z are orthogonal. In this case, $x \times y \times z = x(\overline{y}z) = -x(yz)$ by (6.57). Thus $\operatorname{Re} x \times y \times z = -\langle 1, x(yz) \rangle = \langle x, yz \rangle = \phi(x \wedge y \wedge z)$. Also, $x \times y \times z = -z \times y \times x = z(yx)$ by (6.57). Thus $\overline{x \times y \times z} = (xy)z$. Consequently, $\operatorname{Im} x \times y \times z = \frac{1}{2}[-x(yz) + (xy)z] = \frac{1}{2}[x, y, z]$.

THE EXCEPTIONAL LIE GROUP G_2

Some of the automorphism groups of the normed algebras are *exceptional*. Let

$$(6.64) \qquad \operatorname{Aut}(A) \equiv \{g \in \operatorname{GL}(A) : g(xy) = g(x)g(y) \text{ for all } x, y \in A\}$$

denote the automorphism group of a finite dimensional algebra A. The group G_2 is most naturally defined as the automorphism group of the octonians

$$(6.65) \qquad G_2 \equiv \operatorname{Aut}(\mathbf{O}),$$

$$(6.66) \qquad \tilde{G}_2 \equiv \operatorname{Aut}(\tilde{\mathbf{O}}) \quad (\text{the } split \text{ case}).$$

Before examining G_2 and $\tilde{G}_2$, we discuss the automorphism groups of the other normed algebras besides $\mathbf{O}$ and $\tilde{\mathbf{O}}$. These other automorphism groups will be seen to be orthogonal groups or special orthogonal groups.

Lemma 6.67. *If A is a normed algebra, then* $\mathrm{Aut}(A) \subset O(\mathrm{Im}\, A)$ *the othogonal group of* $\mathrm{Im}\, A$.

Proof: Let $g \in \mathrm{Aut}(A)$. First note that $g(1) = 1$ since $g(x) = g(1)g(x)$ for all $x \in A$. Second, since $x^2 = (\mathrm{Re}\, x)^2 + (\mathrm{Im}\, x)^2 + 2(\mathrm{Re}\, x)(\mathrm{Im}\, x)$ and $(\mathrm{Im}\, x)^2 = -\|\mathrm{Im}\, x\| \in \mathrm{Re}\, A$, it follows that $x^2 \in \mathrm{Re}\, A$ if and only if x is either real or pure imaginary. Thus, if $x \in \mathrm{Im}\, A$, then $g(x)^2 = g(x^2) = x^2 g(1) \in \mathrm{Re}\, A$, so that $g(x)$ is either real or pure imaginary. Thus, $\mathrm{Aut}(A) \subset \mathrm{GL}(\mathrm{Im}\, A)$ and $\overline{g(x)} = g(\overline{x})$. Therefore,

$$\|g(x)\| = g(x)\overline{g(x)} = g(x)g(\overline{x}) = g(x\overline{x}) = g(\|x\|) = \|x\|g(1) = \|x\|.$$

This proves $g \in O(\mathrm{Im}\, A)$. ∎

Note that this proof also shows that $\mathrm{Aut}(\mathbf{O}(+)) \subset O(\mathbf{O}(+))$.

Corollary 6.68. $\mathrm{Aut}(\mathbf{C}) \cong \mathbf{Z}_2$ *and* $\mathrm{Aut}(\mathbf{L}) \cong \mathbf{Z}_2$ *each* $\mathbf{Z}_2$ *consist of the identity and conjugation.*

Lemma 6.69. *If $g \in \mathrm{Aut}(A)$, then g fixes the associative 3-form ϕ for A.*

Proof: Here $\phi(x, y, z) \equiv \langle x, yz \rangle$ for all $x, y, z \in \mathrm{Im}\, A$. Suppose $g \in \mathrm{Aut}(A)$. Then

$$\langle g(x), g(y)g(z) \rangle = \langle g(x), g(yz) \rangle = \langle x, yz \rangle$$

since g is orthogonal. ∎

If $A \equiv \mathbf{H}$, then the associative form ϕ is the unit volume form on $\mathrm{Im}\, \mathbf{H}$ corresponding to the orientation determined by $\{i, j, k\}$. Thus, if $g \in \mathrm{Aut}(\mathbf{H}) \subset (\mathrm{Im}\, \mathbf{H})$, then, by Lemma 6.69, $\det g = 1$.
  Let

$$C \equiv \begin{pmatrix} 1 & 0 \\ -0 & -1 \end{pmatrix} \; (conjugation),$$

$$J \equiv \begin{pmatrix} 0 & -1 \\ 1 & 0 \end{pmatrix} \; (complex\ structure),$$

and

$$R \equiv \begin{pmatrix} 0 & 1 \\ 1 & 0 \end{pmatrix} \; (reflection\ or\ Lorentz\ structure)$$

denote the standard orthonormal basis for $\mathrm{Im}\, M_2(\mathbf{R})$. Then ϕ (the associative form) is the unique unit volume form determined by the orientation $\{C, J, R\}$, since $\phi(C \wedge J \wedge R) = \langle C, JR \rangle = -\langle C, C \rangle = C^2 = 1$. Thus, if $g \in \mathrm{Aut}(M_2(\mathbf{R})) \subset O(\mathrm{Im}\, M_2(\mathbf{R}))$, then $\det g = 1$, because of Lemma 6.69.

Proposition 6.70.

$$\text{Aut}(\mathbf{H}) = \text{SO}(\text{Im}\,\mathbf{R}) = \text{SO}(3,0).$$
$$\text{Aut}(M_2(\mathbf{R})) = \text{SO}(\text{Im}\,M_2(\mathbf{R})) = \text{SO}(1,2).$$

Proof: See Problem 6.7. ∎

This completes our discussion of the automorphism groups of the normed algebras except for G_2 and $\tilde{G}_2$. Sometimes, in order to avoid repetition, results will only be stated for G_2. However, these results will be stated in such a way that no modifications will be necessary for $\tilde{G}_2$ other than replacing $\mathbf{O}$ by $\tilde{\mathbf{O}}$.

The 4-form ψ on $\text{Im}\,\mathbf{O}$ defined by

$$(6.71) \qquad \psi(x,y,z,w) = \frac{1}{2}\langle x, y(\overline{z}w) - w(\overline{z}y)\rangle \quad \text{for all } x,y,z,w \in \text{Im}\,\mathbf{O}$$

is called the *coassociative* 4-*form* for $\mathbf{O}$. If any two of the four variables $x,y,z,w \in \text{Im}\,\mathbf{O}$ are set equal, then $\psi(x,y,z,w) = 0$. (See Problem 6.8). Thus

$$(6.72) \qquad\qquad \psi \in \Lambda^4(\text{Im}\,\mathbf{O})^*$$

is a skew tensor of degree 4.

Lemma 6.73. *If* $g \in \text{Aut}(\mathbf{O})$, *then* g *fixes the coassociative 4-form* $\psi \in \Lambda^4(\text{Im}\,\mathbf{O})^*$.

Proof: Apply the facts that $g \in O(\text{Im}\,\mathbf{O}), g \in \text{Aut}(\mathbf{O})$, and $g(\overline{x}) = \overline{g(x)}$ to the definition of ψ. ∎

One can choose an orthonormal basis $e_1,\ldots,e_7$ for $\text{Im}\,\mathbf{O}$, with dual basis $\omega_1,\ldots,\omega_7$ for $(\text{Im}\,\mathbf{O})^*$, so that direct computation yields

$$(6.74) \quad \phi = \omega_{123} - \omega_{156} - \omega_{426} - \omega_{453} - \omega_{147} - \omega_{257} - \omega_{367}$$

and

$$(6.75) \quad \psi = \omega_{4567} - \omega_{4237} - \omega_{1537} - \omega_{1267} - \omega_{2536} - \omega_{1436} - \omega_{1425},$$

where $\omega_{ijk} \equiv \omega_i \wedge \omega_j \wedge \omega_k$, etc..

Therefore,

Lemma 6.76. $\phi \wedge \psi = 7\omega_{1234567}$.

Corollary 6.77.

(a) $G_2 \subset SO(\text{Im}\,\mathbf{O}) = SO(7)$,

(b) $\tilde{G}_2 \subset SO(\text{Im}\,\tilde{\mathbf{O}}) = SO(3,4)$.

Proof: $g \in G_2 \equiv \text{Aut}(\mathbf{O})$ has $\det g = 1$ since g fixes ϕ and ψ and $\phi \wedge \psi$ is a nonzero volume form. The proof for $\tilde{G}_2$ is analogous. ∎

Remark 6.78. The unit volume element $\frac{1}{7}\phi \wedge \psi$ provides an orientation for $\text{Im}\,\mathbf{O}$. Using this orientation the Hodge star operator $\star : \Lambda\,\text{Im}\,\mathbf{O} \to \Lambda\,\text{Im}\,\mathbf{O}$ is well-defined. One can easily check that

$$(6.79) \qquad\qquad\qquad \psi = \star\phi.$$

Similarly, $\psi = \star\phi$ in the split case $\tilde{\mathbf{O}}$.

As noted above, if $g \in G_2$, then $g^*\phi = \phi$. The converse is more difficult.

Theorem 6.80 (Bryant).

$$(6.81) \qquad\qquad G_2 = \{g \in GL(\text{Im}\,\mathbf{O}) : g^*\phi = \phi\}.$$

Proof: Assume for the moment that

$$(6.82) \qquad \text{if } g \in GL(\text{Im}\,\mathbf{O}) \text{ satisfies } g^*\phi = \phi, \text{ then } g \in O(\text{Im}\,\mathbf{O}).$$

Then, for all $x, y, z \in \text{Im}\,\mathbf{O}$,

$$\langle g(x), g(y \times z) \rangle = \langle x, y \times z \rangle = \phi(x \wedge y \wedge z) = (g^*\phi)(x \wedge y \wedge z)$$
$$= \phi(g(x) \wedge g(y) \wedge g(z)) = \langle g(x), g(y) \times g(z) \rangle.$$

This proves that g preserves the cross product:

$$(6.83) \qquad g(y \times z) = g(y) \times g(z) \quad \text{for all } y, z \in \text{Im}\,\mathbf{O}.$$

Therefore, by Problem 11, $g \in \text{Aut}(\mathbf{O}) \equiv G_2$. It remains to prove (6.82). The identity (Problem 14 (c))

$$(6.84) \qquad\qquad (x \lrcorner \phi) \wedge (x \lrcorner \phi) \wedge \phi = 6\|x\|\lambda,$$

where λ is the standard volume element for $\text{Im}\,\mathbf{O}$, can be polarized to yield

$$(6.85) \qquad (x \lrcorner \phi) \wedge (y \lrcorner \phi) \wedge \phi = 6\langle x, y \rangle \lambda \quad \text{for all } x, y \in \text{Im}\,\mathbf{O}.$$

Now suppose $g \in \mathrm{GL}(\mathrm{Im}\,\mathbf{O})$ satisfies $g^*\phi = \phi$. Applying g^* to (6.85) yields

$$(g^{-1}(x) \lrcorner \phi) \wedge (g^{-1}(y) \lrcorner \phi) \wedge \phi = 6\langle x, y \rangle (\det g)\lambda,$$

while replacing x by $g^{-1}(x)$ and y by $g^{-1}(y)$ in (6.85) yields

$$(g^{-1}(x) \lrcorner \phi) \wedge (g^{-1}(y) \lrcorner \phi) \wedge \phi = \langle g^{-1}(x), g^{-1}(y) \rangle \lambda.$$

Therefore,

(6.86) $\langle g(x), g(y) \rangle = (\det g)^{-1} \langle x, y \rangle$ for all $x, y \in \mathrm{Im}\,\mathbf{O}$.

Therefore, g is either conformal ($\det g > 0$) or anticonformal ($\det g < 0$).

Since the signature of $\mathbf{O}$ is *not* split, g cannot be anticonformal (see Problem 2.3). Thus, g is conformal with conformal factor $\lambda = (\det g)^{-1} > 0$. Recall (Problem 2.9(d)) that if g is a conformal transformation with conformal factor λ, then

(6.87) $$(\det g)^2 = \lambda^n,$$

where n is the dimension. Substituting $\lambda = (\det g)^{-1}$ and $n \equiv 7$, we obtain

$$(\det g)^9 = 1 \quad \text{or} \quad \det g = 1.$$

Thus, we have shown that

(6.88)   if $g \in \mathrm{GL}(\mathrm{Im}\,\mathbf{O})$ satisfies $g^*\phi = \phi$, then $g \in \mathrm{SO}(\mathrm{Im}\,\mathbf{O})$,

completing the proof of Theorem 6.80. ∎

This theorem describes G_2 as a level set of the function from $\mathrm{End}(\mathrm{Im}\,\mathbf{O})$ to $\Lambda^3(\mathrm{Im}\,\mathbf{O})^*$ that maps $A \in \mathrm{End}(\mathrm{Im}\,\mathbf{O})$ to $A^*\phi \in \Lambda^3(\mathrm{Im}\,\mathbf{O})^*$. The linearization of this map at the identity $I \in \mathrm{End}(\mathrm{Im}\,\mathbf{O})$ is the map sending $A \in \mathrm{End}(\mathrm{Im}\,\mathbf{O})$ to

$$D(A) \equiv \frac{d}{dt}(I + tA)^*\phi|_{t=0}.$$

Now

(6.89)
$$D(A)(x \wedge y \wedge z) = \frac{d}{dt}\phi((x + tAx) \wedge (y + tAy) \wedge (z + tAz))\bigg|_{t=0}$$
$$= \phi((Ax) \wedge y \wedge z) + \phi(x \wedge (Ay) \wedge z) + \phi(x \wedge y \wedge (Az)).$$

Thus, $D(A)$ is just the action of A on $\phi \in \Lambda^3(\mathrm{Im}\,\mathbf{O})^*$, considering A as a *derivation*.

Lemma 6.90. *The linear map* (6.89),

$$D : \mathrm{End}(\mathrm{Im}\,\mathbf{O}) \to \Lambda^3(\mathrm{Im}\,\mathbf{O})^*,$$

is surjective with 14-dimensional kernel.

Because of the implicit function theorem, we have several intereting consequences.

Corollary 6.91. G_2 *is a closed 14-dimensional submanifold of* $\mathrm{End}(\mathrm{Im}\,\mathbf{O})$ *implicitly defined by* $G_2 = \{A \in \mathrm{End}(\mathrm{Im}\,\mathbf{O}) : A^*\phi = \phi\}$. *The tangent space* $\mathfrak{g}_2$ *to* G_2 *at the identity is* $\ker D$.

Corollary 6.92. *The orbit of* ϕ *under* $\mathrm{GL}(\mathrm{Im}\,\mathbf{O})$ *is an open subset of* $\Lambda^3(\mathrm{Im}\,\mathbf{O})^*$.

Proof of Lemma 6.90: Since $\dim_{\mathbb{R}} \mathrm{End}(\mathrm{Im}\,\mathbf{O}) = 49$ and $\dim \Lambda^3(\mathrm{Im}\,\mathbf{O})^* = 35$, the map D must have a kernel of at least 14 dimensions; and the kernel is exactly 14-dimensional if and only if D is surjective. Therefore, it suffices to prove that

$$(6.93) \qquad\qquad \dim \ker D \le 14.$$

First, we derive another formula for $D(A)$ in the special case where $A \equiv u \otimes \alpha \in (\mathrm{Im}\,\mathbf{O}) \otimes (\mathrm{Im}\,\mathbf{O})^* \cong \mathrm{End}(\mathrm{Im}\,\mathbf{O})$ is simple, i.e., $Ax = \alpha(x)u$, where $\alpha \in (\mathrm{Im}\,\mathbf{O})^*, u \in \mathrm{Im}\,\mathbf{O}$.

$$\begin{aligned}
D(\alpha \otimes u)(x \wedge y \wedge z) &= \alpha(x)\phi(u \wedge y \wedge z) \\
&\quad + \alpha(y)\phi(x \wedge u \wedge z) + \alpha(z)\phi(x \wedge y \wedge z) \\
&= (u \lrcorner \phi)(\alpha(x)y \wedge z - \alpha(y)x \wedge z + \alpha(z)x \wedge y) \\
&= (\alpha \wedge (u \lrcorner \phi))(x \wedge y \wedge z).
\end{aligned}$$

That is,

$$(6.94) \qquad\qquad D(u \otimes \alpha) = \alpha \wedge (u \lrcorner \phi).$$

Consider the linear map

$$B : \Lambda^3(\mathrm{Im}\,\mathbf{O})^* \to S^2(\mathrm{Im}\,\mathbf{O})^*$$

defined by

$$(6.95) \qquad \star B(\beta)(x, y) \equiv (x \lrcorner \phi) \wedge (y \lrcorner \phi) \wedge \beta \quad \text{for } x, y \in \mathrm{Im}\,\mathbf{O}.$$

Assume, for the moment, that

(6.96) $$\frac{1}{2}BD(A) = A + A^* + \text{trace } A$$

has been verified. Then, if $A \in \ker D$,

(6.97) $$A + A^* + \text{trace } A = 0.$$

Taking the trace, $2 \text{ trace } A + 7 \text{ trace } A = 0$, so that

(6.98) $$A + A^* = 0 \quad \text{if } A \in \ker D.$$

That is,

(6.99) $$\ker D \subset \text{Skew}(\text{Im } \mathbf{O}).$$

Let $L \equiv \{A_u : u \in \text{Im } \mathbf{O}\}$, where

$$A_u x \equiv u \times x \equiv \frac{1}{2}(ux - xu) \quad \text{for all } x \in \text{Im } \mathbf{O}.$$

Then L is a 7-dimensional subspace of the 21-dimensional vector space $\text{Skew}(\text{Im } \mathbf{O})$. It is straightforward to verify that

(6.100) $$D(A_u)(x \wedge y \wedge z) = -\frac{3}{2}\langle u, [x, y, z]\rangle$$

(see Problem 6.15). The associator $[x, y, z]$ takes on all values in $\text{Im } \mathbf{O}$. Therefore, $D(A_u) = 0$ implies $u = 0$. That is,

$$(\ker D) \cap L = \{0\}.$$

To complete the proof of the Lemma 6.90, the identity (6.97) must be verified. It suffices to assume $A = u \otimes \alpha$ is simple. Then

$$\frac{1}{2}BD(u \otimes \alpha) = \frac{1}{2}B(\alpha \wedge (u \rfloor \phi),$$

applied to x $x \in \text{Im } \mathbf{O}$ is equal to

(6.101) $$\frac{1}{2}(x \rfloor \phi) \wedge (x \rfloor \phi) \wedge \alpha \wedge (u \rfloor \phi).$$

However, $A + A^* + \text{trace } A$ applied to x $x \in \text{Im } \mathbf{O}$ is equal to

(6.102) $$\alpha(u)\|x\| + 2\alpha(x)\langle u, x\rangle.$$

120 Problems

Finally, (6.101) and (6.102) are equal; see Problem 6.14(b). ∎

PROBLEMS

1. Suppose A is an algebra with an inner product. Assume 1 is not null, and use this fact to define conjugation. Show that if A has the two properties

   (a) $\|x\| = x\,\overline{x}$, $\overline{xy} = \overline{y}\,\overline{x}$,

   (b) any subalgebra generated by two elements is associative, then A is normed.

2. Verify the *Moufang identities*:

$$(xyx)z = x(y(xz)) \quad (L_{xyx} = L_x L_y L_x)$$
$$z(xyx) = ((zx)y)x \quad (R_{xyx} = R_x R_y R_x)$$
$$(xy)(zx) = x(yz)x$$

   for all x, y, z in $\mathbf{O}$ or $\tilde{\mathbf{O}}$.

   Hint: First show that the difference of the two sides vanishes if any two of the variables are equal.

3. Suppose A is a normed algebra.

   (a) Show that each nonnull element of A has a unique left and right inverse.

   (b) Given $x, y \in A$ with $\|x\| \neq 0$, show that the equations $xw = y$ and $wx = y$ can be uniquely solved for w with $w = \overline{x}y/\|x\|$ and $w = y\overline{x}/\|x\|$ respectively.

   Note that (a) does not automatically imply (b), since (a) is true for $\mathbf{O}(+)$ but not (b).

4. A complex algebra with unit, equipped with a nondegenerate complex symmetric bilinear form $\langle\,,\,\rangle$ satisfying $\|xy\| = \|x\|\,\|y\|$, where $\|x\| = \langle x, x \rangle$, is called a *complex normed algebra*.

   (a) Suppose $A, \langle\,,\,\rangle$ is a normed algebra. Show that $A \otimes_{\mathbf{R}} \mathbf{C}$ is a complex normed algebra, where $\langle\,,\,\rangle$ is extended from A to $A \otimes_{\mathbf{R}} \mathbf{C}$ to be complex bilinear.

   (b) Consider the following four complex normed algebras:

$$\mathbf{C}, \quad \mathbf{C} \oplus \mathbf{C}, \quad M_2(\mathbf{C}), \quad \mathbf{O} \otimes_{\mathbf{R}} \mathbf{C}.$$

   Show that each complex normed algebra is isomorphic and isometric to one of these four.

5. Assuming that each prime number p can be written as the sum of four squares, $p = n_1^2 + n_2^2 + n_3^2 + n_4^2$ ($n_1, n_2, n_3, n_4 \in \mathbf{N}$), show that each natural number $n \in \mathbf{N}$ can be written as the sum of four squares.

6. Show that, for all $x, y, z \in \mathbf{O}$,
   (a) $x \times y = \frac{1}{2}[x, y] - (x_1 y' - y_1 x')$,
   (b) $x \times y \times z = \langle x', y'z' \rangle + \frac{1}{2}[x, y, z] + \frac{1}{2}x_1[z, y] + \frac{1}{2}y_1[x, z] + \frac{1}{2}z_1[y, x]$.

7. Use the Cayley–Dickson process to complete the proof of Proposition 6.70.

8. Show that the coassociative form ψ, defined by (6.71), is skew.

9. (a) Let S^6 denote the unit sphere in $\mathrm{Im}\,\mathbf{O}$. Show that G_2 acts transitively on S^6 with isotropy subgroup SU_3:

$$G_2/\mathrm{SU}_3 \cong S^6 \subset \mathrm{Im}\,\mathbf{O}.$$

(b) Let $V_{7,2}$ denote the Stiefel manifold of ordered pairs of orthonormal vectors in $\mathrm{Im}\,\mathbf{O}$. Show that G_2 acts transitively on $V_{7,2}$ with isotropy subgroup SU_2:

$$G_2/\mathrm{SU}_2 \cong V_{7,2}.$$

(c) Let $V_{7,3}(\phi = 0)$ denote the collection of ordered orthonormal triples $e_1, e_2, e_3 \in \mathrm{Im}\,\mathbf{O}$ with $\phi(e_1 \wedge e_2 \wedge e_3) = 0$ (i.e., $e_3 \perp e_1 e_2$). Show that G_2 acts transitively on $V_{7,3}(\phi = 0)$ with no isotropy:

$$G_2 \cong V_{7,3}(\phi = 0).$$

10. Consider the algebra homomorphism

$$\phi : \mathbf{H} \otimes_{\mathbf{R}} \mathbf{H} \to M_4(\mathbf{R})$$

induced on $\mathbf{H} \otimes_{\mathbf{R}} \mathbf{H}$ by defining ϕ on simple tensors to be

$$\phi(p \otimes q)(x) \equiv px\overline{q} \quad \text{for all } x \in \mathbf{R}^4 \cong \mathbf{H}.$$

Let $\mathbf{H} \otimes_{\mathbf{R}} \mathbf{H}$ have the inner product defined by $\langle a \otimes b, c \otimes d \rangle = \langle a, c \rangle \langle b, c \rangle$, and let $M_4(\mathbf{R})$ have the inner product defined by $\langle A, B \rangle \equiv \frac{1}{4}$ trace AB^t.

Prove that ϕ preserves inner products and hence is an algebra isomorphism (cf. Remark 1.33).

Hint: Show that trace $\phi(a \otimes b) = 0$ if b equals i, j, or k.

11. Show that

$$G_2 = \{g \in \mathrm{GL}(\mathrm{Im}\,\mathbf{O}) : g(x \times y) = g(x) \times g(y) \quad \text{for all } x, y \in \mathrm{Im}\,\mathbf{O}\}.$$

Hint: $x' \times y' = x'y' + \langle x', y' \rangle$ for all $x', y' \in \text{Im}\,\mathbf{O}$.

12. Suppose $u \in \text{Im}\,\mathbf{O}$ and $\|u\| = 1$. Show that, for all $x, y, z \in \mathbf{O}$,
    (a) $(xu) \times (yu) = u(x \times y)\overline{u}$,
    (b) $(xu) \times (yu) \times (zu) = (x \times y \times z)u$.

13. A *cross product* on an inner product space $V, \langle\,,\,\rangle$ is usually defined to be a bilinear map $x \times y$ from $V \times V$ to V that satisfies not only
    (a) $\|x \times y\| = \|x \wedge y\|$
    but also
    (b) $x \times (x \times y) = -\|x\|y + \langle x, y \rangle x$.
    Adopt this (stronger) standard definition. Given a cross product on an inner product space $V, \langle\,,\,\rangle$ define a product on $\mathbf{R} \oplus V$ by

    $$(x_0 + x)(y_0 + y) = x_0 y_0 + x_0 y + y_0 x + \langle x, y \rangle + x \times y$$
    for all $x_0, y_0 \in \mathbf{R}$ and $x, y \in V$.

    Define $\langle x_0 + x, y_0 + y \rangle \equiv x_0 y_0 + \langle x, y \rangle$. Show that $\mathbf{R} \oplus V, \langle\,,\,\rangle$ is a normed algebra.

14. Let λ denote the standard volume element on $\text{Im}\,\mathbf{O}$.
    (a) Show that $(x \lrcorner \phi) \wedge (x \lrcorner \phi) \wedge (x \lrcorner \phi) = 6\|x\|(x \lrcorner \lambda)$.
    Hint: Set $x \equiv e_1 \equiv i$ and note that $\omega \equiv x \lrcorner \phi$ is the Kähler form on $[i]^\perp$ under the complex structure right multiplication by i, while $x \lrcorner \lambda$ is the unit volume element on $[i]^\perp \cong \mathbf{C}^3$.
    (b) Show that $(u \lrcorner \phi) \wedge (x \lrcorner \phi) \wedge (x \lrcorner \phi) = 2\left(\|x\|u + 2\langle x, u \rangle x\right) \lrcorner \lambda$.
    (c) Show that $(x \lrcorner \phi) \wedge (x \lrcorner \phi) \wedge \phi = 6\|x\|\lambda$.
    Hint: Use (a) and $x \wedge (x \lrcorner \lambda) = \|x\|\lambda$.

15. (a) Given a 3-form $\phi \in \Lambda^3(\mathbf{R}^7)^*$ in 7-variables and a volume element λ on $\mathbf{R}^7$, let

    $$(x \lrcorner \phi) \wedge (x \lrcorner \phi) \wedge \phi = \langle x, x \rangle_\phi \lambda$$

    define a real bilinear form on $\mathbf{R}^7$. If ϕ is nondegenerate (i.e., $\langle\,,\,\rangle_\phi$ is nondegenerate), exhibit either an octonian structure with $\mathbf{R}^7 \cong \text{Im}\,\mathbf{O}$, or a split octonian structure with $\mathbf{R}^7 \cong \text{Im}\,\tilde{\mathbf{O}}$, so that $\phi(x \wedge y \wedge z) = \langle x, yz \rangle_\phi$ is the associative 3-form.
    (b) (Gureirch [8]) Show that $GL(7, \mathbf{R})$ acting on $\Lambda^3(\mathbf{R}^7)^*$ has two open orbits with isotropy G_2 and $\tilde{G}_2$ (at the associative three form ϕ and the split associative three form $\tilde{\phi}$, respectively).

16. Suppose N is one of the normed algebras $\mathbf{C}, \mathbf{H}$, or $\mathbf{O}$. Let $\text{Herm}(2, N) \equiv \{A \in M_2(N) : \overline{A}^t = A\}$. Define $\mathbf{P}^1(N) \equiv \{A \in \text{Herm}(2, N) : A^2 =$

A and trace $A = 1\}$. Given a unit vector $a = (a_1, a_2) \in N^2$, define

$$[a] \equiv a^t\,\overline{a} = \begin{pmatrix} |a_1|^2 & a_1\overline{a}_2 \\ a_2 a_1 & |a_2|^2 \end{pmatrix}.$$

(a) Show that $[a] \in \mathbf{P}^1(N)$, and that $[b] = [a]$ if and only if $b = a\lambda$ for some $\lambda \in N$.

(b) Let $S(N^p)$ denote the unit sphere in N^p and verify that

$$S(N) \to S(N^2) \to \mathbf{P}^1(N)$$

is a fibering of spheres by spheres. That is,

$$S^1 \to S^{\,3} \to \mathbf{P}^1(\mathbf{C}) \cong S^2 \quad \text{(Hopf fibration)},$$
$$S^3 \to S^{\,7} \to \mathbf{P}^1(\mathbf{H}) \cong S^4,$$
$$S^7 \to S^{15} \to \mathbf{P}^1(\mathbf{O}) \cong S^8.$$

(c) Show that the Hopf fibration

$$S^1 \to S^3 \to S^2$$

can also be defined by sending

$$x \in S^3 \subset \mathbf{H} \quad \text{to} \quad x\,i\,\overline{x} \in S^2 \subset \mathrm{Im}\,\mathbf{H}.$$

Note: Setting $x \equiv z + jw$ with $z, w \in \mathbf{C}$ yields a formula for an isomorphism from

$$\mathbf{P}^1(\mathbf{C}) \equiv \mathbf{C}^2 - \{0\}/\sim \quad \text{to} \quad S^2 \subset \mathrm{Im}\,\mathbf{H} \cong \mathbf{R}^3$$

in terms of quaternion multiplication. Namely,

$$[(z, w)] \quad \text{is mapped to} \quad \frac{(z + jw)i(\overline{z} - jw)}{|z|^2 + |w|^2} \in \mathbf{H} \cong \mathbf{C} \oplus j\mathbf{C}.$$

17. Consider the representations of G_2 induced on $\Lambda^k(\mathrm{Im}\,\mathbf{O})^*$ by the standard action of G_2 on $\mathrm{Im}\,\mathbf{O}$. Show that these representations decompose as

   (a) $\Lambda^2(\mathrm{Im}\,\mathbf{O})^* = \Lambda_7^2 \oplus \Lambda_{14}^2$,
   (b) $\Lambda^3(\mathrm{Im}\,\mathbf{O})^* = \Lambda_1^3 \oplus \Lambda_7^3 \oplus \Lambda_{27}^3$,

where

$$\Lambda_7^2 = \{\omega_u : u \in \operatorname{Im} \mathbf{H}\},$$
$$\omega_u = u \lrcorner \phi \text{ or equivalently } \omega_u(x, y) \equiv \langle u, x \times y \rangle,$$
$$\Lambda_4^2 \cong \mathfrak{g}_2,$$
$$\Lambda_7^3 = \{u \lrcorner \phi : u \in \operatorname{Im} \mathbf{H}\},$$
$$\Lambda_1^3 = \operatorname{span} \phi,$$

and

$$\Lambda_1^3 \oplus \Lambda_{27}^3 = \{D(A) : A \in \operatorname{Sym}(\operatorname{Im} \mathbf{O})\},$$

with D the linear map defined in (6.89) or (6.94).

7. Calibrations

Throughout this chapter, we restrict attention to $\mathbf{R}^n$ with the standard positive definite inner product $\langle\,,\,\rangle$. The concept of a "calibration" has been alluded to in the previous discussion of Proposition 4.15. This result states that a straight line segment in $\mathbf{R}^n$ minimizes length. The "Fundamental Theorem" of the theory of "calibrations" is a straightforward generalization of this proposition replacing curves by higher dimensional submanifolds in $\mathbf{R}^n$. In this chapter, it is assumed that the reader is familiar with topics from advanced calculus, such as submanifolds of $\mathbf{R}^n$ and Stokes' Theorem.

THE FUNDAMENTAL THEOREM

Again, we identify an oriented p-dimensional linear subspace of $\mathbf{R}^n$ with the element $\xi \equiv e_1 \wedge \ldots \wedge e_p \in \Lambda^p \mathbf{R}^n$ where $e_1, \ldots, e_p$ is an oriented orthonormal basis for the p-dimension subspace. Thus, $G(p, \mathbf{R}^n) \equiv \{\xi \in \Lambda^p \mathbf{R}^n : \xi = e_1 \wedge \cdots \wedge e_p$ for some $e_1, \ldots, e_p$ orthonormal in $\mathbf{R}^n\}$ denotes the grassmannian of oriented p-dimensional subspaces of $\mathbf{R}^n$.

The dual space of $\Lambda^p \mathbf{R}^n$ is $\Lambda^p(\mathbf{R}^n)^*$, the space of p-forms (with constant coefficients). If $e_1, \ldots, e_n$ is an orthonormal basis for $\mathbf{R}^n$ and $\alpha^1, \ldots, \alpha^n$ is the dual basis for $(\mathbf{R}^n)^*$ (for example $e_1 \equiv \partial/\partial x^1, \ldots, e_n \equiv \partial/\partial x^n$ the standard basis and $\alpha^1 \equiv dx^1, \ldots, \alpha^n \equiv dx^n$ the standard dual basis), then each p-form ϕ on an open subset of $\mathbf{R}^n$ can be expressed as $\phi = \sum'_{|I|=p} \phi_I \alpha^I$, where $I = (i_1, \ldots, i_p)$ is a multi-index of length

$|I| = p$, $\alpha^I = \alpha^{i_1} \wedge \cdots \wedge \alpha^{i_p}$, and $\sum'$ denotes summation over strictly increasing multi-indices, i.e., $i_1 < \cdots < i_p$. The coefficients ϕ_I are functions on the open subset of $\mathbf{R}^n$, given in terms of ϕ by $\phi_I = \phi(e_I)$.

Definition 7.1. *A p-form ϕ on an open subset U of $\mathbf{R}^n$ is said to be a calibration if*

(a) $d\phi = 0$ on U, and,
(b) for each fixed point $x \in U$, the form $\phi_x \in \Lambda^p(\mathbf{R}^n)$ satisfies $\phi_x(\xi) \le 1$ for all $\xi \in G(p, \mathbf{R}^n)$, with the contact set

$$G(\phi) \equiv \{\xi \in G(p, \mathbf{R}^n) : \phi(\xi) = 1\}$$

nonempty.

Condition (b) says that the maximum of ϕ on the compact set $G(p, \mathbf{R}^n)$ is one.

Recall that a p-form ϕ can be integrated over an oriented p-dimensional submanifold M of $\mathbf{R}^n$. By using a partition of unity (so that one can assume ϕ is supported in a coordinate chart), this integral $\int_M \phi$ can be defined by pulling back ϕ (with a parameterization map) to an open subset of $\mathbf{R}^p$ and then performing ordinary $\mathbf{R}^p$ integration.

Using the inner product $\langle\,,\,\rangle$ on $\mathbf{R}^n$ this integral can be expressed in terms of volume measure on M. For each point $x \in M$, let $\lambda_M \in \lambda^p(\mathbf{R}^n)^*$ denote the unit volume form for M, and let $\overrightarrow{M}$ denote the unit volume element for M; that is, $\overrightarrow{M} \equiv e_1 \wedge \cdots \wedge e_p$ and $\lambda_M \equiv \alpha^1 \wedge \cdots \wedge \alpha^p$, where $e_1, \ldots, e_p$ is an oriented orthonormal basis for $T_x M$ and $\alpha^1, \ldots, \alpha^p$ is an oriented orthonormal basis for $T_x^* M$. Then

$$(7.2) \qquad \int_M \phi = \int \phi(\overrightarrow{M}) \lambda_M,$$

i.e., $\int_M \phi$ equals the integral of the scalar valued function $\phi(\overrightarrow{M})$ over M with respect to volume measure over M.

Definition 7.3. *A closed oriented submanifold M of $\mathbf{R}^n$ is said to be (area) volume minimizing if, for each relatively compact open subset U of M with smooth boundary ∂U,*

$$(7.4) \qquad \mathrm{vol}(U) \le \mathrm{vol}(V)$$

for all other compact oriented p-dimensional submanifolds V with the same boundary as U, i.e., $\partial U = \partial V$.

A calibration ϕ can be used to distinguish a class of oriented submanifolds M; namely, those with $\overrightarrow{M} \in G(\phi)$ for each point $x \in M$. Equivalently, ϕ restricted to M equals λ_M, the unit oriented volume form on M.

The Fundamental Theorem 7.5. *Suppose ϕ is a calibration of degree p on $\mathbf{R}^n$. Each closed oriented p-dimensional submanifold M distinguished by ϕ, i.e.,*

$$\phi(\vec{M}) = 1 \quad \text{for all points } x \in M$$

is volume minimizing.

Proof:

$$\text{vol}(U) = \int_U \phi = \int_V \phi = \int_V \phi(\vec{V})\lambda_V$$

$$\leq \int_V \lambda_V \equiv \text{vol}(V),$$

where

$$\int_U \phi - \int_V \phi = \int_{U-V} \phi = \int_{U-V} d\psi = \int_{\partial(U-V)} \psi = 0,$$

because of Stokes' Theorem and the fact that:

(7.6)   if $d\phi = 0$, then $\phi = d\psi$ for some $p-1$ form ψ on $\mathbf{R}^n$. ∎

THE KÄHLER CASE

Consider $\mathbf{C}^n$ equipped with the standard positive definite $\mathbf{C}$-hermitian inner product h. Then $h = \langle\,,\,\rangle - i\omega$, where $\langle\,,\,\rangle$ is the standard positive definite real inner product on $\mathbf{R}^{2n}(\cong \mathbf{C}^n)$ and

(7.7) $\omega \equiv \dfrac{i}{2}\, dz' \wedge d\bar{z}' + \cdots + \dfrac{i}{2}\, dz^n \wedge d\bar{z}^n$

is the standard symplectic inner product on $\mathbf{R}^{2n}(\cong \mathbf{C}^n)$, which is also called the standard Kähler form on $\mathbf{C}^n$. Note that ω may be viewed as the sum of the complex axis lines

(7.8) $\lambda^j \equiv \dfrac{i}{2}\, dz^j \wedge d\bar{z}^j = dx^j \wedge dy^j, \quad j = 1, \ldots, n.$

Since forms of degree 2 commute under wedge product,

(7.9) $\dfrac{1}{p!}\, \omega^p = \sum'_{|I|=p} \lambda^I$

is the sum of the complex axis p-planes λ^I. The unitary group $U(n)$ fixes h, and hence ω. Therefore, $U(n)$ also fixes $\frac{1}{p!}\omega^p$. Also note that this form is d-closed since it has constant coefficients.

Theorem 7.10. *The 2p-form $\phi \equiv \frac{1}{p!}\omega^p$ on $\mathbf{R}^{2n}(\cong \mathbf{C}^n)$ is a calibration. The contact set equals the complex grassmannian. That is,*

$$(7.11) \qquad G(\phi) = G_{\mathbf{C}}(p, \mathbf{C}^n) \subset G_{\mathbf{R}}(2p, \mathbf{C}^n).$$

Since $d\phi = 0$, the theorem is equivalent to

Theorem 7.12 (Wirtinger's Inequality).

$$\frac{1}{p!}\ \omega^p(\xi) \leq 1 \quad \text{for all } \xi \in G_{\mathbf{R}}(2p, \mathbf{C}^n),$$

with equality if and only if $\xi \in G_{\mathbf{C}}(p, \mathbf{C}^n)$.

Corollary 7.13. *Each complex submanifold of $\mathbf{C}^n$ is volume minimizing.*

First Proof (Federer): The case $p = 1$ is particularly easy. Let u, v denote an oriented orthonormal basis for ξ, i.e., $\xi \equiv u \wedge v$. Then

$$(7.14) \qquad \omega(\xi) = \omega(u, v) = \langle Ju, v \rangle \leq |Ju|\,|v| = 1$$

by the C.–S. inequality. Moreover, equality occurs in (7.14) if and only if $v = Ju$. That is, precisely when $\xi = u \wedge Ju$ is a complex line (with the orientation induced from the complex structure on the line).

The general case reduces to the $p = 1$ case. Suppose $\xi \in G_{\mathbf{R}}(2p, \mathbf{C}^n)$ and let $P \equiv \operatorname{span} \xi$ denote the real $2p$-subspace corresponding to ξ. Restriction of ω to P yields a 2-form $\omega_P \in \Lambda^2 P^*$ on P. This 2-form can be put in canonical form (see Problem 4.2) with respect to the real inner product on P inherited from $\mathbf{R}^{2n} \cong \mathbf{C}^n$:

$$(7.15) \qquad \omega_P \equiv \lambda_1 \alpha^1 \wedge \alpha^2 + \lambda_2 \alpha^3 \wedge \alpha^4 + \cdots + \lambda_r \alpha^{2r-1} \wedge \alpha^{2-r}$$

where $\lambda_1 \geq \lambda_2 \geq \ldots \geq \lambda_r > 0$ and $\alpha^1, \ldots, \alpha^p$ is an orthonormal basis for P^*. Let $e_1, \ldots, e_p$ denote the dual basis for P. Then, by (7.14), for each $j = 1, \ldots, r$,

$$(7.16) \qquad \lambda_j = \omega(e_{2j-1}, e_{2j}) \leq 1,$$

with equality if and only if $e_{2j-1} \wedge e_{2j}$ represents a complex line. Now $\phi \equiv \frac{1}{p!}\omega^p$ restricted to P is the same as $\frac{1}{p!}\omega_P^p$, which equals $\lambda_1 \ldots \lambda_p \alpha^1 \wedge \cdots \wedge \alpha^p$. Therefore,

$$(7.17) \qquad \phi(\xi) = \lambda_1 \ldots \lambda_p.$$

Combined with (7.16), this completes the proof. ∎

A second proof of Wirtinger's Inequality utilizes a canonical form for a real $2p$-plane in $\mathbf{C}^n$ under the action of the unitary group $U(n)$.

Lemma 7.18 (Harvey–Lawson). *Each oriented $2p$-dimensional real subspace $\xi \in G_{\mathbf{R}}(2p, \mathbf{C}^n)$ of $\mathbf{R}^{2n} \cong \mathbf{C}^n$ can be expressed in terms of a unitary basis $e_1, \ldots, e_n$ as*

(a) *for $2p \leq n$,*

$$(7.19) \qquad \begin{aligned} \xi \equiv &e_1 \wedge (\cos\theta_1 \, Je_1 + \sin\theta_1 \, e_2) \wedge \ldots \\ &\wedge e_{2p-1} \wedge (\cos\theta_p \, Je_{2p-1} + \sin\theta_p \, e_{2p}), \end{aligned}$$

where the angles satisfy

$$0 \leq \theta_1 \leq \cdots \leq \theta_{p-1} \leq \frac{\pi}{2} \quad \text{and} \quad \theta_{p-1} \leq \theta_p \leq \pi;$$

(b) *for $2p \geq n$,*

$$(7.19') \qquad \begin{aligned} \xi \equiv \; &e_1 \wedge Je_1 \wedge \ldots \wedge e_r \wedge Je_r \\ &\wedge e_{r+1} \wedge (\cos\theta_1 \, Je_{r+1} + \sin\theta_1 \, e_{r+2}) \\ &\wedge \cdots \wedge e_{n-1} \wedge (\cos\theta_q \, Je_{n-1} + \sin\theta_q \, e_n), \end{aligned}$$

where the angles satisfy

$$0 \leq \theta_1 \leq \cdots \leq \theta_{q-1} \leq \frac{\pi}{2} \quad \text{and} \quad \theta_{q-1} \leq \theta_q \leq \pi.$$

Here $r \equiv 2p - n$ and $q \equiv n - p$.

Second Proof of Wirtinger: The Kähler form can be expressed as

$$(7.20) \qquad \omega = \alpha^1 \wedge J\alpha^1 + \cdots + \alpha^n \wedge J\alpha^n$$

for any orthonormal basis $\alpha^1, J\alpha^1, \ldots, \alpha^n, J\alpha^n$ dual to the orthonormal basis $e_1, Je_1, \ldots, e_n, Je_n$ (i.e., $e_1, \ldots, e_n$ is a unitary basis for $\mathbf{C}^n$). Then, by (7.19) and (7.20),

$$(7.21) \qquad \frac{1}{p!} \, \omega^p(\xi) = \cos\theta_1 \cdots \cos\theta_p, \quad 2p \leq n,$$

which is less than or equal to one and equal to one if and only if each $\theta_j = 0$, $j = 1, \ldots, p$. That is, if and only if $\xi = e_1 \wedge Je_1 \wedge \cdots \wedge e_p \wedge Je_p \in G_{\mathbf{C}}(p, \mathbf{C}^n)$. ∎

Proof of Lemma 7.18: The proof is by induction on p. First, maximize $\langle Ju, v \rangle$ over all orthonormal pairs of vectors $u, v \in \text{span } \xi$ and set $\cos\theta_1 \equiv \langle Ju, v \rangle$ equal to the maximum value with $0 \leq \theta_1 \leq \pi/2$. Since the function $f(\psi) \equiv \langle Ju, (\cos\psi)v + (\sin\psi)w \rangle$ has maximum value at $\psi = 0$ for each $w \in$

span ξ with $w \perp \text{span}\{u, v\}$, the derivative $f'(0) = 0$ vanishes. However, $f'(0) = \langle Ju, w \rangle$, so that $w \perp Ju$. Similarly, $w \perp Jv$. Now define $e_1 \equiv u$. Then $v = \cos\theta_1 \, Je_1 + \sin\theta_1 \, z$, z a unit vector orthogonal to both e_1 and Je_1. Define $e_2 \equiv z$. As noted above, if $w \in \text{span}\,\xi$ and $w \perp \{u, v\}$, then $w \perp \{Ju, Jv\}$. Therefore $w \perp \{e_1, Je_1, e_2, Je_2\}$.

Consequently, $\xi = e_1 \wedge (\cos\theta_1 Je_1 + \sin\theta_1 e_2) \wedge \eta$ with $\eta \in G_{\mathbf{R}}(2p-2, V)$ where $V = \{e_1, Je_1, e_2, Je_2\}^\perp$. In the special case $\theta_1 = 0$, the proof must be modified: $\xi \equiv e_1 \wedge Je_1 \wedge \eta$ with $\eta \in G_{\mathbf{R}}(2p-2, V)$, where $V = \{e_1, Je_1\}^\perp$. Also note that in the last step of the induction, v cannot be replaced by $-v$ so that one must allow the possibility of $\cos\theta_p$ negative, or $\theta_{p-1} \leq \theta_p \leq \pi$. ∎

THE SPECIAL LAGRANGIAN CALIBRATION

Consider $\mathbf{C}^n$ with the standard positive definite **C**-hermitian inner product $h = \langle \, , \, \rangle - i\omega$, exactly as in the last section. However, also assume that $\mathbf{C}^n$ is equipped with a complex volume form

$$(7.22) \qquad\qquad dz \equiv dz^1 \wedge \cdots \wedge dz^n$$

of norm $|dz|^2 = |dz^1|^2 \cdots |dz^n|^2 = 2^n$.

In this section, we shall examine the real n-form

$$(7.23) \qquad\qquad \phi \equiv \operatorname{Re} dz \equiv \frac{1}{2}(dz + \overline{dz}).$$

Note that $\mathrm{SU}(n)$ fixes this form ϕ. Of course, $d\phi = 0$, since ϕ has constant coefficients on $\mathbf{C}^n \cong \mathbf{R}^{2n}$. In the coordinates $z^j \equiv x^j + iy^j$, $j = 1, \ldots, n$ for $\mathbf{C}^n$,

$$(7.24) \qquad \phi = \operatorname{Re}\left(dx^1 + idy^1\right) \wedge \cdots \wedge \left(dx^n + idy^n\right)$$

can be expanded out and is the sum of the 2^{n-1} real axis n-planes of the form $\pm dx^I \wedge dy^J$ with $I \cup J = \{1, \ldots, n\}$ and $|J|$ even.

Definition 7.25. *A real oriented n-plane $\xi \in G_{\mathbf{R}}(n, \mathbf{C}^n)$ is said to be special Lagrangian if $\xi \equiv A(e_1 \wedge \cdots \wedge e_n)$ for some $A \in \mathrm{SU}(n)$, where $e_1, \ldots, e_n$ is the standard unitary basis for $\mathbf{C}^n$, i.e., $\xi_0 \equiv e_1 \wedge \cdots \wedge e_n$ represents the standard oriented subspace $\mathbf{R}^n \subset \mathbf{C}^n = \mathbf{R}^n \oplus i\mathbf{R}^n$. Let $\mathrm{SLAG} \subset G_{\mathbf{R}}(n, \mathbf{C}^n)$ denote the subset of special Lagrangian n-planes. An oriented n-dimensional submanifold M of $\mathbf{C}^n$ is said to be special Lagrangian if $\overrightarrow{M} \in \mathrm{SLAG}$ for each point $x \in M$.*

Theorem 7.26. *The n-form $\phi \equiv \operatorname{Re} dz$ is a calibration on $\mathbf{C}^n$, and the contact set $G(\phi) = \mathrm{SLAG}$ consists of all the special Lagrangian n-planes.*

Corollary 7.27. *Each special Lagrangian submanifold M of $\mathbf{C}^n$ is area minimizing.*

Remark 7.28. For each fixed angle θ, consider the n-form

$$(7.29) \qquad\qquad \phi_\theta \equiv \operatorname{Re} e^{-i\theta} dz.$$

For example, if $\theta = \pi/2$, then $\phi = \operatorname{Im} dz$. Each such ϕ_θ is also a calibration referred to as *the special Lagrangian calibration of phase* θ.

The proof of Theorem 7.26 is via several lemmas that clarify the notion of special Lagrangian subspace. First, the weaker notion of Lagrangian is investigated. A subspace L of dimension n in $\mathbf{C}^n \cong \mathbf{R}^{2n}$ is said to be *Lagrangian* if L is totally null with respect to the Kähler (symplectic) form ω. However, there are many equivalent ways of expressing this fact. Let $\perp$ denote orthogonality with respect to $\langle\,,\,\rangle \equiv \operatorname{Re} h$.

Lemma 7.30. *Suppose L is an n-dimensional real subspace of $\mathbf{C}^n, h$. The following are equivalent.*

(a) $\omega(u, v) = 0$ *for all* $u, v \in L$.
(b) *The 1-form* $\sum_{j=1}^n y^j dx^j$ *restricted to L is d-closed.*
(c) *If* $u \in L$, *then* $Ju \perp L$.
(d) $L = \{Ax : x \in \mathbf{R}^n \subset \mathbf{C}^n \cong \mathbf{R}^n \oplus i\mathbf{R}^n\}$ *for some* $A \in U(n)$ *unitary.*
(e) $|dz(u_1 \wedge \ldots \wedge u_n)| = 1$ *if* $u_1, \cdots, u_n$ *is an orthonormal basis for L.*

Further, if $L = \operatorname{graph} f$ *can be graphed over* $\mathbf{R}^n \subset \mathbf{R}^n \oplus i\mathbf{R}^n$, $f : \mathbf{R}^n \to \mathbf{R}^n$, *then the following can be added to the equivalence.*

(f) $L = \operatorname{graph} \nabla\phi$ *is the graph of a gradient* $y = (\nabla\phi)(x)$ *of a scalar valued function* $\phi(x)$.

Proof: (cf. Problem 5.8) Condition (a) has been taken as the definition of L Lagrangian, i.e.,

$$\omega = \sum dx^j \wedge dy^j$$

restricted to L should vanish. Since $d\alpha = -\omega$ with

$$\alpha = \sum_{j=1}^n y^j dx^j,$$

and restriction to L commutes with exterior differentiation, (b) is equivalent to (a).

If $L = \operatorname{graph} f \equiv \{(x, f(x)) : x \in \mathbf{R}^n\}$, then,

$$\alpha|_L = \sum_{j=1}^n f_j(x) dx^j.$$

Condition (b) says that this 1-form is d-closed, i.e.,

$$\frac{\partial f_j}{\partial x^i} = \frac{\partial f_i}{\partial x^j}.$$

This symmetry is true if and only if

$$f_i(x) = \frac{\partial \phi}{\partial x^i}(x), \quad i = 1, \ldots, n,$$

for some scalar-valued function ϕ on $\mathbf{R}^n$. Thus, (b) and (f) are equivalent.

To prove that (a) and (c) are equivalent use the formula

$$\langle Ju, v \rangle = \omega(u, v).$$

Since unitary transformations preserve $\langle\,,\,\rangle$, J, and ω, and since (c) is obviously valid for $L = \mathbf{R}^n \subset \mathbf{C}^n$, the condition (d) implies condition (c). Conversely, if L satisfies (c), then an orthonormal basis $u_1, \ldots, u_n$ for L can be used to define a unitary map $A \in U(n)$ sending the standard basis $e_1, \ldots, e_n$ for $\mathbf{R}^n$ into $u_1, \ldots, u_n$ so that (d) follows.

Let $u_j \equiv Ae_j$ denote the image of e_j under any complex linear transformation A. Then

$$(7.31) \qquad (dz)(u_1 \wedge \cdots \wedge u_n) = (dz)(A(e_1 \wedge \cdots \wedge e_n)) = \det_{\mathbf{C}} A.$$

If (d) is valid, then $\xi = A(e_1 \wedge \cdots \wedge e_n)$ with A unitary, so that $|\det_{\mathbf{C}} A| = 1$, which verifies (e).

To prove that (e) implies (c) and thus complete the proof of Lemma 7.30, we shall prove the following lemma. ∎

Lemma 7.32. $|(dz)(\xi)| \leq 1$ *for all* $\xi \in G_{\mathbf{R}}(n, \mathbf{C}^n)$, *with equality if and only if* $L = \mathrm{span}\,\xi$ *is Lagrangian.*

Proof: Assume $\lambda \xi = u_1 \wedge \cdots \wedge u_n$ with $\lambda \in \mathbf{R}$ and $u_1, \ldots, u_n$ any oriented basis for $L = \mathrm{span}\,\xi$ (not necessarily orthogonal). Define $A \in \mathrm{GL}(n, \mathbf{C})$ to be the complex linear map sending the basis $e_1, \ldots, e_n$ for $\mathbf{R}^n$ to $u_1, \ldots, u_n$. Then, exactly as in (7.31),

$$(7.33) \qquad\qquad \lambda\, dz(\xi) = \det_{\mathbf{C}} A.$$

However,

$$(7.34) \qquad
\begin{aligned}
|\det_{\mathbf{C}} A|^2 &= \det_{\mathbf{R}} A = |A(e_1 \wedge Je_1 \wedge \cdots \wedge e_n \wedge Je_n)| \\
&= |A(e_1 \wedge \cdots \wedge e_n \wedge Je_1 \wedge \cdots \wedge Je_n)| = \lambda^2 |\xi \wedge J\xi|.
\end{aligned}$$

Therefore,

$$(7.35) \qquad |(dz)(\xi)|^2 = |\xi \wedge J\xi|.$$

The classical Hadamard inequality (see Corollary 9.33 in Part II for a proof) states that $|w_1 \wedge \cdots \wedge w_k|^2 \leq |w_1|^2 \cdots |w_k|^2$ with equality if and only if $w_1, \ldots, w_k$ are orthogonal. In particular,

$$(7.36) \qquad |\xi \wedge J\xi| \leq |u_1| \cdots |u_n| \, |Ju_1| \cdots |Ju_n|$$

with equality if and only if

$$u_1, \ldots, u_n, Ju_1, \ldots, Ju_n$$

are orthogonal (i.e., if and only if $u_1, \ldots, u_n$ is a unitary basis for $\mathbf{C}^n$).

Now assume that $u_1, \ldots, u_n$ is an oriented orthonormal basis for ξ. Combining (7.35) and (7.36) yields the proof of Lemma 7.32, since L is Lagrangian if and only if $u_1, \ldots, u_n, Ju_1, \ldots, Ju_n$ is orthonormal (by (c) above). ∎

Corollary 7.37. *Suppose* span ξ *is Lagrangian*, $\xi \in G_{\mathbf{R}}(n, \mathbf{C}^n)$. *Then*

$$(7.38) \qquad (dz)(\xi) = \det{}_{\mathbf{C}} A |\det{}_{\mathbf{C}} A|^{-1},$$

where A is any complex linear map sending the standard basis $e_1, \ldots, e_n$ for $\mathbf{R}^n$ into an oriented basis $u_1, \ldots, u_n$ for ξ.

Proof: By (7.33), $\lambda(dz)(\xi) = \det{}_{\mathbf{C}} A$ with $\lambda > 0$. Since $|(dz)(\xi)| = 1$, the corollary is immediate. ∎

Proof of Theorem 7.26. Consider $\xi \in G_{\mathbf{R}}(n, \mathbf{C}^n)$. By Lemma 7.32, $\phi(\xi) = \mathrm{Re}\,(dz)(\xi) \leq |(dz)(\xi)| \leq 1$ so that ϕ is a calibration. If equality $\phi(\xi) = 1$ holds, then $|(dz)(\xi)| = 1$, so that ξ must be Lagrangian. Therefore, by condition (d) of Lemma 7.30, $\xi = A(e_1 \wedge \cdots \wedge e_n)$ for some unitary map A. Corollary 7.37 implies that $\det{}_{\mathbf{C}} A = 1$. Therefore, $\xi \in$ SLAG is a special Lagrangian. Conversely, if $\xi \in$ SLAG (by definition $\xi = A(e_1 \wedge \cdots \wedge e_n)$ some $A \in SU(n)$), then Corollary 7.37 implies that $(dz)(\xi) = 1$ so that $\phi(\xi) = 1$. ∎

Corollary 7.37 provides several useful criterion for a manifold to be special Lagrangian.

134   *The Special Lagrangian Differential Equation*

Corollary 7.39. *Suppose $\xi \in G_{\mathbf{R}}(n, \mathbf{C}^n)$ and span ξ is Lagrangian.*

(a) *$\pm\xi$ is special Lagrangian if and only if $\operatorname{Im}(dz)(\xi) = 0$.*

(b) *ξ is special Lagrangian if and only if for any (some) complex linear map A sending the standard basis $e_1, \ldots, e_n$ for $\mathbf{R}^n$ into an oriented basis for ξ:*

$$(7.40) \qquad \det{}_{\mathbf{C}} A > 0 \quad (\text{real and positive}).$$

THE SPECIAL LAGRANGIAN DIFFERENTIAL EQUATION

In pursuing the analogues between the Kähler (form) calibration and the special Lagrangian calibration, one question naturally comes to mind. What is the special Lagrangian analogue of the Cauchy–Riemann equations? More precisely, suppose

$$(7.41) \qquad M \equiv \operatorname{graph} f \equiv \{x + if(x) : x \in U^{\text{open}} \subset \mathbf{R}^n\}$$

is the graph of a smooth function $f : U \to \mathbf{R}^n$. Then the question becomes: What is the differential equation(s) imposed on f by requiring that M be a special Lagrangian submanifold?

The n vectors

$$(7.42) \qquad u_j \equiv e_j + i\,\frac{\partial f}{\partial x^j}, \; j = 1, \ldots, n,$$

form a basis for the tangent space to M at each point. Thus, one answer to this question is the equation

$$
(7.43) \qquad
\begin{aligned}
&\phi\left(\left(e_1 + i\,\frac{\partial f}{\partial x^1}\right) \wedge \cdots \wedge \left(e_n + i\,\frac{\partial f}{\partial x^n}\right)\right) \\
&= \left|\left(e_1 + i\,\frac{\partial f}{\partial x^1}\right) \wedge \cdots \wedge \left(e_n + i\,\frac{\partial f}{\partial x^n}\right)\right|.
\end{aligned}
$$

This answer is not very useful. In fact, in the Kähler case $\mathbf{C}^2, \omega$ a similar analysis yields the equation

$$
(7.44) \qquad
\begin{aligned}
&\omega\left(e_1 + \frac{\partial u}{\partial x^1}\,e_2 + \frac{\partial v}{\partial x^1}\,Je_2, Je_1 + \frac{\partial u}{\partial y^1}\,e_2 + \frac{\partial v}{\partial y^1}\,Je_2\right) \\
&= \left[1 + |\nabla u|^2 + |\nabla v|^2 + \left(\frac{\partial u}{\partial x^1}\frac{\partial v}{\partial y^1} - \frac{\partial v}{\partial x^1}\frac{\partial u}{\partial y^1}\right)^2\right]^{\frac{1}{2}},
\end{aligned}
$$

which is *not* the elegant Cauchy–Riemann equations:

$$(7.45) \qquad \frac{\partial u}{\partial x^1} = \frac{\partial v}{\partial y^1}, \quad \frac{\partial u}{\partial y^1} = \frac{\partial v}{\partial x^2}, \quad \text{with } f = u + iv.$$

After this unproductive analysis, we can make a fresh start by using Corollary 7.39(a). Since

$$\lambda \vec{M} = \left(e_1 + i\, \frac{\partial f}{\partial x^1} \right) \wedge \cdots \wedge \left(e_n + i\, \frac{\partial f}{\partial x^n} \right),$$

with $\lambda \in \mathbf{R}$, this corollary states: Either M or $-M$ (M with orientation reversed) is special Lagrangian if and only if

$$(7.46) \qquad\qquad M \text{ is Lagrangian}$$

and

$$(7.47) \qquad (\operatorname{Im} dz) \left(\left(e_1 + i\, \frac{\partial f}{\partial x^1} \right) \wedge \cdots \wedge \left(e_n + i\, \frac{\partial f}{\partial x^n} \right) \right) = 0.$$

The first condition, M Lagrangian, can be combined with the (first order) partial differential equation (7.47) by using a *potential* (or *generating*) *function* ϕ.

Lemma 7.48. *Suppose $M \equiv$ graph $f \equiv \{x + if(x) : x \in U^{\text{open}} \subset \mathbf{R}^n\}$ and U is simply connected. Then M is Lagrangian if and only if*

$$(7.49) \qquad\qquad f = \nabla \phi$$

is the gradient of a scalar function ϕ on U.

The proof of this lemma is exactly the same as the proof that parts (a) and (f) of Lemma 7.30 are equivalent.

Now, if $f = \nabla \phi$, then $\partial f / \partial x^i$ is the ith column of the *Hessian matrix*

$$(7.50) \qquad\qquad \operatorname{Hess}(\phi) \equiv \left(\frac{\partial^2 \phi}{\partial x^i \partial x^j} \right).$$

That is, $\partial f / \partial x^i = \operatorname{Hess}(\phi)(e_i)$.

Therefore, (with λ real)

$$(7.51) \qquad \begin{aligned} \lambda \vec{M} &= \left(e_1 + i\, \frac{\partial f}{\partial x^i} \right) \wedge \cdots \wedge \left(e_n + i\, \frac{\partial f}{\partial x^n} \right) \\ &= (1 + i\, \operatorname{Hess}(\phi))(e_1 \wedge \cdots \wedge e_n), \end{aligned}$$

so that

$$(7.52) \qquad (dz)(\lambda \overrightarrow{M}) = \det{}_{\mathbf{C}}(I + i \text{ Hess}(\phi)).$$

Theorem 7.53. *Suppose* $M \equiv$ *graph* $f \equiv \{x + if(x) : x \in U \subset \mathbf{R}^n \subset \mathbf{C}^n\}$ *is the graph of a smooth function over a simply connected domain* U *in* $\mathbf{R}^n$. *Then* M *(with one of the two possible orientations) is special Lagrangian if and only if*

$$(7.54) \qquad f = \nabla \phi \quad \text{for some potential } \phi \text{ on } U,$$

and either of the following equivalent conditions hold:

$$(7.55) \qquad \text{Im } \det{}_{\mathbf{C}}(I + i \text{ Hess}(\phi)) = 0 \quad \text{on } U,$$

or

$$(7.56) \qquad \sum_{k=0}^{[(n-1)/2]} (-1)^k \alpha_{2k+1} \text{Hess}(\phi) = 0 \quad \text{on } U.$$

Remark 7.57. The equation (7.56) will be referred to as the *special Lagrangian differential equation*. Here $\alpha_j(A)$ denotes the jth elementary symmetric function of a symmetric matrix A (such as $A \equiv \text{Hess}(\phi)$). In terms of the eigenvalues $\lambda_1, \ldots, \lambda_n$ of A, recall that

$$(7.58) \qquad \alpha_j(A) = \sum{}'_{|I|=j} \lambda_{i_1} \cdots \lambda_{i_j}.$$

If $n = 2$, then the special Lagrangian differential equation is not a new equation but is just $\Delta \phi = \text{trace Hess}(\phi) = 0$. The first nonclassical interesting case is $n = 3$. In this case, the equation is

$$(7.59) \qquad \Delta \phi = MA(\phi),$$

where $\Delta \phi \equiv \text{trace Hess}(\phi)$ and $MA(\phi) \equiv \det \text{Hess}(\phi)$ is the *Monge–Ampere operator* on ϕ.

Proof of Theorem 7.53: The theorem is derived from the characterization of M as special Lagrangian given by (7.46) and (7.47). By Lemma 7.48, M is Lagrangian if and only if there exists a potential function ϕ as in (7.54). The equation (7.55) has already been verified; see (7.52). Thus, it remains to show that (7.55) and (7.56) are the same equation. Let A denote the symmetric matrix $\text{Hess}(\phi)$ at a fixed point $x \in U$. The matrix

A can be put in canonical form with respect to the orthogonal group $O(n)$. This standard result is often referred to as the *Principle Axis Theorem*. Its proof is very similar to the proof of Problem 4.2 (a canonical form for $O(n)$ acting on $\text{Skew}(n, \mathbf{R}) \cong \Lambda \mathbf{R}^n$), and hence is omitted. The result says: Given

$$A \in \text{Sym}(n, \mathbf{R}) \subset M_n(\mathbf{R})$$

a symmetric matrix, there exists $g \in O(n)$ such that

$$(7.60) \qquad gAg^t = D$$

is a diagonal matrix with nonzero entries the eigenvalues $\lambda_1, \ldots, \lambda_n$ of A.

Now $\det_{\mathbf{C}}(I + i\,\text{Hess}(\phi)) = \det_{\mathbf{C}}(g(I + i\,\text{Hess}(\phi))g^t) = \det_{\mathbf{C}}(I + iD)$, and hence (7.55) is equivalent to

$$(7.61) \qquad \text{Im} \prod_{J=1}^{n}(1 + i\lambda_j) = 0,$$

which is the same as (7.56).

Theorem 7.62. *Suppose M is a connected oriented n-dimensional real submanifold, of an open subset U of $\mathbf{C}^n$, which is defined implicitly by smooth functions on U:*

$$(7.63) \qquad M \equiv \{z \in U : f_1(z) = c_1, \ldots, f_n(z) = c_n\}.$$

Then M (with one of the two possible orientations) is special Lagrangian if and only if

$$(7.64) \qquad \sum_{k=1}^{n}\left(\frac{\partial f_p}{\partial \overline{z}_k}\frac{\partial f_q}{\partial z_k} + \frac{\partial f_p}{\partial z_k}\frac{\partial f_q}{\partial \overline{z}_k}\right) = 0 \quad \text{for each } p, q,$$

and

$$(7.65) \qquad \text{Im}\left(\det_{\mathbf{C}}\left(i\,\frac{\partial f_q}{\partial \overline{z}^p}\right)\right) = 0$$

for all points $z \in M$.

Proof: The normal n-plane to M is Lagrangian if and only if the tangent n-plane to M is Lagrangian (c.f., condition (c) of Lemma 7.30). This normal n-plane is spanned by the vectors

$$(7.66) \qquad u_p \equiv \frac{1}{2}\left(\frac{\partial f_p}{\partial x} + i\,\frac{\partial f_p}{\partial y}\right) = \frac{\partial f_p}{\partial \overline{z}}.$$

Note that

$$(7.67) \qquad Ju_q = \frac{1}{2}\left(i\,\frac{\partial f_q}{\partial x} - \frac{\partial f_q}{\partial y} \right) = i\,\frac{\partial f_q}{\partial \overline{z}}.$$

Therefore, M is Lagrangian if and only if

$$(7.68) \qquad \langle u_p, Ju_q \rangle = 0 \quad \text{for each } 1 \le p, q \le n.$$

Equivalently,

$$(7.68') \qquad \operatorname{Re} \sum_{k=1}^{n} \frac{\partial f_p}{\partial z_k}\frac{\partial f_q}{\partial \overline{z}_k} = 0,$$

which is Equation (7.62).

Now assume M is Lagrangian. Then

$$Ju_p \equiv i\,\frac{\partial f_p}{\partial \overline{z}} \quad p = 1, \ldots, n,$$

is a basis for the tangent space. The complex matrix

$$A \equiv \left(i\frac{\partial f_p}{\partial \overline{z}_q} \right)$$

maps the standard basis $e_1, \ldots, e_n$ for $\mathbf{R}^n$ to the basis $Ju_1, \ldots, Ju_n$ for $\vec{M}$, i.e., $\lambda\,\vec{M} = Ju_1 \wedge \cdots \wedge Ju_n$ with $\lambda > 0$. Therefore, $\operatorname{Im}(dz)(\lambda\,\vec{M}) = \lambda\operatorname{Im}\det_{\mathbf{C}} A$ by (7.31). Thus, A satisfies (7.65) if and only if $\pm\,\vec{M}$ is special Lagrangian by Corollary 7.39.

EXAMPLES OF SPECIAL LAGRANGIAN SUBMANIFOLDS

Rather than discuss general classes of examples, specific examples are discused in this Section. The first example is invariant under the action of the $n-1$ torus $= \{\operatorname{diag}(e^{i\theta_1}, \ldots, e^{i\theta_n}) : \theta_1 + \theta_2 + \cdots + \theta_n = 0\}$ contained in $\mathrm{SU}(n)$.

Theorem 7.69. *Let M_c denote the locus of the equations:*

$$(7.70) \qquad |z_j|^2 - |z_1|^2 = c_j, \quad j = 2, \ldots, n,$$

and

$$(7.71) \qquad \operatorname{Re} z_1 \cdots z_n = c_1 \quad \text{if } n \text{ is even,}$$

or

(7.71') $\operatorname{Im} z_1 \cdots z_n = c_1$ if n is odd.

Then M_c (with the correct orientation) is a special Lagrangian submanifold.

This provides perhaps the simplest example of an area minimizing object (with a singularity at the origin) that is not a *real-analytic variety*, that is, not the zero set of a family of real analytic functions. The key property distinguishing a real analytic zero set is that if it contains a half-ray then it must contain the full line. This is because a real-analytic function of one variable t that vanishes for $t > 0$ must vanish identically.

Corollary 7.72. *Suppose n is odd and let*

$$M^+ \equiv \left\{ \left(re^{i\theta_1}, \ldots, re^{i\theta_n} \right) : \theta_1 + \ldots + \theta_n = 0 \text{ and } r > 0 \right\}$$

$$M^- \equiv \left\{ \left(re^{i\theta_1}, \cdots, re^{i\theta_n} \right) : \theta_1 + \cdots + \theta_n = \pi \text{ and } r > 0 \right\}.$$

Then the cones M^+ and M^- are both special Lagrangian, and hence both volume minimizing. Note that M^+ and M^- are disjoint cones with one the image of the other under the map $-\operatorname{Id}$, so that neither M^+ nor M^- is a real-analytic variety.

Proof: Set $c_1 = \cdots = c_n = 0$ in Theorem 7.69 and note that $M_0 = M^+ \cup M^-$ on $\mathbf{C}^n - \{0\}$. ∎

Proof of Theorem 7.69: Set $f_1(z) = \operatorname{Re} i^n z_1 \cdots z_n$ and $f_q(z) \equiv |z_q|^2 - |z_1|^2$ for $q = 2, \ldots, n$. Then (see Problem 2a)

(7.73) $\det_{\mathbf{C}} \left(i \, \dfrac{\partial f_q}{\partial \bar{z}_p} \right) = (-1)^n \displaystyle\sum_{j=1}^{n} \dfrac{1}{|z_j|^2} \, |z_1|^2 \cdots |z_n|^2.$

Now verify (see Problem 2a) the conditions (7.64) and (7.65) of Theorem 7.62. ∎

The next example provides a useful tool for the study of pairs of minimizing planes in euclidean space; see the following section entitled *Angle Theorem*.

Given $a_1, \ldots, a_n \geq 0$, let

(7.74) $P(a, y) \equiv \dfrac{1}{y^2} \left[(1 + a_1 y^2) \cdots (1 + a_p y^2) - 1 \right].$

Now for each $k = 1, \ldots, n$, and each $y \in \mathbf{R}$, define

$$(7.75) \qquad z_k(y) \equiv r_k(y)e^{i\theta_k(y)},$$

with

$$(7.76) \qquad r_k(y) \equiv \sqrt{a_k^{-1} + y^2} \,,$$

and

$$(7.77) \qquad \theta_k(y) \equiv a_k \int_0^y \frac{dy}{(1 + a_k y^2)\sqrt{P(a, y)}} \,.$$

It is convenient to use the calibration $\phi \equiv \operatorname{Im} dz_1 \wedge \cdots \wedge dz_n$ rather than $\operatorname{Re} dz$.

Theorem 7.78. *The manifold*

$$M_a \equiv \left\{ w \in \mathbf{C}^n : w_k = t_k z_k(y) \text{ with } y \in \mathbf{R}, \ t \in \mathbf{R}^n \text{ and } \sum_{k=1}^{n} t_k^2 = 1 \right\}$$

is $(\phi \equiv \operatorname{Im} dz)$ special Lagrangian (if correctly oriented).

This example (see Lawlor [11]), introduced by Lawlor and completed by Harvey, is used in the proof of the Angle Theorem 7.134.

Remark 7.79. Let Γ denote the curve

$$(7.80) \qquad \Gamma \equiv \{\theta(y) : y \in \mathbf{R}\},$$

defined by (7.77). Let

$$(7.81) \qquad P_\theta \equiv e^{i\theta} \cdot \mathbf{R}^n \equiv \left\{ \left(t_1 e^{i\theta_1}, \ldots, t_n e^{i\theta_n}\right) : t \in \mathbf{R}^n \right\}$$

denote the n-plane in $\mathbf{C}^n$ obtained by rotating $\mathbf{R}^n$ by $e^{i\theta}$. The key property of the special Lagrangian manifold M_a is that each nontrivial intersection (i.e., $\theta \in \Gamma$) of M_a with P_θ is the boundary of a bounded open subset of P_θ. In fact, for $\theta \in \Gamma$,

$$(7.82) \qquad M_a \cap P_\theta = \left\{ \left(t_1 e^{i\theta_1}, \cdots, t_n e^{i\theta_n}\right) : \sum_{j=1}^{n} \frac{t_j^2}{r_j^2(y)} = 1 \right\}$$

is an ellipsoid.

Proof of Theorem 7.78: In addition to giving a proof of Theorem 7.78, we shall sketch the proof that the submanifolds M_a are the only possible special Lagrangian submanifolds with each nontrivial intersection $M_a \cap P_\theta$ a compact hypersurface of P_θ, thus removing some of the mystery of this example.

If M is to consist of the union of hypersurfaces in a family of the n-planes P_θ, and M is required to be Lagrangian, then M must be of the form (set $w_j \equiv R_j e^{i\theta_j}$)

$$(7.83) \qquad M(\Gamma, h) \equiv \left\{ w \in \mathbf{C}^n : \sum_{j=1}^n R_j^2 \frac{d\theta_j}{ds} = \frac{dh}{ds} \right\}$$

for some curve Γ in the θ-space and some function $h(s)$ defined on Γ (here $\theta(s)$ parametrizes Γ). This can be reformulated as a general fact about Lagrangian submanifolds with degenerate (one-dimensional) projection onto one of the Lagrangian axis planes, using the alternate symplectic coordinates $p_j \equiv \frac{1}{2} R_j^2$ and $q_j \equiv \theta_j$, $j = 1, \cdots, n$. (Note that $dp_j \wedge dq_j \equiv dx_j \wedge dy_j$. See Harvey-Lawson [9] for the proof of (7.83) (cf. Problem 3).

Note that for $\theta \in \Gamma$, $E_\theta \equiv M(\Gamma, h) \cap P_\theta$ is an ellipsoid with radii $r_1, \cdots, r_n$ defined by

$$(7.84) \qquad \frac{1}{r_j^2} = \frac{d\theta_j}{dh} \quad \text{if} \quad \frac{d\theta_j}{dh} > 0,$$

and that otherwise the hypersurface $M(\Gamma, h) \cap P_\theta$ is not compact.

The vectors

$$(7.85) \qquad V_k \equiv (-t_k z_1, 0, \ldots, 0, t_1 z_k, 0, \ldots, 0), \quad k = 2, \ldots, n,$$

with $t_k \equiv R_k / r_k$ and $z_k \equiv r_k e^{i\theta_k}$, provide tangent vectors to the ellipsoid

$$E_\theta \equiv \left\{ w \in \mathbf{C}^n : w_k = t_k z_k, \ \sum t_k^2 = 1 \right\}$$

since $N \equiv (t_1 z_1, \ldots, t_n z_n)$ is normal to E_θ. In addition,

$$(7.86) \qquad V_1 \equiv (t_1 \dot{z}_1, \ldots, t_n \dot{z}_n)$$

is also tangent to M.

Let L denote the complex linear map sending e_j to V_j, $j = 1, \ldots, n$, when $e_1, \ldots, e_n$ is the standrad basis for $\mathbf{R}^n \subset \mathbf{C}^n$. Since M is Lagrangian, M is $\phi \equiv \operatorname{Im} dz$ special Lagrangian if and only if

$$(7.87) \qquad \det_{\mathbf{C}} L = i\lambda, \quad \text{with } \lambda > 0,$$

because of Corollary 7.37.

One can compute that (Problem 2(b))

$$\det_{\mathbf{C}} L = t_1^{n-1} \left(\frac{\overline{z}_1 \dot{z}_1 t_1^2}{|z_1|^2} + \cdots + \frac{\overline{z}_n \dot{z}_n t_n^2}{|z_n|^2} \right) (z_1 \cdots z_n). \tag{7.88}$$

Thus, M is special Lagrangian if and only if each $\overline{z}_k \dot{z}_k$ is of the form

$$\overline{z}_k \dot{z}_k = f_k i \, \overline{z}_1 \cdots \overline{z}_n, \text{ with } f_k \text{ real and } > 0. \tag{7.89}$$

In polar coordinates $z_k = r_k e^{i\theta_k}$, with $\psi \equiv \sum_{k=1}^n \theta_k$, Equation (7.89) becomes

$$r_k \dot{r}_k = f_k r_1 \cdots r_n \sin \psi \tag{7.90}$$

and

$$r_k^2 \dot{\theta}_k = f_k r_1 \cdots r_n \cos \psi. \tag{7.91}$$

Since $r_k^2 (d\theta_k/dh) = 1$, Equation (7.91) implies that all the f_k must be equal. Reparametrizing, we may assume $f_k = 1$, yielding the system (with $\psi \equiv \sum \theta_k$)

$$r_k \dot{r}_k = r_1 \cdots r_n \sin \psi \tag{7.92}$$

and

$$r_k^2 \dot{\theta}_k = r_1 \cdots r_n \cos \psi. \tag{7.93}$$

Let $c_j \equiv z_j(0) > 0$, i.e., $r_j(0) = c_j > 0$ and $\theta_j(0) = 0$, $j = 1, \ldots, n$ denote the initial conditions at time $t = 0$.

It remains to solve this system (7.92), (7.93). First, note that if $r(t), \theta(t)$ is a solution, then $r(t) \equiv r(-t)$, $\theta(t) \equiv -\theta(-t)$ is also a solution. Hence, we need only consider $t \geq 0$.

Since $r_k \dot{r}_k$ is independent of k.

$$r_k^2 = c_k^2 + u \quad \text{and} \quad u(0) = 0 \tag{7.94}$$

for some function u. Also (7.92) and (7.93) imply that $r_1 \cdots r_n \cos \psi$ has derivative zero so that

$$r_1 \cdots r_n \cos \psi = c_1 \cdots c_n. \tag{7.95}$$

Remark. Alternatively, note that the solution curves to (7.92), (7.93) provide the flow lines of the Hamiltonian vector field associated with the Hamiltonian function $H \equiv r_1 \cdots r_n \cos \psi = \operatorname{Re} z_1 \cdots z_n$ (it is interesting to compare with Theorem 7.69).

Since $\dot{\theta}(0) = c_1 \cdots c_n / c_k^2 > 0$, the functions $\theta_k(t)$ and $\psi(t)$ must be strictly increasing for t small. Since $\dot{u} = 2 r_k \dot{r}_k = 2 r_1 \cdots r_n \sin \psi$, u is also strictly increasing for $t > 0$ small. Now, in order for $\dot{r}$ to vanish, $\sin \psi = 0$ or $\cos \psi = 1$, i.e., $r_1 \cdots r_n = c_1 \cdots c_n$ by (7.95). This proves that $\dot{r}_k$ never vanishes, and hence u is strictly increasing for all $t > 0$. Assume $t > 0$ and define y by $u \equiv y^2$. Since $\dot{u} = 2 y \dot{y} = 2 r_k \dot{r}_k = 2 r_1 \cdots r_n \sin \psi$, we have

$$(7.96) \qquad y \dot{y} = r_1 \cdots r_n \sin \psi.$$

Substituting $\cos \psi \equiv c_1 \cdots c_n / r_1 \cdots r_n$ into (7.96) and (7.93) yields the system

$$(7.97) \qquad \dot{y} = \sqrt{\frac{(c_1^2 + y^2) \cdots (c_n^2 + y^2) - c_1^2 \cdots c_n^2}{y^2}},$$

$$(7.97') \qquad \dot{\theta}_k = c_1 \cdots c_n / \left(c_k^2 + y^2 \right), \quad k = 1, \ldots, n.$$

Eliminating t and replacing c_k^2 by $1/a_k$ yields

$$(7.98) \qquad \frac{d\theta_k}{dy} = \frac{a_k}{(1 + a_k y^2) \sqrt{P(a, y)}}$$

as desired. ∎

Remark 7.99. Equation (7.97) implies that

$$(7.100) \qquad t = \int_0^y \frac{dy}{\sqrt{Q(y)}},$$

where

$$(7.101) \qquad Q(y) \equiv y^{-2} \left[(c_1^2 + y^2) \cdots (c_n^2 + y^2) - c_1^2 \cdots c_n^2 \right].$$

If $n = 2$, the integral diverges as $y \to \infty$, and the system is solved for all time; while if $n \geq 3$, the integral converges and the system is solved for $t \in [-T, T]$, where the terminal time T is given by

$$(7.102) \qquad T = \int_0^\infty \frac{dy}{\sqrt{Q(c, y)}}.$$

Note that as c_1 approaches $+\infty$, the terminal time T approaches zero.

ASSOCIATIVE GEOMETRY

In this section, consider the 3-form ϕ on $\operatorname{Im} \mathbf{O} = \mathbf{R}^7$ defined by

$$(7.103) \qquad \phi(x, y, z) \equiv \langle x, yz \rangle \quad \text{for all } x, y, z \in \operatorname{Im} \mathbf{O}$$

that was examined in Chapter 6. Recall that $\phi \in \Lambda^3(\operatorname{Im} \mathbf{O})^*$, since ϕ vanishes when any two of the variables x, y, z are set equal. The analogue of the Wirtinger inequality for ϕ is most elegantly described by an equality.

Theorem 7.104 (Associator Equality). *For all $x, y, z \in \operatorname{Im} \mathbf{O}$,*

$$(7.105) \qquad \phi(x \wedge y \wedge z)^2 + \frac{1}{4}|[x, y, z]|^2 = |x \wedge y \wedge z|^2.$$

Proof: Recall that the triple cross product satisfies

$$(7.106) \qquad\qquad |x \times y \times z| = |x \wedge y \wedge z|,$$
$$(7.107) \qquad\qquad \operatorname{Re} x \times y \times x = \phi(x, y, z),$$
$$(7.108) \qquad\qquad \operatorname{Im} x \times y \times z = \frac{1}{2}[x, y, z].$$

The *associator equality* (7.105) then follows.

Definition 7.109. *If $\xi \in G_{\mathbf{R}}(3, \operatorname{Im} \mathbf{O})$ is the canonically oriented imaginary part of any quaternion subalgebra of $\mathbf{O}$, then ξ is said to be an associative 3-plane. Let ASSOC denote the set of all associative 3-planes. A real oriented 3-manifold $M \subset \operatorname{Im} \mathbf{O}$ with $\vec{M} \in$ ASSOC at each point $x \in M$ is said to be an associative submanifold of $\operatorname{Im} \mathbf{O}$.*

Given $\xi \equiv u_1 \wedge u_2 \wedge u_3$ with u_1, u_2, u_3 an oriented orthonormal basis for $\xi \in G_{\mathbf{R}}(3, \operatorname{Im} \mathbf{O})$, Artin's Theorem 6.39 implies that the associator $[u_1, u_2, u_3]$ vanishes if and only if $\operatorname{span}\{1, u_1, u_2, u_3\}$ is a subalgebra of $\mathbf{O}$, algebra isomorphic to $\mathbf{H}$.

Consequently, given $\xi \in G_{\mathbf{R}}(3, \operatorname{Im} \mathbf{O})$,

$$(7.110) \qquad \pm\xi \in \text{ASSOC if and only if } [u_1, u_2, u_3] = 0$$

for some (or equivalently all) basis u_1, u_2, u_3 for span ξ. Therefore, as a consequence of the associator equality we have that ϕ is a calibration and that

$$(7.111) \qquad \pm\xi \in \text{ASSOC if and only if } \phi(\xi) = \pm 1.$$

This proves the next result.

Theorem 7.112. *The associative form* $\phi \in \Lambda^3(\operatorname{Im}\mathcal{O})^*$ *is a calibration with the contact set*

$$G(\phi) = \text{ASSOC}.$$

Corollary 7.113. *Each associative submanifold of* $\operatorname{Im}O \cong \mathbf{R}^7$ *is volume minimizing.*

Definition 7.114. *An oriented 4-plane* $\eta \in G_{\mathbf{R}}(4, \operatorname{Im}O)$ *is said to be coassociataive if the 3-plane* $\eta^\perp$ *is associative. The 4-form* $\psi \equiv \star\phi \in \Lambda^4(\operatorname{Im}O)^*$ *is called the coassociative calibration.*

Theorem 7.115. *The coassociative 4-form* $\psi \in \Lambda^4(\operatorname{Im}O)^*$ *is a calibration, and any real 4-dimensional submanifold of* $\operatorname{Im}O$ *whose normal 3-plane (at each point) belongs to* ASSOC *is volume minimizing.*

This theorem can be deduced from Theorem 7.112 (see Problem 4).

Interesting systems of partial differential equations arise in associative geometry (see Problem 5) and coassociative geometry, which in some ways are analogous to the Cauchy–Riemann equations. However, we turn now to examples.

A particularly interesting example is the Lawson–Osserman coassociative cone.

Theorem 7.116. *The cone*

$$(7.117) \qquad M \equiv \left\{ \frac{\sqrt{5}}{2} \frac{xi\,\overline{x}}{|x|} + xi : x \in \mathbf{H} \right\}$$

on the graph of the Hopf map

$$\eta(x) \equiv \frac{\sqrt{5}}{2} \frac{xi\overline{x}}{|x|},$$

where η *maps the unit 3-sphere* $S^3 \subset \mathbf{H}$ *to the 2-sphere* S^2 *of radius* $\sqrt{5}/2$ *in* $\operatorname{Im}\mathbf{H}$, *is coassociative, and hence volume minimizing.*

Remark 7.118. The function $\eta : \mathbf{H} \to \operatorname{Im}\mathbf{H}$ defined by

$$\eta(x) \equiv \frac{\sqrt{5}}{2} \frac{xi\overline{x}}{|x|}$$

is Lipschitz continuous but not of class C^1 (continuously differentiable). This makes the fact that its graph is volume minimizing particularly interesting, since a classical result of C. B. Morry says that if the graph of a C^1-function is volume minimizing, then the function is real analytic.

A 3-*dimensional cone* M in $\mathbf{R}^7 \cong \operatorname{Im} \mathbf{O}$ with vertex at the origin is a real 3-dimensional submanifold of $\mathbf{R}^7 - \{0\}$ of the form

$$M \equiv \{ tx : t \in \mathbf{R}^+ \text{ and } s \in L \},$$

where L is a 2-dimensional submanifold of the unit 6-sphere $S^6 \subset \operatorname{Im} \mathbf{O}$. The real surface $L \subset S^6$ is called the *link* of the cone M. Associative cones M have an elegant description in terms of their links L given by Corollary 7.121 below.

Lemma 7.119. *At each point $x \in S^6 \subset \operatorname{Im} \mathbf{O}$, left multiplication by x, denoted L_x, is a real linear map from the tangent space $T_x S^6$ into itself and $L_x^2 = -1$ so that L_x is an almost complex structure on $T_x S^6$.*

Proof: If $u \in T_x S^6$, then $ux + xu = 2\langle u, x \rangle = 0$. Therefore, $ux \in \operatorname{Im} \mathbf{O}$. Since $\langle xu, x \rangle = \langle u, 1 \rangle = 0$, this proves $xu \in T_x S^6$. ∎

Lemma 7.120. *A two plane $\eta \in T_x S^6$ is almost complex if and only if the three plane $x \wedge \eta$ is associative.*

Proof: $\eta \equiv u \wedge L_x u$ implies $x \wedge \eta = x \wedge u \wedge xu$ is associative. Conversely, if $x \wedge u \wedge v$ is associative then $v = xu$ (if x, u, v are orthonormal). ∎

Corollary 7.121. *A 3-dimensional cone in $\operatorname{Im} \mathbf{O}$ is associative if and only if the link L is an almost-complex curve in $S^6 \subset \operatorname{Im} \mathbf{O}$.*

R. Bryant proved that each compact Riemann surface occurs as the link on some associative cone. The proof, using the Newlander–Nirenberg Theorem, is beyond the scope of this book.

Theorem 7.122 (Bryant). *Given a compact Riemann surface S there exists a almost-complex embedding of S in the 6-sphere $S^6 \in \operatorname{Im} \mathbf{O}$.*

THE ANGLE THEOREM

This beautiful result uses a general class of calibrations as well as special Lagrangian geometry in its solution. But before presenting the result, the motivation will be discussed in a heuristic manner.

A standard method for understanding a singularity (of, for example, a real analytic variety M) is to study the associated "tangent cone." At a point of M, a "tangent cone" is obtained by considering the intersection of M with successively smaller balls (centered at the point in question) and then enlarging each ball with a rescaling from the small ball to the unit ball. In the limit, one obtains a cone in the unit ball, which can be considered a cone in $\mathbf{R}^n$, "the tangent cone to M at the point."

Thus, it is important to have a good understanding of tangent cones. In associative geometry, Corollary 7.121 reduces the study of (tangent) cones to the study of complex curves in $S^6 \subset \mathrm{Im}\,\mathbf{O}$.

In the study of general area minimizing cones, one of the simplest questions one can ask is:

$$(7.123) \qquad \begin{array}{l} \text{When is the union of a pair of oriented } p\text{-dimensional} \\ \text{planes } = \xi, \eta \in G_{\mathbf{R}}(p, \mathbf{R}^n) \text{ minimizing?} \end{array}$$

This question has a remarkably simple answer, as conjectured by Frank Morgan. This answer involves certain *characterizing angles* $\theta_1, \ldots, \theta_p$ which distinguish one pair of planes ξ, η from another such pair ξ', η' (see (7.133$'$) below).

Lemma 7.124. *For each pair of oriented planes* $\xi, \eta \in G_{\mathbf{R}}(p, \mathbf{R}^{2p})$, *there exists an orthonormal basis* $e_1, \ldots, e_{2p}$ *for* $\mathbf{R}^{2p}$ *and angles*

$$(7.125) \qquad 0 \leq \theta_1 \leq \cdots \leq \theta_p, \quad \theta_p \leq \pi - \theta_{p-1},$$

so that

$$(7.126) \qquad \xi = e_1 \wedge \cdots \wedge e_p,$$

$$(7.127) \qquad \eta = \left(\cos\theta_1\, e_1 + \sin\theta_1\, e_{p+1} \right) \wedge \cdots \wedge \left(\cos\theta_p\, e_p + \sin\theta_p\, e_{2p} \right).$$

Remark. The statement and proof are easily adapted to the more general case where span $\xi = \mathbf{R}^m \times \{0\}$ and $\eta \in G_{\mathbf{R}}(p, \mathbf{R}^m \times \mathbf{R}^n)$. For example, if $m \leq p \leq n$, then η can be expressed as

$$\eta = \left(\cos\theta_1\, e_1 + \sin\theta_1\, f_1 \right) \wedge \cdots \wedge \left(\cos\theta_m\, e_m + \sin\theta_m\, f_m \right) \wedge f_{m+1} \wedge \cdots \wedge f_p.$$

Proof of Lemma 7.124: Choose $e_1 \in P \equiv$ span ξ and $u_1 \in Q \equiv$ span η, both unit vectors, so that $\langle e_1, u_1 \rangle \equiv \cos\theta_1$ is maximized. (Note that $0 \leq \langle e_1, u_1 \rangle \leq 1$ so we may choose $0 \leq \theta_1 \leq \pi/2$.)

Since $\langle e_1, u_1 \rangle$ is maximized, the orthogonal projection of u_1 onto P must be equal to $\cos\theta_1\, e_1$. In particular,

$$(7.128) \qquad u_1 = \cos\theta_1\, e_1 + \sin\theta_1\, e_{p+1}$$

defines a unit vector $e_{p+1} \in P^\perp$. The equation (7.128) says that

(7.129)   if $e \in P$ and $e \perp e_1$, then $e \perp u_1$.

By symmetry,

(7.130)   if $u \in Q$ and $u \perp u_1$, then $u \perp e_1$.

Next choose $e_2 \in P, e_2 \perp e_1$ and $u_2 \in Q, u_2 \perp u_1$ so that $\langle e_2, u_2 \rangle \equiv \cos \theta_2$ is maximized, with $0 \le \theta_1 \le \theta_2 \le \pi/2$. Because of (7.130), the projection of u_2 onto P only has an e_2 component. Similarly, the projection of u_2 onto $P^\perp$ must be orthogonal to e_{p+1}. Thus,

(7.131) $u_2 = \cos \theta_2 e_2 + \sin \theta_2 e_{p+2}$

defines $e_{p+2} \in P^\perp, e_{p+2} \perp e_{p+1}$. Continue in this manner. At the last step, choose $e_p \in P$ and $u_p \in Q$ so that $e_1, \ldots, e_p$ is an oriented orthonormal basis for ξ and $u_1, \ldots, u_p$ is an orthonormal basis for η. Since the angle θ_p defined by $\cos \theta_p = \langle e_p, u_p \rangle$ must satisfy $|\cos \theta_p| \le \cos \theta_{p-1}$, the angle θ_p can be chosen so that $\theta_{p-1} \le \theta_p \le \pi - \theta_{p-1}$. ∎

Intuitively, the union of the pair ξ, η is volume minimizing if and only if ξ and $-\eta$ are not too close together. (Picture two large disks very close together, almost parallel, and with opposite orientation. Then a narrow strip will have the same boundary and smaller area.)

If $\theta_1, \ldots, \theta_p$ are the characterizing angles for the pair ξ, η, then the characterizing angles for the pair $\xi, -\eta$ are given by

(7.132) $\psi_1 \equiv \theta_1, \ldots, \psi_{p-1} \equiv \theta_{p-1}$,   and $\psi_p \equiv \pi - \theta_p$.

To prove this fact, replace u_p by $-u_p$ in the proof of Lemma 7.124. Thus, $-u_p = -\cos \theta_p e_p - \sin \theta_p e_{2p} = \cos \psi_p e_p - \sin \psi_p e_{2p}$. Finally, replace e_{2p} by $-e_{2p}$ and note that $\psi_p \le \pi - \psi_{p-1}$.

The next theorem states that the union of ξ and η is *not* volume minimizing if and only if $\xi, -\eta$ are "close together" in the precise sense that

(7.133) $\psi_1 + \cdots + \psi_p < \pi$,

or, in terms of the characterizing angles $\theta_1, \ldots, \theta_p$ for ξ and η,

(7.133′) $\theta_1 + \cdots + \theta_{p-1} < \theta_p$.

This result was conjectured by Frank Morgan.

Angle Theorem 7.134 (Lawlor–Nance). *The union of a pair* $\xi, \eta \in G_{\mathbf{R}}(p, \mathbf{R}^{2p})$ *of oriented p-dimensional subspaces of* $\mathbf{R}^p$ *is volume minimizing if and only if the characterizing angles satisfy the angle criterion:*

$$(7.135) \qquad \theta_p \leq \theta_1 + \cdots + \theta_{p-1} \quad (\text{or equivalently } \psi_1 + \cdots + \psi_p \geq \pi).$$

Proof: First, assume that ξ, η do not satisfy the angle criterion (7.135), or equivalently, assume that the characterizing angles $\psi_1, \ldots, \psi_p$ for the pair $\xi, -\eta$ satisfy (7.133) (i.e, ξ and $-\eta$ are "close"). We must show that $\xi \cup \eta$ is not volume minimizing.

It is convenient to make ξ and $-\eta$ symmetrical about a fixed p-plane. Let $\bar{\theta}_j$ denote $\psi_j/2$, $j = 1, \cdots, p$, and pick a new orthonormal basis $e_1, \ldots, e_{2p}$ so that

$$(7.136) \qquad \xi = \xi(\bar{\theta}) \quad \text{and} \quad -\eta = \xi(-\bar{\theta}),$$

where $\xi(\theta)$ is defined to be

$$(7.137) \quad \xi(\theta) \equiv (\cos\theta_1 \ e_1 + \sin\theta_1 \ e_{p+1}) \wedge \cdots \wedge (\cos\theta_p \ e_p + \sin\theta_p \ e_{2p})$$

The hypothesis (7.133) now takes the form

$$(7.138) \qquad \sum_{j=1}^{p} \bar{\theta}_j < \frac{\pi}{2}.$$

Identify $\mathbf{R}^{2p}$ with $\mathbf{C}^p$ by identifying the e_j axis in $\mathbf{R}^{2p}$ with the x_j axis in $\mathbf{C}^p$ ($j = 1, \ldots, p$), and identifying the e_{p+j} axis in $\mathbf{R}^{2p}$ with the y_j axis in $\mathbf{C}^p$ ($j = 1, \ldots, p$). Assume, for the moment, that for some choice of the constants $a \equiv (a_1, \ldots, a_n)$, the special Lagrangian manifold M_a constructed in Theorem 7.78 has nontrivial intersection with $P_{\bar{\theta}} \equiv$ span $\xi(\bar{\theta})$. Then one can show that the portion $\overline{M}$ of M cut off by $P_{\bar{\theta}} \equiv$ span ξ and $P_{-\bar{\theta}} \equiv$ span η satisfies

$$(7.139) \quad \begin{aligned} \mathrm{vol}(\overline{M}) &= \int_{\overline{M}} \mathrm{Im}\, dz \\ &= \int_{U_{\bar{\theta}}} \mathrm{Im}\, dz - \int_{U_{-\bar{\theta}}} \mathrm{Im}\, dz < \ \mathrm{vol}(U_{\bar{\theta}}) + \ \mathrm{vol}(U_{-\bar{\theta}}), \end{aligned}$$

where U_θ denotes the solid ellipsoid in $P_\theta \equiv$ span $\xi(\theta)$, with the boundary the ellipsoid $M_a \cap P_\theta$. Thus $U_{\bar{\theta}} - U_{-\bar{\theta}}$, and hence $\xi \cup \eta$ is not area minimizing.

Note that

(7.140) $(dz)(\xi(\theta)) = e^{i\theta_1} \cdots e^{i\theta_p}$.

Therefore, both $(\operatorname{Im} dz)(\xi(\bar{\theta}))$ and $-(\operatorname{Im} dz)(\xi(-\bar{\theta}))$ equal $\sin \sum \bar{\theta}_j$, which is strictly less than one by the hypothesis (7.138). This proves the strict inequality in (7.139).

It remains to prove that the constants $a_1, \ldots, a_n \geq 0$ can be chosen so that $M_a \cap P_{\bar{\theta}}$ is nonempty. Fix a value of $y > 0$ and consider the function $\theta(y)$ defined by (7.98) as a function $\theta(a)$ of the parameters a, i.e., set

(7.141) $$\theta(a) \equiv a_k \int_0^y \frac{dy}{(1 + a_k y^2) \sqrt{P(a, y)}} \, ,$$

with

(7.142) $$P(a, y) \equiv y^{-2} \left[\left(1 + a_1 y^2\right) \cdots \left(1 + a_n y^2\right) - 1 \right] .$$

It remains to prove that the map $a \mapsto \theta(a)$ is surjective onto the set $\{\theta : \sum \theta_j < \pi/2 \text{ with } 0 \leq \theta_j, \ j = 1, \ldots, n\}$. The proof is by induction on n. If $n = 1$, then $\theta(a) = \arctan(\sqrt{a}\, y)$ surjects onto $0 \leq \theta < \pi/2$. Note that $a \mapsto \theta(a)$ extends continuously to the closed positive quadrant since setting any one of the variables a_j equal to zero in the formula for θ simply results in the formula for $\theta(a)$ in fewer variables.

Because of (7.93), the inverse image of the hyperplane (in the positive quadrant)

(7.143) $$H_\psi \equiv \left\{ \theta : \sum \theta_j = \psi \text{ and } \theta_j \geq 0, \ j = 1, \ldots, n \right\}$$

under the map $\theta(a)$ is contained in the surface

$$S_\psi \equiv \left\{ a : (1 + a_1 y^2) \cdots (1 + a_n y^2) = \cos^{-2} \psi \right\} .$$

It suffices to show that

(7.144) $\theta : S_\psi \to H_\psi$ is surjective for $0 < \psi < \pi/2$.

The proof of (7.144) is obtained by standard topological methods using the notion of degree of a map. By the induction hypothesis, the degree of the boundary map (θ restricted to the boundary), $\theta : \partial S_\psi \to \partial H_\psi$, can be shown to be one. Now a standard topological result says that the map $\theta : S_\psi \to H_\psi$ must have the same degree one and hence is surjective.

To prove the remaining half of the theorem, assume that the characterizing angles $\theta_1, \ldots, \theta_p$ for ξ, η satisfy the angle criterion (7.135).

Assume that the orthonormal basis $e_1, \cdots, e_{2p}$ of Lemma 7.124 is the standard basis for $\mathbf{R}^{2p}$. Let $z \equiv (x, y)$ with $x, y \in \mathbf{R}^p$ denote coordinates. Following D. Nance [13], consider the p-form

$$(7.145) \qquad \phi_u \equiv \mathrm{Re}\,(dx^1 + u_1 dy^1) \wedge \cdots \wedge (dx^p + u_p dy^p) \in \Lambda^p \left(\mathbf{R}^{2p}\right)^*$$

for each p-tuple $u_1, \ldots, u_p \in S^2 \subset \mathrm{Im}\,\mathbf{H}$ of unit imaginary quaternions. Note that if each $u_k = i$ for $k = 1, \ldots, p$, then ϕ is the special Lagrangian calibration on $\mathbf{C}^p = \mathbf{R}^{2p}$.

Proposition 7.146. *Each ϕ_u is a calibration. Moreover,*

$$(7.147) \quad \xi(\theta) \equiv (\cos\theta_1\, e_1 + \sin\theta_1\, e_{p+1}) \wedge \cdots \wedge (\cos\theta_p\, e_p + \sin\theta_p\, e_{2p})$$

belongs to $G(\phi_u)$ if and only if the quaterion product $v_1 \ldots v_p$ equals 1, where $v_j = \cos\theta_j + \sin\theta_j\, u_j \in \mathbf{H}$, $j = 1, \cdots, p$.

Proof: Note that

$$(7.148) \qquad \phi_u(\xi(\theta)) = \mathrm{Re}\,v_1 \cdots v_p \leq |v_1 \cdots v_p| = |v_1| \cdots |v_p| = 1,$$

with equality if and only if

$$(7.149) \qquad\qquad\qquad v_1 \cdots v_p = 1.$$

It remains to show that the maximum of ϕ_u over the full grassmannian cannot be larger than the maximum of ϕ_u over the special p-planes $\xi(\theta)$. This is a consequence of Frank Morgan's Torus Lemma, which is discussed below.

First we use this proposition to calibrate both ξ and $\xi(\theta)$ with ϕ_u. We must find unit imaginary quaternions $u_1, \ldots, u_p$ so that $v_1 \cdots v_p = 1$, where $v_p = \cos\theta_j + \sin\theta_j\, u_j$, $j = 1, \ldots, p$. If we choose $v_1, \ldots, v_p$ to be of the form

$$(7.150) \qquad\qquad v_1 \equiv w_1\overline{w}_2, \; v_2 \equiv w_2\overline{w}_3, \; \ldots, \; v_p = w_p\overline{w}_1,$$

with $w_1, \ldots, w_p$ unit quaternions then $v_1 \cdots v_p = 1$ is automatic. The only other requirement on $v_1, \ldots, v_p$ is that

$$(7.151) \qquad\qquad \mathrm{Re}\,v_j = \cos\theta_j, \quad j = 1, \ldots, p.$$

The u_j's are defined by

(7.152) $$u_j \equiv \frac{\mathrm{Im}\,v_j}{|\mathrm{Im}\,v_j|}, \quad j = 1, \ldots, p.$$

Note that with v_j given by (7.150) the final condition (7.151) is that

(7.153) $\langle w_1, w_2 \rangle = \cos\theta_1, \ \ldots, \ \langle w_p, w_1 \rangle = \cos\theta_p.$

For simplicity, suppose $w_1, \ldots, w_p \in S^2$ are unit imaginary quaternions. Then the problem is reduced to the following standard problem in spherical trigonometry: Given real numbers $0 \leq \theta_1 \leq \cdots \leq \theta_p$, $\theta_p \leq \pi - \theta_{p-1}$, satisfying the condition $\theta_p \leq \theta_1 + \cdots + \theta_{p-1}$, construct p disjoint points $w_1, \ldots, w_p$ on the 2-sphere so that the spherical distance between the points equals the θ's. That is,

(7.154) $d(w_1, w_2) = \theta_1, \ d(w_2, w_3) = \theta_3, \ \ldots, \ d(w_p, w_1) = \theta_p.$

A picture of a p-sided spherical polygon is helpful (the sides are arcs of great circles). ■

The useful torus lemma can be described as follows. Let $\mathbf{T}$ denote the subset of $G_{\mathbf{R}}(p, \mathbf{R}^{2p})$ consisting of all p-planes of the form

$$\xi(\theta) = (\cos\theta_1\, e_1 + \sin\theta_1\, e_{p+1}) \wedge \cdots \wedge (\cos\theta_p\, e_p + \sin\theta_p\, e_{2p}).$$

Given a calibration ϕ, let $G_{\mathbf{T}}(\phi) = \{\xi(\theta) \in \mathbf{T} : \phi(\xi(\theta)) = 1\}$ denote the *torus contact set* as opposed to the (full) contact set

$$G(\phi) = \{\xi \in G_{\mathbf{R}}(p, \mathbf{R}^{2p}) : \phi(\xi) = 1\}$$

of Definition 7.1. In coordinates, $e_j = \partial/\partial x^j$ and $dx^j = e_j^*\ j = 1, \cdots, p$, while $e_{p+j} = \partial/\partial y^j$ and $dy^j = e_{p+j}\ j = 1, \cdots, p$. A form $\phi \in \Lambda^p(\mathbf{R}^{2p})^*$ in the span of the forms $dx^I \wedge dy^J$, where I and J are disjoint and $I \cup J = \{1, \ldots, p\}$ is called a *torus form*.

Torus Lemma 7.155 (Morgan). *A torus form $\phi \in \Lambda^p(\mathbf{R}^{2p})^*$ is a calibration if and only if*

$$\phi(\xi(\theta)) \leq 1 \quad \text{for all } \xi(\theta) \in \mathbf{T}.$$

Proof: The proof is by induction on p. Renormalize ϕ so that $M = 1$ is the maximum value of ϕ on $G(p, \mathbf{R}^{2p})$. Choose $\xi \in G(p, \mathbf{R}^{2p})$ to be a maximum point, i.e., $\phi(\xi) = 1$. We must show that the torus contact set $G_{\mathbf{T}}(\phi)$ is nonempty.

Put ξ in canonical form (see the remark following Lemma 7.124) with respect to $\mathbf{R}^{2p} = \mathbf{R}^2 \times \mathbf{R}^{2p-2}$, with $\mathbf{R}^2 \times \{0\}$ the span of e_1 and e_{p+1}. That is,

$$\xi \equiv (\cos\theta_1 \ u_1 + \sin\theta_1 \ f_1) \wedge (\cos\theta_2 \ u_2 + \sin\theta_2 \ f_2) \wedge f_3 \wedge \cdots \wedge f_p,$$

where $0 \le \theta_1 \le \theta_2 \le \pi/2$; u_1, u_2 is an orthonormal basis for $\mathbf{R}^2 \times \{0\}$ and $f_1, \ldots, f_{2p-2}$ is an orthonormal basis for $\{0\} \times \mathbf{R}^{2p-2}$. Since ϕ is a torus form, $\phi = e_1 \wedge \Phi + e_{p+1} \wedge \psi$ with Φ and ψ torus forms on $\mathbf{R}^{2p-2}$. Expressing e_1 and e_{p+1} as linear combinations of u_1 and u_2 yields

$$\phi = u_1^* \wedge \alpha + u_2^* \wedge \beta,$$

with α and β torus forms on $\mathbf{R}^{2p-2}$. Now

$$\phi(\xi) = a \cos\theta_1 \sin\theta_2 + b \sin\theta_1 \cos\theta_2,$$

where $a = \alpha(f_2 \wedge \cdots \wedge f_p) = \phi(u_1 \wedge f_2 \wedge \cdots \wedge f_p)$ and $b = -\beta(f_1 \wedge f_3 \wedge \cdots \wedge f_p) = \phi(f_1 \wedge u_2 \wedge f_3 \wedge \cdots \wedge f_p)$. Therefore,

$$1 = \phi(\xi) \le \sqrt{a^2 \cos^2\theta_1 + b^2 \sin^2\theta_1} \le \max\{|a|, |b|\} \le 1,$$

so that we must have equality. Therefore $\theta_2 = \pi/2 - \theta_1$, so that with $\theta \equiv \theta_1$,

$$\xi = (\cos\theta \ u_1 + \sin\theta \ f_1) \wedge (\sin\theta \ u_2 + \cos\theta \ f_2) \wedge f_3 \wedge \cdots \wedge f_p.$$

In particular, $a = 1$ and/or $b = 1$ so that by the induction hypothesis either the torus form α or the torus form β has attained its maximum value 1 on the $p-1$ torus, say $\alpha(\eta) = 1$. Then $\phi(u_1 \wedge \eta) = \alpha(\eta) = 1$, proving that ϕ attains its maximum value 1 on the p-torus $\mathbf{T}$. ∎

The Torus Lemma has several interesting reinterpretations and consequences. Note that the space of torus forms is the dual space of span $\mathbf{T}$, so the Torus Lemma can be stated as

(7.156a) $$\max_{\xi \in \mathbf{T}} \phi(\xi) = \max_{\xi \in G(p, \mathbf{R}^{2p})} \phi(\xi),$$

for all $\phi \in (\text{span } \mathbf{T})^*$.

Given a set A, let ch A denote the convex hull of A. Let P denote orthogonal projection onto the linear subspace span $\mathbf{T}$ of the vector space $\Lambda^p \mathbf{R}^{2p}$. The Torus Lemma implies

(7.156b) $P(\mathrm{ch}(G(p, \mathbf{R}^{2p}))) = \mathrm{ch}\,\mathbf{T}$

and

(7.156c) $\mathrm{ch}\,\mathbf{T} = \mathrm{ch}(G(p, \mathbf{R}^{2p})) \cap \mathrm{span}\,\mathbf{T}.$

In fact, both (7.156b) and (7.156c) are equivalent to the statement (7.156a) of the Torus Lemma.

Assuming (7.156a) we prove (7.156b). Of course,

$$\mathrm{ch}\,\mathbf{T} \subset P(\mathrm{ch}\,(G(p, \mathbf{R}^{2p}))).$$

Now suppose $\xi \in G(p, \mathbf{R}^{2p})$. If $P\xi \notin \mathrm{ch}\,\mathbf{T}$, then by the Hahn–Banach Theorem there exists $\phi \in (\mathrm{span}\,\mathbf{T})^*$ with $\phi(\eta) \leq 1$ for all $\eta \in \mathrm{ch}\,\mathbf{T}$ but with $\phi(\xi) > 1$. This is impossible by (7.156a). Next note that (7.156b) implies (7.156c) and that (7.156c) implies (7.156a).

For a torus calibration the torus contact set $G_{\mathbf{T}}(\phi)$ actually determines the full contact set.

Corollary 7.157. *Given a torus calibration ϕ and $\xi \in G(p, \mathbf{R}^{2p})$,*

$$\xi \in G(\phi) \text{ if and only if } P\xi \in \mathrm{ch}(G_{\mathbf{T}}(\phi)).$$

Proof: If $P\xi \in \mathrm{ch}(G_{\mathbf{T}}(\phi))$, then $P\xi = \sum \lambda_j \xi_j$ with each $\xi_j \in G_{\mathbf{T}}(\phi)$ and $\sum \lambda_j = 1$. Therefore, $\phi(\xi) = \phi(P\xi) = \sum \lambda_j \phi(\xi_j) = 1$. Conversely, given $\xi \in G(\phi), \phi(P\xi) = \phi(\xi) = 1$.

By (7.156b), $P\xi \in \mathrm{ch}\,\mathbf{T}$. It remains to show that

$$\{\eta \in \mathrm{ch}\,\mathbf{T} : \phi(\eta) = 1\} \subset \mathrm{ch}\,(G_{\mathbf{T}}(\phi)).$$

Since the extreme points of a convex set generate the convex set via convex combinations, it suffices to show that the extreme points of $\{\eta \in \mathrm{ch}\,\mathbf{T} : \phi(\eta) = 1\}$ are contained in $\mathrm{ch}(G_{\mathbf{T}}(\phi))$. However, if η is such an extreme point, then one can easily verify that η is also extreme in $\mathrm{ch}\,\mathbf{T}$. Since the extreme points of $\mathrm{ch}\,\mathbf{T}$ are just the elements of $\mathbf{T}$ this proves that $\eta \in G_{\mathbf{T}}(\phi)$.

An alternate method for determining $G(\phi)$ from $G_{\mathbf{T}}(\phi)$, due to Frank Morgan, is described next.

Suppose $\omega \in \Lambda^2 \mathbf{R}^4$ can be written as $\omega = \omega^1 \wedge \omega^2 + \omega^3 \wedge \omega^4$, where $\omega^1, \omega^2, \omega^3, \omega^4$ is an orthonormal basis. Let u_1, u_2, u_3, u_4 denote the dual basis. Then ω has contact set $G(\omega) = \{\xi \in G_{\mathbf{R}}(2, \mathbf{R}^4) : \xi = u \wedge Ju \in G_{\mathbf{C}}(1, \mathbf{C}^2)\} \cong \mathbf{CP}^1$ with respect to the complex structure on $\mathbf{R}^4$ defined by $Je_1 = e_2$, $Je_3 = e_4$.

More generally if $\psi \in \Lambda^p(\mathbf{R}^m)^*$ is of the form $\psi = \omega \wedge \omega^5 \wedge \cdots \wedge \omega^{p-2}$, with $\omega \equiv \omega^1 \wedge \omega^2 + \omega^3 \wedge \omega^4$ as before, then $G(\psi) = G(\omega) \wedge u^5 \wedge \cdots \wedge u^{p-2}$ will also be referred to as a $\mathbf{CP}^1$-contact set.

A subset $B \subset G(p, \mathbf{R}^n)$ with the following property will be called $\mathbf{CP}^1$-*closed*.

If K is any $\mathbf{CP}^1$−contact set containing two points of B,

then the whole $(\mathbf{CP}^1$−contact) set K is a subset of B.

Proposition 7.158. *The contact set $G(\phi)$ of a calibration $\phi \in \Lambda^p(\mathbf{R}^n)^*$ is $\mathbf{CP}^1$-closed.*

This is a consequence of the "First Cousin Principle"; see Problem 8. Note that an arbitrary intersection of $\mathbf{CP}^1$ closed sets is again $\mathbf{CP}^1$-closed.

Proposition 7.159. *Suppose $\phi \in \Lambda^p(\mathbf{R}^{2p})^*$ is a torus calibration. The contact set $G(\phi)$ is the smallest $\mathbf{CP}^1$-closed set containing the torus contact set $G_{\mathbf{T}}(\phi)$.*

Proof: Given any set B with the two properties; B is larger than $G_{\mathbf{T}}(\phi)$ and B is $\mathbf{CP}^1$-closed, we must show that $G(\phi)$ is a subset of B. Again, the proof is by induction on p. Suppose $\xi \in G(\phi)$, i.e., $\phi(\xi) = 1$. We must show that $\xi \in B$. We proceed exactly as in the proof of the Torus Lemma to show that

$$\xi = (\cos\theta \ u_1 + \sin\theta \ f_1) \wedge (\sin\theta \ u_2 + \cos\theta \ f_2) \wedge f_3 \wedge \cdots \wedge f_p.$$

Case 1: If $\theta = 0$, then $\xi = u_1 \wedge f_2 \wedge \cdots \wedge f_p$ and $\phi(\xi) = \alpha(f_2 \wedge \cdots \wedge f_p) = 1$, so that $\eta_0 = f_2 \wedge \cdots \wedge f_p \in G(\alpha)$.

Let $B' \equiv \{\eta \in G(p-1, \mathbf{R}^{2p-2}) : u_1 \wedge \eta \in B\}$. Then B' contains $G_{\mathbf{T}}(\alpha)$ and B' is $\mathbf{CP}^1$-closed. Therefore, by the induction hypothesis B' contains $G(\alpha)$. Since $\eta_0 \in G(\alpha) \subset B'$, this proves that $\xi = u_1 \wedge \eta_0 \in B$.

Case 2: If $0 < \theta < \pi/2$, then $a = b = 1$ and both $u_1 \wedge f_2 \wedge \cdots \wedge f_p$ and $f_1 \wedge u_2 \wedge f_3 \wedge \cdots \wedge f_p$ belong to $G(\phi)$. As in the previous case, by induction both belong to B. Both also belong to the $\mathbf{CP}^1$-contact set $K \equiv G((u_1 \wedge f_2 + f_1 \wedge u_2) \wedge f_3 \wedge \cdots \wedge f_p)$. Therefore, the hypothesis on B implies that B contains all of K. Since $\xi \in K$ the proof is complete. ∎

GENERALIZED NANCE CALIBRATIONS AND COMPLEX STRUCTURES

The Nance calibrations ϕ_u, defined in (7.145), provide a large class of "torus calibrations" using the quaternions, or more precisely, a collection of unit imaginary quaternions $u_1, \ldots, u_p$. This construction can be generalized as follows.

Consider the vector space $M_{2n}(\mathbf{R})$ of matrices with positive definite inner product given by

$$\langle A, B \rangle = \frac{1}{2n} \text{ trace } AB^t.$$

Let $P : M_{2n}(\mathbf{R}) \to \text{span } 1$ denote the orthogonal projection onto the span of the identity matrix $1 \in M_{2n}(\mathbf{R})$. The orthogonal complex structures on $\mathbf{R}^{2n}$ are given by

$$\text{Cpx}(\mathbf{R}^{2n}) \equiv \left\{ J \in M_{2n}(\mathbf{R}) : J^t = J^{-1} \text{ and } J^2 = -1 \right\}.$$

Definition 7.160. *The p-form $\phi_u \in \Lambda^p(\mathbf{R}^{2p})^*$ defined by*

$$(7.161) \qquad \phi_u \equiv P\left[(dx^1 + u^1 dy^1) \wedge \cdots \wedge (dx^p + u^p dy^p) \right],$$

where $u_1, \ldots, u_p \in \text{Cpx}(\mathbf{R}^{2n})$, is called a generalized Nance form.

Proposition 7.162. *Each generalized Nance form ϕ_u is a calibration. Moreover,*

$$(7.163) \qquad \xi(\theta) \in G(\phi_u) \text{ if and only if } A_1 \cdots A_p = 1,$$

where $A_j = \cos\theta_j + \sin\theta_j \ u_j \in M_{2n}(\mathbf{R})$, $j = 1, \ldots, p$.

Proof: If $u \in \text{Cpx}(\mathbf{R}^{2n}) \subset M_{2n}(\mathbf{R})$ and $\theta \in \mathbf{R}$, we set

$$A \equiv \cos\theta + \sin\theta \ u.$$

Then

$$A^t = \cos\theta - \sin\theta \ u,$$

and

$$AA^t = \cos^2\theta + \sin^2\theta = 1,$$

so that $A \in O(2n)$ is also an orthgonal transformation. If $u_1, \ldots, u_p \in \text{Cpx}(\mathbf{R}^{2n})$ and $\theta_1, \ldots, \theta_p \in \mathbf{R}$ are given, define $A_j \equiv \cos\theta_j + \sin\theta_j \ u_j$, $j = 1, \ldots, p$. Then each $A_j \in O(2n)$ and hence the product $A_1 \cdots A_p \in O(2n)$.

Therefore, $A_1 \cdots A_p$ is a unit vector in $M_{2n}(\mathbf{R})$. To complete the proof, note that

$$\phi_u(\xi(\theta)) \equiv P[(\cos\theta_1 + \sin\theta_1 \ u_1) \cdots (\cos\theta_p + \sin\theta_p \ u_p)]$$
$$= \langle A_1 \cdots A_p, 1 \rangle \le 1,$$

with equality if and only if $A_1 \cdots A_p = 1$; and then apply the Torus Lemma. ∎

In order to understand why this construction—using complex structures $\mathrm{Cpx}(\mathbf{R}^{2n})$ rather than unit imaginary quaternions $S^2 \subset \mathrm{Im}\,\mathbf{H}$—generalizes the Nance construction, we examine the set $\mathrm{Cpx}(\mathbf{R}^{2n})$ of orthogonal complex structures in more detail.

Since each complex structure determines a canonical orientation on $\mathbf{R}^{2n}$, the set $\mathrm{Cpx}(n)$ naturally divides into two disjoint sets of complex structures

$$(7.164) \qquad \mathrm{Cpx}(n) = \mathrm{Cpx}^+(n) \cup \mathrm{Cpx}^-(n),$$

with $\mathrm{Cpx}^+(n)$ the subset of $\mathrm{Cpx}(n)$ consisting of complex structures that induce a given orientation on $\mathbf{R}^{2n}$, and $\mathrm{Cpx}^-(n)$ the complex structures that induce the opposite orientation on $\mathbf{R}^{2n}$.

The group $O(2n)$ acts on $\mathrm{Cpx}(n)$ by

$$(7.165) \qquad J \in \mathrm{Cpx}(n) \text{ maps to } gJg^{-1} \in \mathrm{Cpx}(n) \text{ for each } g \in O(2n).$$

The isotropy subgroup of $O(2n)$ at a particular $J \in \mathrm{Cpx}(n)$ is just

$$(7.166) \qquad U(n) = \mathrm{GL}(n, \mathbf{C}) \cap O(2n),$$

where $\mathrm{GL}(n, \mathbf{C}) = \{A \in \mathrm{GL}(2n, \mathbf{R}) : JA = AJ\}$.

Given a complex structure $J \in O(2n)$ on $\mathbf{R}^{2n}$, there exists a (complex) unitary basis $e_1, \ldots, e_n$ for $\mathbf{C}^n = \mathbf{R}^{2n}$. Thus,

$$(7.167) \qquad e_1, \ldots, e_n, Je_1, \ldots, Je_n$$

is an (real) orthonormal basis for $\mathbf{R}^{2n}$. With respect to this orthonormal basis the matrix for the complex structure J is

$$(7.168) \qquad J = \begin{pmatrix} 0 & 1 \\ -1 & 0 \end{pmatrix},$$

where $\mathbf{R}^n = V \oplus JV$ with $V \equiv \mathrm{span}_{\mathbf{R}}\{e_1, \ldots, e_n\}$ induces the 2×2 blocking of J. This proves that $O(2n)$ acts transitively on $\mathrm{Cpx}(n)$. Since $O(2n)$ has two connected components and $U(n)$ is connected, the quotient $O(2n)/U(n)$ has two connected components. In summary we have the following proposition.

Proposition 7.169. *The set* $\mathrm{Cpx}(n)$ *of orthogonal complex structures on* $\mathbf{R}^{2n}$ *has two connected components* $\mathrm{Cpx}^+(n)$ *and* $\mathrm{Cpx}^-(n)$, *with*

$$(7.170) \qquad \begin{aligned} \mathrm{Cpx}(n) &\cong O(2n)/U(n), \\ \mathrm{Cpx}^{\pm}(n) &\cong \mathrm{SO}(2n)/U(n). \end{aligned}$$

In low dimensions, the space of complex structures $\mathrm{Cpx}(n)$ can be represented in terms of the quaternions and octonians. For example, $\mathbf{R}_u$, right multiplication by a unit imaginary quaternion, is an orthogonal complex structure on $\mathbf{H} \cong \mathbf{R}^4$.

Proposition 7.171.

$$\mathrm{Cpx}^+(2) = \{ L_u : u \in S^2 \subset \mathrm{Im}\,\mathbf{H} \} \cong S^2$$

and

$$\mathrm{Cpx}^-(2) = \{ R_u : u \in S^2 \subset \mathrm{Im}\,\mathbf{H} \} \cong S^2,$$

where $\mathbf{H}$ *has the standard orientation.*

Proof: First note that, for each $u \in S^2 \subset \mathrm{Im}\,\mathbf{H}$, $L_u \in \mathrm{Cpx}^+(2)$. This is because the complex structure L_i induces the standard orientation on $\mathbf{H}$. Hence, each L_u also induces the standard orientation on $\mathbf{H}$, since S^2 is connected.

Now suppose $J \in \mathrm{Cpx}^+(2)$ is given and identify $\mathbf{R}^4 = \mathbf{H}$. Define $e_1 \equiv J1$ and choose a unit vector $e_2 \perp 1, e_1$. Then, because of the Cayley–Dickson process, $1, e_1, e_2$ and $e_3 = e_1 e_2$ is an oriented orthonormal basis for $\mathbf{H}$. Now Je_2 must equal $\pm e_3$. However, since $1, J1, e_2, Je_2$ must induce the standard orientation on $\mathbf{H}$, $Je_2 = e_3$. This proves $J = L_{e_1}$.

Similarly, $\mathrm{Cpx}^-(2) = \{ R_u : u \in S^2 \subset \mathrm{Im}\,\mathbf{H} \} \cong S^2$. ∎

Alternatively, Proposition 7.171 can be proved using the fact (7.170) that each $J \in \mathrm{Cpx}^+(2)$ is of the form $J = g L_i g^{-1}$ with $g \in \mathrm{SO}(4)$. Because of (1.34′), $g(x) = p x \bar{q}$ for some unit vectors $p, q \in \mathbf{H}$. Thus $J = L_{p i \bar{p}}$.

Replacing the quaternions $\mathbf{H}$ by the octonians $\mathbf{O}$, each R_u with $u \in S^6 \subset \mathrm{Im}\,\mathbf{O}$ determines a (6-sphere of) complex structures on $\mathbf{R}^8 = \mathbf{O}$ (compatible with the standard orientation on $\mathbf{O}$). However, $\mathrm{Cpx}^+(\mathbf{O}) \cong \mathrm{SO}(8)/U(4)$ has dimension 12, so that S^6 is too small.

Definition 7.172. *Given* $\xi = u \wedge v \in G(2, \mathbf{O})$, *let* $J_\xi \equiv \frac{1}{2}(R_v R_{\bar{u}} - R_u R_{\bar{v}})$.

First, note that J_ξ is well-defined (independent of the choice of u, v with $\xi = u \wedge v$) since the right hand side vanishes when $u = v$. Second, the basic octonian formula

$$R_u R_{\bar{v}} + R_v R_{\bar{u}} = 2\langle u, v \rangle$$

implies that, for u, v orthonormal, $J_\xi = R_v R_{\overline{u}} = -R_u R_{\overline{v}}$, so that

$$(7.173) \qquad J_\xi^2 = -R_v R_{\overline{u}} R_u R_{\overline{v}} = -1.$$

This proves that $J_\xi \in \mathrm{Cpx}(4)$.

Proposition 7.174.

$(n = 3) \qquad \mathrm{Cpx}^+(3) = \left\{ J_{u \wedge ui} : u \in S^6 \subset \mathrm{Im}\,\mathbf{O} \right\} \cong \mathbf{P}^6(\mathbf{R})$

$(n = 4) \qquad \mathrm{Cps}^+(4) = \left\{ J_\xi : \xi \in G(2, \mathbf{O}) \right\} \cong G(2, \mathbf{O}).$

The proof will be given in Chapter 14 *Low Dimensions*.

Remark 7.175. Consider the case $n = 4$. Given $\xi = u \wedge v \in G(2, \mathbf{O})$ with u, v orthonormal.

$$(7.176) \qquad J_\xi u = v \quad \text{and} \quad J_\xi v = -u.$$

Thus, in the plane $\mathrm{span}\{u, v\}$, J_ξ is just counterclockwise rotation by $\pi/2$. Suppose $x \in \mathrm{span}\{u, v\}^\perp$. Then $v(\overline{u}x) = -v(\overline{x}u) = x(\overline{v}u)$, so that

$$(7.177) \qquad J_\xi x = R_{u \times v} x \quad \text{if } x \in \mathrm{span}\{u, v\}^\perp,$$

where $u \times v = \frac{1}{2}(\overline{v}u - \overline{u}v)$. In summary, the complex structure J_ξ is

(1) counterclockwise rotation by $\pi/2$ on the 2-plane span ξ, and
(2) right multiplication by $u \times v$ on $(\mathrm{span}\ \xi)^\perp$, where u, v is any oriented orthonormal basis for ξ.

PROBLEMS

1. Use the canonical form given by Lemma 7.18 to give a second proof of Theorem 7.26:

$$(\mathrm{Re}\,dz)(\xi) \leq 1 \quad \text{for all } \xi \in G_{\mathbf{R}}(n, \mathbf{C}^n),$$

with equality if and only if $\xi \in \mathrm{SLAG}$.

2. (a) Complete the proof of Theorem 7.69, computing

$$\det{}_{\mathbf{C}} \left(i\, \frac{\partial f_q}{\partial z_p} \right),$$

and then verifying conditions (7.64) and (7.65) so that Theorem 7.62 is applicable.

(b) Compute the complex determinant $\det_{\mathbf{C}} L$ of the linear map L used in the proof of Theorem 7.78.

3. (HL [9]—assumes knowledge of harmonic functions on surfaces in $\mathbf{R}^3$, minimal surfaces, and normal bundles.)

Given a surface S in $\mathbf{R}^3$ and a smooth function F on a neighborhood of S, let

$$A(S, f) \equiv \{(x, tn_x + \nabla F(x)) \in \mathbf{R}^6 : x \in S, \ t \in \mathbf{R}\}.$$

Here n_x is the unit normal to S at $x \in S$ and $f \equiv F|_S$.

(a) Show that $A(S, f)$ depends only on $f \equiv F|_S$ and not on the ambient extension F of f.

(b) Show that if S is a minimal surface and f is an harmonic function on S, then $A(S, f)$ is a special Lagrangian 3-manifold in $\mathbf{R}^6 \cong \mathbf{C}^3$.

4. Deduce the Coassociative Theorem 7.115 from the Associative Theorem 7.112.

5. Show that $M \equiv \{x + f(x)\, e : x \in U \subset \operatorname{Im} \mathbf{H}\}$, the graph of $f : U \to \mathbf{H}$ over an open domain U in $\operatorname{Im} \mathbf{H}$, correctly oriented, is associative if and only if

$$\left[i + \frac{\partial f}{\partial x^1}\, e, \ j + \frac{\partial f}{\partial x^2}\, e, \ k + \frac{\partial f}{\partial x^3}\, e \right] = 0 \quad \text{on } U.$$

6. (a) When $n = 2$, explicitly solve (using elementary functions) the system of differential equations (7.92), (7.93) for $z_1(t), z_2(t)$.

(b) Find a real orthogonal coordinate change in $\mathbf{R}^4 \cong \mathbf{C}^2$ so that the special Lagrangian manifold M_a described in Theorem 7.78 equals $\{w \in \mathbf{C}^2 : a_1 w_1^2 - a_2 w_2^2 = 1\}$.

7. (a) At the point

$$r\left(\frac{\sqrt{5}}{2}\, i + e \right), \quad r \in \mathbf{R},$$

compute a basis for the normal 3-plane to the cone M defined by (7.117) (the cone on the graph of the Hopf map). Verify that this normal 3-plane is associative.

(b) Show that M is fixed by $HU(1)$ acting on $\mathbf{O} = \mathbf{H} \oplus \mathbf{H}e$, with $p \in HU(1)$ sending $a + be$ to $pa\bar{p} + (b\bar{p})e$.

(c) Prove Theorem 7.116.

8. (a) (The First Cousin Principle) Suppose $\phi \in \Lambda^p(\mathbf{R}^n)^*$ is a calibration and $\phi(e_1 \wedge \cdots \wedge e_p) = 1$, where $e_1, \ldots, e_n$ is an orthonormal basis. Prove that $\phi(\xi) = 0$ for all *first cousins*:

$$\xi = e_i \wedge (e_j \rfloor (e_1 \wedge \cdots \wedge e_p)), \quad j = 1, \ldots, p \text{ and } i = p+1, \ldots, n$$

of $e_1 \wedge \ldots \wedge e_p$.

(b) Prove Proposition 7.158.

9. Consider a 7-torus $M \equiv \operatorname{Im} \mathbf{O}/\Lambda$ where Λ is a rank 7 lattice in $\mathbf{R}^7 \cong \operatorname{Im} \mathbf{O}$ equipped with the associative calibration $\phi \in \Lambda^3(M)$. Prove that the lattice may be chosen so that M has a compact associative submanifold.

8. Matrix Algebras

This chapter collects information about matrix algebras that will be needed in our discussions of spinors and pinors. More knowledgeable readers may wish to treat this as a reference chapter rather than reading it in detail.

The matrix algebras

$$M_N(\mathbf{R}), \ M_N(\mathbf{C}), \ M_N(\mathbf{H})$$

are examined in this chapter. The algebras $M_N(\mathbf{R})$ and $M_N(\mathbf{H})$ will always be considered real algebras, while the algebra $M_N(\mathbf{C})$ may be considered a real algebra or a complex algebra.

REPRESENTATIONS

At first, consider all three of the algebras $M_N(\mathbf{R}), \ M_N(\mathbf{C}), \ M_N(\mathbf{H})$ to be real algebras.

Definition 8.1. *Given a real algebra A and a vector space V over F ($\equiv$ $\mathbf{R}, \mathbf{C},$ or $\mathbf{H}$), a real linear map*

$$(8.2) \qquad\qquad \rho : A \to \mathrm{End}_F(V)$$

sending 1 to the identity on V that satisfies

$$(8.3) \qquad\qquad \rho(ab) = \rho(a)\rho(b) \quad \text{for all } a, b \in A,$$

*is called an F-representation of A. Two F-representations, $\rho_1 : A \to$
$\text{End}_F(V_1)$ and $\rho_2 : A \to \text{End}_F(V_2)$ of the real algebra A are, equivalent if
there exists an invertible F-linear map*

$$(8.4) \qquad\qquad f : V_1 \to V_2$$

satisfying

$$(8.5) \qquad\qquad \rho_2(a) = f \circ \rho_1(a) \circ f^{-1} \quad \text{for all } a \in A.$$

Such an operator f is called an intertwining operator.

The kernel of a representation ρ is a two-sided ideal in A (which is
proper since $1 \notin \ker \rho$).

Lemma 8.6. *Each two-sided ideal in one of the matrix algebras $M_N(\mathbf{R})$,
$M_N(\mathbf{C})$, or $M_N(\mathbf{H})$ is either $\{0\}$ or the entire matrix algebra.*

The proof is a straightforward calculation, left as an exercise. An
algebra with this property is said to be *simple*.

Corollary 8.7. *Each representation of $M_N(\mathbf{R})$, $M_N(\mathbf{C})$, or $M_N(\mathbf{H})$ is
injective.*

Examples (The Standard Representations)

$$(8.8) \qquad\qquad \rho(a)x \equiv ax \quad \text{for } a \in M_N(F) \text{ and } x \in F^N$$

*defines an F-representation of $M_N(F)$ called the standard representation
of $M_N(F)$. Here F is either $\mathbf{R}, \mathbf{C}$, or $\mathbf{H}$.*

Given a matrix $a \in M_N(\mathbf{C})$, let $\bar{a}$ denote the conjugate matrix. The
$\mathbf{C}$-representation of the real algebra $M_N(\mathbf{C})$, defined by

$$(8.9) \qquad\qquad \bar{\rho}(a)z \equiv \bar{a}z \quad \text{for } a \in M_N(\mathbf{C}) \text{ and } z \in \mathbf{C}^N,$$

is called the *conjugate representation* of $M_N(\mathbf{C})$.

Note that each $\mathbf{H}$-representation, or each $\mathbf{C}$-representation of A, is
automatically an $\mathbf{R}$-representation. As $\mathbf{R}$-representations, the standard
representation of $M_N(\mathbf{C})$ and the conjugate representation of $M_N(\mathbf{C})$ are
equivalent; the intertwining operator $C : \mathbf{C}^N \to \mathbf{C}^N$ is conjugation on $\mathbf{C}$
extended to $\mathbf{C}^N$ by acting on each component. However, since $\rho(i) = i$
and $\bar{\rho}(i) = -i$, the intertwining operator *cannot* be chosen to be complex linear. Therefore, these two representations are not equivalent as
$\mathbf{C}$-representations of the real algebra $M_N(\mathbf{C})$.

Definition 8.10. *Suppose* $\rho_1 : A \to \operatorname{End}_F(V_1)$ *and* $\rho_2 : A \to \operatorname{End}_F(V_2)$ *are two F-representations of A. Then* $\rho \equiv \rho_1 \oplus \rho_2 : A \to \operatorname{End}_F(V_1 \oplus V_2)$, *defined by* $\rho(a) \equiv \rho_1(a) \oplus \rho_2(a)$, *is also an F-representation of A. An F-representation of this form* $\rho = \rho_1 \oplus \rho_2$ *is said to be F-reducible. If an F-representation* ρ *of A is not reducible, then it is called F-irreducible.*

Theorem 8.11. *Suppose F is* **R**, **C**, *or* **H**. *The standard representation of* $M_N(F)$ *on* F^N *is the only irreducible* **R**-*representation of* $M_N(F)$, *up to equivalence.*

Proof: Let I denote the subspace of $M_N(F)$ consisting of those matrices with nonzero entries confined to the first column, and identify I with F^N. Note that I is a left ideal in $M_N(F)$. Also note that

$$(8.12) \qquad \begin{array}{l} \text{if } b \in I \text{ (nonzero) is given, then each } a \in I \text{ can} \\ \text{be written as } a = cb \text{ for some } c \in M_N(F). \end{array}$$

In particular, if K is a left ideal contained in I, then either $K = \{0\}$ or $K = I$. (Thus, I is called a *minimal* left ideal.)

Let E denote the matrix with all entries zero except for the entry 1 in the position first row–first column. Since ρ is injective (Corollary 8.7), $\rho(E) \neq 0$. Suppose $v_1 \in V$ satisfies $\rho(E)v_1 \neq 0$. Define $f : I \cong F^N \to V$ by

$$(8.13) \qquad f(b) = \rho(b)v_1 \quad \text{for all } b \in I.$$

First, we show that f is injective. Let

$$J \equiv \{ a \in M_N(F) : \rho(a)v_1 = 0 \}.$$

Note that J, and hence $\ker f = I \cap J$, is a left ideal in $M_N(F)$. Since $E \in I$, but $E \notin J$, the left ideal $I \cap J$ is proper in I. However, since I is a minimal left ideal, this proves that $I \cap J = \{0\}$, i.e., f is injective.

Now define $V_1 \equiv \operatorname{image} f \equiv \{ \rho(b)v_1 : b \in I \}$. Note that $\rho(a)V_1 \subset V_1$ for each matrix a, since $\rho(a)\rho(b)v_1 = \rho(ab)v_1$. Let $\rho_1(a) \equiv \rho(a)|_{V_1}$. Then

$$(8.14) \qquad \rho_1(a) = f \circ a \circ f^{-1}.$$

This is because for each $u \equiv \rho(b)v_1 \in V_1$, $\rho(a)u = \rho(a)\rho(b)v_1 = \rho(ab)v_1 = f(ab) = f(af^{-1}(u))$. This proves that ρ_1 is equivalent to the standard **R**-representation of $M_N(F)$ on $F^N \cong I$. If $V_1 = V$ the proof is complete.

Next, we prove that

$$(8.15) \qquad \text{if } \rho(E)V \subset V_1, \text{ then } V_1 = V.$$

If $\rho(E)V \subset V_1$, then for any matrices a and b,

$$\rho(aEb)V = \rho(aE)\rho(b)V \subset \rho(aE)V = \rho(a)\rho(E)V \subset \rho(a)V)_1 \subset V_1.$$

Since the two-sided ideal in $M_N(F)$ generated by E is all of $M_N(\mathbf{F})$, this proves that $\rho(1)V \subset V_1$ or $V = V_1$.

Now choose v_2 so that $\rho(E)v_2 \notin V_1$. Exactly as above, define $V_2 \equiv \{\rho(a)v_2 : a \in I\}$, note that $\rho(a)V_2 \subset V_2$, and prove that $\rho_2(a) \equiv \rho(a)|_{V_2}$ defines an $\mathbf{R}$-representation ρ_2 of $M_N(F)$ equivalent to the standard representation.

Next, since $\rho(E)v_2 \notin V_1$ it follows that $V_1 \cap V_2 = \{0\}$. If $V_1 \cap V_2 \neq \{0\}$, say $\rho(a_1)v_1 = \rho(a_2)v_2$ with $a_1, a_2 \in I - \{0\}$, then by (8.12) $E = ca_2$ for some matrix c. Therefore, $\rho(E)v_2 = \rho(ca_2)v_2 = \rho(c)\rho(a_2)v_2 = \rho(c)\rho(a_1)v_1 = \rho(ca_1)v_1 \in V_1$ since $ca_1 \in I$.

Continuing in this manner, we obtain $V_1 \oplus \cdots \oplus V_k = V$, so that $\rho_1 \oplus \cdots \oplus \rho_k = \rho$ is reducible unless $V = V_1$. $\blacksquare$

Corollary 8.16. *The real algebra $M_N(\mathbf{C})$ has exactly two irreducible $\mathbf{C}$-representations—the standard $\mathbf{C}$-representation and the conjugate $\mathbf{C}$-representation.*

Proof: Suppose $\rho : M_N(\mathbf{C}) \to \mathrm{End}_{\mathbf{C}}(V)$ is any $\mathbf{C}$-representation of $M_N(\mathbf{C})$. The operator $\rho(i) \in \mathrm{End}_{\mathbf{C}}(V)$ has square -1. Therefore, the eigenvalues are $\pm I$ where I is the complex structure on V. Let

$$W_{\pm} \equiv \{x \in V : \rho(i)x = \pm Ix\}$$

denote the eigenspaces, and note that if $\rho(i)x = \pm Ix$, then $\rho(i)\rho(a)x = \rho(a)\rho(i)x = \pm I\rho(a)x$, so that $\rho(a)W_{\pm} \subset W_{\pm}$ for all $a \in M_N(\mathbf{C})$. Let $\rho_{\pm}(a) = \rho(a)|_{W_{\pm}}$. Then

$$\rho = \rho_+ \oplus \rho_- \quad \text{as } \mathbf{C}\text{-representations.}$$

Thus we may assume that either $\rho(i) = I$ or $\rho(i) = -I$ if ρ is an irreducible $\mathbf{C}$-representation. If $\rho = \rho_1 \oplus \rho_2$ is reducible as an $\mathbf{R}$-representation, with $V = V_1 \oplus V_2$, then each subspace V_j is complex with respect to I, since $I = \pm \rho(i)$ and $\rho(i)$ maps V_j to V_j. That is, $\rho = \rho_1 \oplus \rho_2$ is also reducible as a $\mathbf{C}$-representation.

Therefore, we may assume that ρ is $\mathbf{R}$-irreducible, and hence $\mathbf{R}$-equivalent to the standard representation of $M_N(\mathbf{C})$ on $\mathbf{C}^N$ via an intertwining operator $f : \mathbf{C}^N \to V$, which is real linear. Now $f \circ i \circ f^{-1} = \rho(i) = \pm I$, or $f \circ i = \pm I \circ f$. That is, f is $\mathbf{C}$-linear if $\rho(i) = I$ and $\mathbf{C}$-antilinear if $\rho(i) = -I$. $\blacksquare$

Corollary 8.17. *The standard representation of* $M_N(\mathbf{H})$ *on* $\mathbf{H}^N$ *is the only irreducible* $\mathbf{H}$-*representation, up to equivalence.*

Proof: Suppose $\rho : M_N(\mathbf{H}) \to \mathrm{End}_{\mathbf{H}}(V)$ is an $\mathbf{H}$-representation of $M_N(\mathbf{H})$ on a right $\mathbf{H}$-space V. Recall that

$$(8.18) \qquad M_N(\mathbf{H}) \otimes_{\mathbf{R}} \mathbf{H} \cong \mathrm{End}_{\mathbf{R}}(\mathbf{H}^N) \cong M_{4N}(\mathbf{R}),$$

with $(a \otimes \lambda)(x) \equiv ax\overline{\lambda}$ for $a \in M_N(\mathbf{H})$, $x \in \mathbf{H}^N$, and $\lambda \in \mathbf{H}$. Now $\tilde{\rho} : M_{4N}(\mathbf{R}) - \mathrm{End}_{\mathbf{H}}(\mathbf{H}^N) \to \mathrm{End}_{\mathbf{R}}(V)$, defined by

$$(8.19) \qquad \tilde{\rho}(a \oplus \lambda)(v) \equiv \rho(a)v\overline{\lambda}, \quad \text{for all } v \in V,$$

is an $\mathbf{R}$-representation of $M_{4N}(\mathbf{R})$ on V. Thus by Theorem 8.11,

$$\tilde{\rho} = \rho_1 \oplus \cdots \oplus \rho_k, \quad V = V_1 \oplus \cdots \oplus V_k,$$

where each representation $\rho_j : M_{4N}(\mathbf{R}) \cong \mathrm{End}_{\mathbf{R}}(\mathbf{H}^N) \to \mathrm{End}_{\mathbf{R}}(V_j)$ is $\mathbf{R}$-equivalent to the standard representation of $M_{4N}(\mathbf{R})$ via an intertwining operator $f_j : \mathbf{R}^{4N} \cong \mathbf{H}^N \to V_j$. That is,

$$(8.20) \qquad \rho_j(A) = f_j \circ A \circ f_j^{-1}.$$

If $\lambda \in \mathbf{H}$, then right multiplication by λ on $\mathbf{R}^{4N} \cong \mathbf{H}^N$ is given by

$$r_\lambda = 1 \otimes \overline{\lambda},$$

and $\tilde{\rho}(r_\lambda) = \tilde{\rho}(1 \otimes \overline{\lambda})$ is R_λ, right multiplication by λ on V (see (8.19)). Therefore, by (8.20), $f_j \circ r_\lambda = R_\lambda \circ f_j$. That is, each $f_j : \mathbf{H}^N \to V_j$ is equivalent, as an $\mathbf{H}$-representation, to the standard $\mathbf{H}$-representation on $M_N(\mathbf{H})$. ∎

UNIQUENESS OF INTERTWINING OPERATORS

The *center* of an algebra A is, by definition,

$$(8.21) \qquad \mathrm{cen}\, A \equiv \{a \in A : ab = ba \text{ for all } b \in A\}.$$

Let $\mathbf{R} \subset M_N(\mathbf{R})$ and $\mathbf{C} \subset M_N(\mathbf{C})$ denote all scalar multiples of the identity, and let $\mathbf{R} \subset M_N(\mathbf{H})$ denote all real multiples of the identity. The centers of the matrix algebras are listed in the next lemma.

Lemma 8.22.

Algebra	Center
$M_N(\mathbf{R})$	$\mathbf{R}$
$M_N(\mathbf{C})$	$\mathbf{C}$
$M_N(\mathbf{H})$	$\mathbf{R}$

Proof: The proof for $N = 2$ is Problem 1 (note that $\mathbf{H}$ has center $\mathbf{R}$). The general case follows immediately, because, for $a \equiv (a_{ij})$ in the center, each 2×2 submatrix of the form

$$\begin{pmatrix} a_{ii} & a_{ij} \\ a_{ji} & a_{jj} \end{pmatrix}$$

must belong to the center of $M_2(F)$.

Corollary 8.23.

(a) *The intertwining operator for two irreducible $\mathbf{R}$-representations of $M_N(\mathbf{R})$ is unique up to a nonzero scale $\lambda \in \mathbf{R}^*$.*
(b) *The intertwining operator for two equivalent irreducible $\mathbf{C}$-representations of $M_N(\mathbf{C})$ is unique up to a nonzero scale $\lambda \in \mathbf{C}^*$.*
(c) *The interwining operator for two irreducible $\mathbf{H}$-representations of $M_N(\mathbf{H})$ is unique up to a nonzero real scale $\lambda \in \mathbf{R}^*$.*

Proof: (a) Suppose $\rho'(a) = f_1 \circ \rho(a) \circ f_1^{-1} = f_2 \circ \rho(a) \circ f_2^{-1}$, where f_1 and f_2 are both intertwining operators. Then $f_2^{-1} \circ f_1 \in \text{End}_F(V)$ commutes with all $\rho(a) \in \text{End}_F(V)$. Since $\rho : M_N(F) \to \text{End}_F(V)$ is an isomorphism, the corollary follows. ∎

Suppose A is a subalgebra of B. The *centralizer* of A in B is defined to be

$$(8.24) \qquad \{b \in B : ab = ba \text{ for all } a \in A\}.$$

Lemma 8.25.

(a) *The centralizer of $\text{End}_\mathbf{C}(\mathbf{C}^N)$ in $\text{End}_\mathbf{R}(\mathbf{C}^N)$ is $\mathbf{C}$.*
(b) *The centralizer of $\text{End}_\mathbf{H}(\mathbf{H}^N)$ in $\text{End}_\mathbf{R}(\mathbf{H}^N)$ is $\mathbf{H}$-acting on the right as scalar multiplication.*
(c) *The centralizer of $\text{End}_\mathbf{R}(\mathbf{R}^N)$ in $\text{End}_\mathbf{R}(\mathbf{C}^N)$ is $M_2(\mathbf{R})$.*

Proof: (See Problem 2.) ∎

Corollary 8.26.

(a) *The intertwining operator for two irreducible* **R**-*representations of* $M_N(\mathbf{C})$ *is unique up to a nonzero complex scale* $\lambda \in \mathbf{C}^*$.
(b) *The intertwining operator for two irreducible* **R**-*representations of* $M_N(\mathbf{H})$ *is unique up to a nonzero right multiplication* R_λ, $\lambda \in \mathbf{H}^*$.

Proof: Analogous to the proof of Corollary 8.23. ∎

AUTOMORPHISMS

The automorphisms of the algebras $M_N(\mathbf{R})$, $M_N(\mathbf{C})$, and $M_N(\mathbf{H})$ can be computed as a special case of Theorem 8.11 and its two corollaries. If α is a real automorphism of one of these algebras, then

$$(8.27) \qquad \rho(a) \equiv \alpha(a)$$

defines an F-representation of the algebra on F^N. If V is a complex vector space, let $\overline{\mathrm{End}}_{\mathbf{C}}(V)$ denote the space of complex antilinear maps from V to V.

Corollary 8.28 (Automorphisms).

(a) (*Existence*) *Each automorphism* α *of the algebra* $M_N(\mathbf{R})$ *is inner. That is, there exists* $h \in M_N(\mathbf{R})$ *with*

$$(8.29) \qquad \alpha(a) = hah^{-1} \quad \text{for all } a \in M_N(\mathbf{R}).$$

(*Uniqueness*) *The intertwining operator* h *in* (8.29) *is unique up to a real scalar multiple.*
(b) *Suppose* α *is an automorphism of the algebra* $M_N(\mathbf{C})$ *considered as a real algebra. Then* α *is either complex linear or complex antilinear.*
(*Existence*) *If* α *is complex linear, then* α *is inner. That is, there exists* $h \in M_N(\mathbf{C})$ *with*

$$(8.30) \qquad \alpha(a) = hah^{-1} \quad \text{for all } a \in M_N(\mathbf{C}).$$

If α *is complex antilinear, then there exists* $h \in \overline{\mathrm{End}}_{\mathbf{C}}(\mathbf{C}^n)$ *satisfying* (8.30).
(*Uniqueness*) *In both cases the intertwining operator* h *is unique up to complex scalar multiples.*
(c) (*Existence*) *Each automorphism* α *of the real algebra* $M_N(\mathbf{H})$ *is inner. That is, there exists* $h \in M_N(\mathbf{H})$ *with*

(8.31) $\alpha(a) = hah^{-1}$ *for all* $a \in M_N(\mathbf{H})$.

(*Uniqueness*) *The intertwining operator* h *is unique up to a real scalar multiple.*

INNER PRODUCTS

An inner product ε (of one of the seven types presented in Chapter 3) on a vector space V over F determines an antiautomrophism of $\mathrm{End}_F(V)$ called the adjoint (and denoted by a^* or $\mathrm{Ad}(a)$):

(8.32) $\varepsilon(ax, y) = \varepsilon(x, \mathrm{Ad}(a)y)$ for all $x, y \in V$.

The adjoint in each of the seven cases is explicitly given in Problem 2.7. Up to a scalar multiple no other inner product can determine the same adjoint.

Theorem 8.33. *Suppose* V *is an* F-*vector space with* $F \equiv \mathbf{R}, \mathbf{C}$, *or* $\mathbf{H}$. *Given two inner products* ε_1 *and* ε_2 *on* V *that determine the same antiautomorphism of the algebra* $\mathrm{End}_F(V)$ *by taking adjoints,* ε_1 *and* ε_2 *must be of the same type and differ by a constant multiple* $c \neq 0$:

(8.34) $\varepsilon_2(x, y) = c\, \varepsilon_1(x, y)$,

where $c \in \mathbf{R}^*$ *if the type is* $\mathbf{R}$-*symmetric,* $\mathbf{R}$-*skew,* $\mathbf{C}$-*hermitian symmetric,* $\mathbf{H}$-*hermitian symmetric, or* $\mathbf{H}$-*hermitian skew, and* $c \in C^*$ *if the type is* $\mathbf{C}$-*symmetric or* $\mathbf{C}$-*skew.*

Proof: Let $\alpha_1 \equiv \mathrm{Re}\,\varepsilon_1$ and $\alpha_2 \equiv \mathrm{Re}\,\varepsilon_2$. Note that if ε is an inner product on a complex vector space V, then

(8.35) $\langle \varepsilon(x, y), i \rangle = -\mathrm{Re}\,\varepsilon(x, y)i = -\mathrm{Re}\,\varepsilon(x, iy)$;

while if ε is an inner product on a right $\mathbf{H}$-vector space, then

(8.36) $\langle \varepsilon(x, y), u \rangle = \mathrm{Re}\,\varepsilon(x, y)\overline{u} = \mathrm{Re}\,\varepsilon(x, y\overline{u})$

for all scalars $u \in \mathbf{H}$.

   In particular, α_1 and α_2 are real inner products on V. Let $b_j : V \to V^*$ be defined by

$$(b_j(x))(y) \equiv \alpha_j(x, y), \quad j = 1, 2.$$

Given $a \in \text{End}_{\mathbf{R}}(V)$, let $a^* \in \text{End}_{\mathbf{R}}(V^*)$ denote the dual map from V^* to V^*. Let $\text{Ad}_j(a)$ denote the adjoint of a with respect to α_j. That is, $\alpha_j(ax, y) = \alpha_j(x, \text{Ad}_j(a)y)$. Let $b_j^* : V \cong (V^*)^* \to V^*$ denote the map dual to b_j. Note that $\alpha_j(ax, y) = (b_j(ax))(y) = (ax)(b_j^*(y)) = x(a^*b_j^*(y))$, while $\alpha_j(x, \text{Ad}_j(a)y) = (b_j(x))(\text{Ad}_j(a)y) = x(b_j^*\text{Ad}_j(a)(y))$. Therefore,

$$(8.37) \qquad \text{Ad}_j(a) = (b_j^*)^{-1} a^* b_j^*.$$

By hypothesis, $\text{Ad}_1(a) = \text{Ad}_2(a)$ for all $a \in \text{End}_F(V) \subset \text{End}_{\mathbf{R}}(V)$. Thus, by taking the dual of (8.37),

$$b_1 a b_1^{-1} = b_2 a b_2^{-1} \quad \text{for all } a \in \text{End}_F(V).$$

Therefore, $b_2^1 b_1 \in \text{End}_{\mathbf{R}}(V)$ is in the centralizer of $\text{End}_F(V)$, and the theorem follows from Lemma 8.25 as described below.

If $F \equiv \mathbf{R}$, then $b_1 = cb_2$ with $c \in \mathbf{R}^*$, and hence $\varepsilon_1 = c\varepsilon_2$. If $F \equiv \mathbf{C}$, then $b_1 = b_2 c$ with $c \in \mathbf{C}^*$. Therefore,

$$\alpha_1(x, y) = \alpha_1(cx, y),$$

and it follows easily that ε_1 and ε_2 differ by a constant using (8.35). Further, if $\varepsilon_1, \varepsilon_2$ are $\mathbf{C}$-hermitian, then the constant must be real. If $F \equiv \mathbf{H}$, then $b_2^1 b_1 = R_c$, right multiplication by a scalar $c \in \mathbf{H}^*$, because of part (b) of Lemma 8.25. Therefore, $\alpha_1(x, y) = \alpha_2(xc, y)$, and hence $\varepsilon_1(x, y) = \varepsilon_2(xc, y) = \overline{c}\,\varepsilon_2(x, y)$ by Corollary 8.26. Since both ε_1 and ε_2 satisfy $\varepsilon(x\lambda, y) = \overline{\lambda}\varepsilon(x, y)$, the constant $\overline{c}$ must be in the center of $\mathbf{H}$, i.e., $c \in \mathbf{R}^*$. ∎

An F-inner product ε on a vector space V with scalar field F induces an $\mathbf{R}$-symmetric inner product on $\text{End}_F(V)$.

Definition 8.38. *Suppose V, ε is an inner product space with scalar field F and real dimension N. Define*

$$(8.39) \qquad \langle a, b \rangle \equiv \frac{1}{N} \text{trace}_{\mathbf{R}}\, a^* b \quad \text{for all } a, b \in \text{End}_F V,$$

where a^* denotes the ε-adjoint of a.

Lemma 8.40. $\langle\,,\,\rangle$ *is an $\mathbf{R}$-symmetric inner product on $\text{End}_F(V)$, and*

$$(8.41) \qquad \langle ab, c \rangle = \langle b, a^* c \rangle \quad \text{for all } a, b, c \in \text{End}_F(V).$$

Proof: Note that $\langle\,,\,\rangle$ is symmetric if $\text{trace}_{\mathbf{R}}\, a = \text{trace}_{\mathbf{R}}\, a^*$. If V, ε is a real inner product space, then $\text{trace}_{\mathbf{R}}\, a = \text{trace}_{\mathbf{R}}\, a^*$ by Problem 2.7. In general,

$\text{Re}\,\varepsilon$ is an **R**-inner product on V and $(\text{Re}\,\varepsilon)(ax, y) = (\text{Re}\,\varepsilon)(x, a^*y)$, since $\varepsilon(ax, y) = \varepsilon(x, a^*y)$. Therefore, the general case reduces to the real case. To show that $\langle\,,\,\rangle$ is nondegenerate it suffices to prove that if $\text{trace}_{\mathbf{R}}\, ab = 0$ for all $a \in M_N(F)$, then $b = 0$, which follows by considering matrices a with all but one entry zero. Finally, (8.40) is immediate since $(ab)^* = b^*a^*$. ∎

Definition 8.42. *Suppose V, ε is an F-inner product space. Given $x, y \in V$, define the product $x \odot y \in \text{End}_F(V)$ by*

$$(8.43) \qquad\qquad (x \odot y)(z) \equiv x\varepsilon(y, z) \quad \text{for all } z \in V.$$

Note that $x \odot y = 0$ if and only if $x = y = 0$.

Lemma 8.44. *For all $a \in \text{End}_F(V)$,*

(a) $(ax) \odot y = a(x \odot y), \quad x \odot (ay) = (x \odot y)a^*,$
(b) $(x \odot y)(z \odot w) = (x\varepsilon(y, z)) \odot w,$
(c) $(x \odot y)^* = y \odot x$ *if ε is symmetric,*
(d) $(x \odot y)^* = -y \odot x$ *if ε is skew.*

Proof:

(a) $\qquad ((ax) \odot y)(z) = ax\varepsilon(y, z) = a((x \odot y)(z)),$

    and

$$(x \odot (ay))(z) = x\varepsilon(ay, z) = x\varepsilon(y, a^*z) = (x \odot y)(a^*z).$$

(b) $\qquad (x \odot y)(z \odot w)(u) = (x \odot y)(z\varepsilon(w, u))$
$$= x\varepsilon(y, z\varepsilon(w, u)) = x\varepsilon(y, z)\varepsilon(w, u)$$
$$= ((x\varepsilon(y, z)) \odot w)(u).$$

In all cases, but ε **C**-symmetric or **C**-skew, (c) and (d) are proven by noting

$$\varepsilon(u, (x \odot y)^*v) = \varepsilon((x \odot y)(u), v) = \varepsilon(x\varepsilon(y, u), v) = \overline{\varepsilon(y, u)}\varepsilon(x, v)$$
$$= \overline{\varepsilon(x, v)}\varepsilon(y, u) = \overline{\varepsilon(y\varepsilon(x, v), u)} = \overline{\varepsilon((y \odot x)(v), u)}$$
$$= \pm\varepsilon(u, (y \odot x)(v)).$$

The cases ε **C**-symmetric or ε **C**-skew are proven by deleting the conjugations in the previous proof. ∎

Theorem 8.45. *Suppose V, ε is an F-inner product space. Then*

$$\langle 1, x \odot y \rangle = (\dim_{\mathbf{F}} V)^{-1}\,\text{Re}\,\varepsilon(y, x) \quad \text{for each } x, y \in V.$$

Proof: We give the proof in the most difficult case $F \equiv \mathbf{H}$.

Let $(\,,)$ denote a positive definite $\mathbf{H}$-hermitian inner product on V. Choose an $\mathbf{H}$-orthonormal basis with respect to $(\,,)$, say $u_1, \ldots, u)_n \in V$, and let $v_1, \ldots, v_{4n}$ denote the corresponding real orthonormal basis. That is $v_1 \equiv u_1, v_2 \equiv u_1 i, v_3 \equiv u_1, j, v_4 \equiv u_1 k, v_5 \equiv u_2, v_6 \equiv u_2 i$, etc. Then the real dual basis $v_1, \ldots, v_{4n}^*$ is given by $v_j^*(x) = \mathrm{Re}\,(v_j, x), j = 1, \ldots, 4n$. Therefore,

$$
\mathrm{trace}_{\mathbf{R}}\; x \odot y = \sum_{j=1}^{4n} v_j^*\,[x\varepsilon(y, v_j)] = \sum_{j=1}^{4n} \mathrm{Re}\,(v_j, (x\varepsilon(y, v_j))
$$

$$
= \sum_{j=1}^{4n} \mathrm{Re}\,[(v_j, x)\varepsilon(y, v_j)] = \sum_{j=1}^{4n} \mathrm{Re}\,[\varepsilon(y, v_j)(v_j, x)]
$$

$$
= \mathrm{Re}\,\varepsilon\left(y, \sum_{j=1}^{4n} v_j(v_j, x)\right).
$$

However, since $ui(ui, x) = u(u, x)$, etc.,

$$
\sum_{j=1}^{4n} v_j(v_j, x) = 4 \sum_{j=1}^{n} u_j(u_j, x) = 4x,
$$

completing the proof, since $\dim_{\mathbf{R}} v = 4 \dim_{\mathbf{H}} v$. ∎

Corollary 8.46. *Suppose V, ε is an F-inner product space of F-dimension n.*

(8.47) $\qquad \langle x \odot y, a \rangle = \dfrac{1}{n}\,\mathrm{Re}\,\varepsilon(ay, x) \quad$ *for all $a \in \mathrm{End}_F(V)$.*

Proof:

$$
\langle x \odot y, a \rangle = \langle a^*(x \odot y), 1 \rangle = \langle (a^*x) \odot y, 1 \rangle
$$

$$
= \frac{1}{n}\,\mathrm{Re}\,\varepsilon(y, a^*x) = \frac{1}{n}\,\mathrm{Re}\,\varepsilon(ay, x). \quad \blacksquare
$$

Corollary 8.48. *Suppose V, ε is an F-inner product space of F-dimension n. If ε is F-symmetric or F-skew, then*

(8.49) $\qquad\qquad \langle x \odot y, z \odot w \rangle = \dfrac{1}{n}\,\mathrm{Re}\,[\varepsilon(x, z)\varepsilon(y, w)].$

If ε is *F*-hermitian, then

(8.50) $\langle x \odot y, z \odot w \rangle = \dfrac{1}{n} \, \mathrm{Re} \left[\overline{\varepsilon(x,z)} \varepsilon(y,w) \right].$

Proof: We give the proof of (8.50). The proof of (8.49) is exactly the same except for the last equality. By Corollary 8.46,

$$\langle x \odot y, z \odot w \rangle = \frac{1}{n} \, \mathrm{Re} \, \varepsilon((z \odot w)y, x) = \frac{1}{n} \mathrm{Re} \, \varepsilon(z\varepsilon(w,y), x)$$
$$= \frac{1}{n} \, \mathrm{Re} \left[\overline{\varepsilon(x,z)} \varepsilon(y,w) \right]. \quad \blacksquare$$

PROBLEMS

1. Show that $M_2(\mathbf{R})$ has center $\mathbf{R}$, $M_2(\mathbf{C})$ has center $\mathbf{C}$, and $M_2(\mathbf{H})$ has center $\mathbf{R}$.

2. Prove Lemma 8.25.

3. Show that $M_N(F) \times M_N(F)$, $F \equiv \mathbf{R}$ or $\mathbf{H}$, has exactly two irreducible *F*-representations, up to equivalence. Namely, given $(a,b) \in M_N(F) \times M_N(F)$,
$$\rho_+(a,b)(x) = ax \quad \text{for all } x \in F^N,$$
   and
$$\rho_-(a,b)(x) = bx \quad \text{for all } x \in F^N.$$

4. Let V_1 denote $\mathbf{H}^N$ with the usual right scalar multiplication R_λ, $\lambda \in \mathbf{H}$. Let V_2 denote $\mathbf{H}^N$ with right multiplication by $\lambda \in \mathbf{H}$ given by $R_{\mu\lambda\mu^{-1}}$, where $\mu \in \mathbf{H}$ is a fixed unit quaternion. Let ρ_α $(\alpha = 1, 2)$ denote the **R**-representation of $M_N(\mathbf{H})$ on V_α defined by $\rho_\alpha(a)v \equiv av$ for all $v \in V_\alpha$. Find (explicitly) the intertwining operator f for these two **H**-representations.

PART II. SPINORS

9. The Clifford Algebras

We start with V, $\langle\,,\,\rangle$, $\|\;\|$, a euclidean vector space of dimension n and signature p, q, and with the orthogonal group $O(p, q)$ acting on V.

The *Clifford algebra* $\mathrm{Cl}(V)$ is constructed in this chapter. It has considerably more structure than just that of an algebra. In fact, if $\mathrm{alg\,Cl}(V)$ denotes the underlying associative algebra with unit, then each algebra $\mathrm{alg\,Cl}(V)$ is nothing more than one, or occasionally two, copies of a matrix algebra, i.e., $M_N(\mathbf{R}), M_N(\mathbf{C})$, or $M_N(\mathbf{H})$. Some patience is required, as this important fact is not proven until Chapter 11!

The reader is assumed to be familiar with the *exterior algebra*

$$\Lambda V \equiv \sum_{p=0}^{n} \Lambda^p V$$

and the *tensor algebra*

$$\otimes V \equiv \sum_{p=0}^{\infty} \otimes^p V$$

over V. Since the extent of this familiarity may vary, we include a brief reminder of these constructions.

For example, the tensor algebra is defined by $\otimes^p V \equiv F/R$. Here F is the vector space of all real-valued functions on $V \times \cdots \times V$ (p times) that vanish except at a finite number of points. Let $\phi(v_1, \ldots, v_p) \in F$ denote the function that vanishes everywhere except at the single point

177

$(v_1, \ldots, v_p)$, where it has the value one. Thus, F consists of all expressions of the form $\sum_{i=1}^{N} a_i \phi(v_1^i, \ldots, v_p^i)$, with each $a_i \in \mathbf{R}$. These expressions are subject to certain rules. This is made precise by quotienting out by all the rules or expressions one wishes to vanish. Thus, R is defined to be the vector subspace of F generated by

$$\phi\left(v_1, \ldots, v_{i-1}, x + y, v_{i+1}, \ldots, v_n\right)$$
$$- \phi\left(v_1, \ldots, v_{i-1}, x, v_{i+1}, \ldots, v_n\right) - \phi\left(v_1, \ldots, v_{i-1}, y, v_{i+1}, \ldots, v_n\right)$$

and

$$\phi\left(v_1, \ldots, v_{i-1}, cx, v_{i+1}, \ldots, v_n\right) - c\phi\left(v_1, \ldots, v_{i-1}, x, v_{i+1}, \ldots, v_n\right)$$

with $c \in \mathbf{R}$. The equivalence class $\phi(v_1, \ldots, v_n) + R$ is denoted $v_1 \otimes \cdots \otimes v_p \in \otimes^p V$.

The tensor algebra $\otimes V$ is graded, associative, with unit, and has an inner product that we also denote by $\langle \, , \, \rangle$. Let $I(V)$ denote the two-sided ideal in $\otimes V$ generated by all elements $x \otimes x \in \otimes^2 V$ with $x \in V$. The quotient algebra $\Lambda V \equiv \otimes V / I(V)$ is called the *exterior algebra* over V, and the equivalence class of $v_1 \otimes \cdots \otimes v_p$ is denoted $v_1 \wedge \cdots \wedge v_p$. In fact, $\Lambda V = \sum_{p=0}^{n} \Lambda^p V$ is a graded, associative, anticommutative algebra with unit (and with inner product also denoted $\langle \, , \, \rangle$).

The exterior algebra ΛV may also be identified with a vector subspace of $\otimes V$, the space of *skew tensors*. For example, $x \wedge y = \frac{1}{2}[x \otimes y - y \otimes x]$, $x, y \in V$. More generally,

$$(9.1) \quad x_1 \wedge \cdots \wedge x_p = \text{Alt}\left(x_1 \otimes \cdots \otimes x_p\right) \equiv \frac{1}{p!} \sum_{\sigma} \text{sign } \sigma \, x_{\sigma(1)} \otimes \cdots \otimes x_{\sigma(p)}.$$

Elements of $\Lambda^p V$ are called *p*-vectors and those of the form $v_1 \wedge \cdots \wedge v_p$ are said to be *simple* or *decomposable*. The set of all unit (i.e., $\|u\| = \pm 1$) simple *p*-vectors is called the grassmannian of oriented, nondegenerate, *p*-planes in V, and denoted by $G(p, V)$. Of course, when $p = 1$, this is just the unit sphere in V.

Given $w \in V$, let $E_w : \Lambda V \to \Lambda V$, *left exterior multiplication by w*, be defined by

$$E_w u = w \wedge u \quad \text{for all } u \in \Lambda V.$$

The adjoint $I_w : \Lambda V \to \Lambda V$, defined by

$$\langle w \wedge u, v \rangle = \langle u, I_w v \rangle \quad \text{for all } u, v \in \Lambda v,$$

is called *interior multiplication by w* and sometimes is denoted by

$$I_w(u) = w \, \llcorner \, u.$$

Observe that
$$I_w v = \langle w, v \rangle \quad \text{if } v \in V,$$
and that interior multiplication I_w is an antiderivation on ΛV; that is,

$$(9.2) \quad I_w (v_1 \wedge \cdots \wedge v_p) = \sum_{k=1}^{p} (-1)^{k-1} \langle w, v_k \rangle \, v_1 \wedge \cdots \wedge v_{k-1} \wedge v_{k+1} \wedge \cdots \wedge v_p$$

on simple p-vectors, and

$$I_w (u \wedge v) = I_w(u) \wedge v + (-1)^{\deg u} u \wedge I_w(v)$$

for all $u, v \in \Lambda V$ with u of pure degree.

In particular,

$$(9.3) \quad \|w\| u = w \wedge (w \llcorner u) + w \llcorner (w \wedge u) \quad \text{or} \quad \|w\| = E_w \circ I_w + I_w \circ E_w$$

for $w \in V$ and $u \in \Lambda V$. This is called a *chain homotopy for E_w* if $\|w\| \neq 0$.

Definition 9.4. *Given a euclidean vector space $V, \langle \, , \, \rangle$ of signature p, q, the Clifford algebra $\mathrm{Cl}(V)$ is the quotient $\otimes V / I(V)$, where $I(V)$ is the two-sided ideal in $\otimes V$ generated by all elements:*

$$(9.5) \quad x \otimes x + \langle x, x \rangle \quad \text{with } x \in V.$$

If $V \equiv \mathbf{R}(p, q)$, then $\mathrm{Cl}(V)$ will be denoted $\mathrm{Cl}(p, q)$. The multiplication on $\mathrm{Cl}(V)$ will be denoted by a dot.

Polarizing $x \otimes x + \langle x, x \rangle \in I(V)$ yields

$$(9.6) \quad x \otimes y + y \otimes x + 2\langle x, y \rangle \in I(V).$$

The Fundamental Lemma for Clifford algebras follows.

Lemma 9.7. *Suppose $\phi : V \to A$ is a linear map from V into A, an associative algebra with unit A. If*

$$(9.8) \quad \phi(x)\phi(x) = -\|x\| \quad \text{for all } x \in V,$$

then ϕ has a unique "extension" (also denoted ϕ) to an algebra homomorphism of $\mathrm{Cl}(V)$ into A.

$$(9.9) \quad
\begin{array}{ccc}
V & \xrightarrow{\ \phi\ } & A \\
\downarrow & \nearrow & \\
\mathrm{Cl}(V) & \phi &
\end{array}$$

Proof: The fundamental lemma for tensor algebras says that ϕ has a unique extension to $\otimes V$ as an algebra homomorphism, which we also denote by $\phi : \otimes V \to A$.

The hypothesis $\phi(x)\phi(x) = -\|x\|$ is equivalent to requiring that the kernel of $\phi : \otimes V \to A$ contains $I(V)$. Thus ϕ, descends to a well-defined map on the quotient $\mathrm{Cl}(V) \equiv \otimes V / I(V)$. ∎

Frequently, it is more convenient to use Lemma 9.7 in the following form.

Remark 9.10. Suppose $e_1, \ldots, e_n$ is an orthonormal basis for V and $\phi(e_j) \in A$ satisfy

(a) $\phi(e_j)^2 = -\|e_j\|$, and
(b) $\phi(e_i)\phi(e_j) + \phi(e_j)\phi(e_i) = 0$ for all $i \neq j$.

Then ϕ has a unique linear extension $\phi : V \to A$ that satisfies $\phi(x)^2 = -\|x\|$, and hence Lemma 9.7 is applicable.

Proposition 9.11. *The subspace ΛV of $\otimes V$ is isomorphic (as a vector space, not as an algebra!) to $\mathrm{Cl}(V) \equiv \otimes V / I(V)$ under the quotient map. In particular, the natural map*

$$V \mapsto \mathrm{Cl}(V)$$

is an inclusion.

Moreover, if $x \in V$ and $u \in \Lambda V \cong \mathrm{Cl}(V)$, then (Clifford product)

(9.12) $$x \cdot u = x \wedge u - x \,\llcorner\, u.$$

That is, Clifford multiplication is an enhancement of exterior minus interior multiplication.

Proof: The quotient map restricted to ΛV maps onto $\mathrm{Cl}(V)$; that is, each tensor can be expressed modulo $I(V)$ as a skew tensor. This is automatic for tensors of degree zero and one. Each 2-tensor $x \otimes y$ is the sum of a symmetric 2-tensor and a skew 2-tensor. The symmetric part belongs to $I(V)$ modulo a tensor of degree zero. The proof is completed by induction.

To prove that ΛV injects into $\mathrm{Cl}(V)$, consider the map $\phi : V \to \mathrm{End}(\Lambda V)$ defined to be exterior minus interior multiplication. Given $x \in V$,

$$\phi(x) \equiv E_x - I_x \in \mathrm{End}(\Lambda V).$$

Note that

$$[\phi(x) \circ \phi(x)](u) = (E_x - I_x)(E_x - I_x)(u)$$
$$= -(E_x \circ I_x + I_x \circ E_x)(u) = -\|x\|u$$

by (9.3). That is, $\phi(x) \circ \phi(x) = -\|x\|$, so by the Fundamental Lemma, ϕ extends to an algebra homomorphism $\phi : \mathrm{Cl}(V) \to \mathrm{End}(\Lambda V)$. Now consider the linear map $\psi : \mathrm{End}(\Lambda V) \to \Lambda V$ defined by evaluation at $1 \in \Lambda V$. The map $\psi \circ \phi : \mathrm{Cl}(V) \to \Lambda V$ is a right inverse to the natural map from ΛV to $\mathrm{Cl}(V)$. To prove this, choose $u \in \Lambda V$ of the form

$$(9.13) \qquad u = u_1 \wedge \cdots \wedge u_p, \quad \text{with } u_1, \ldots, u_p \text{ orthonormal.}$$

Let $[u]$ denote the corresponding element of $\mathrm{Cl}(V)$. Since

$$u = \frac{1}{p!} \sum_\sigma \text{sign } \sigma \, u_{\sigma(1)} \otimes \cdots \otimes u_{\sigma(p)},$$

$$\phi([u]) = \frac{1}{p!} \sum_\sigma \text{sign } \sigma \phi\left(u_{\sigma(1)}\right) \circ \cdots \circ \phi\left(u_{\sigma(p)}\right)$$

$$= \frac{1}{p!} \sum_\sigma \text{sign } \sigma \left(E_{u_{\sigma(1)}} - I_{u_{\sigma(1)}}\right) \circ \cdots \circ \left(E_{u_{\sigma(p)}} - I_{u_{\sigma(p)}}\right).$$

Since $u_1, \ldots, u_p$ is orthonormal, evaluating at $1 \in \Lambda V$ yields

$$\frac{1}{p!} \sum_\sigma \text{sign } \sigma u_{\sigma(1)} \wedge \cdots \wedge u_{\sigma(p)} = u_1 \wedge \cdots \wedge u_p.$$

Because V has a basis with each basis vector of the form (9.13), this proves that the composite

$$\Lambda V \to \mathrm{Cl}(V) \overset{\phi}{\to} \mathrm{End}(\Lambda V) \overset{\psi}{\to} \Lambda V$$

is the identity map on ΛV.

Finally, we prove (9.12). Note that $\phi(x \cdot u) = \phi(x) \circ \phi(u)$ and hence,

$$(\psi \circ \phi)(x \cdot u) = \phi(x)(\phi(u)(1)) = \phi(x)(u) = (E_x - I_x)(u),$$

since $\phi(u)(1) = (\psi \circ \phi)(u) = u.$ ∎

THE CLIFFORD AUTOMORPHISMS

An important special case of the Fundamental Lemma of Clifford algebras is where the target algebra A is also a Clifford algebra.

Proposition 9.14. *Suppose* $V_1, \langle\, , \rangle_1$ *and* $V_2, \langle\, , \rangle_2$ *are two inner product spaces. Each linear map* $f : V_1 \to V_2$ *that preserves the inner products has a unique extension (also denoted* f*) to an algebra homomorphism*

$$(9.15) \qquad\qquad f : \mathrm{Cl}(V_1) \to \mathrm{Cl}(V_2).$$

Proof: $f(x) \cdot f(x) = -\langle f(x), f(x) \rangle_2 = -\langle x, x \rangle_1 = -\|x\|_1$ for all $x \in V_1$, so that f satisfies (9.8). ∎

Remark 9.16. Each linear map $f : V_1 \to V_2$ also has a unique extension to an algebra homomorphism

$$(9.17) \qquad\qquad f : \Lambda V_1 \to \Lambda V_2.$$

The extensions (9.15) and (9.17) are the same under the canonical identifications $\mathrm{Cl}(V_1) \cong \Lambda V_1$ and $\mathrm{Cl}(V_2) \cong \Lambda V_2$.

Remark 9.18. Suppose that $V_3, \langle\, , \rangle_3$ is another inner product space and $g : V_2 \to V_3$ is a linear map preserving the inner products then the composite $h = g \circ f : V_1 \to V_3$ has a unique extension H to an algebra homomorphism of $\mathrm{Cl}(V_1)$ to $\mathrm{Cl}(V_3)$. By the uniqueness part of Proposition 9.14, h must be the composite of the extension of f with the extension of g.

In order to emphasize the fact that $\mathrm{Cl}(V)$ is much more than an algebra, let alg $\mathrm{Cl}(V)$ denote $\mathrm{Cl}(V)$ considered just as an associative algebra with unit. In particular, $\mathrm{Aut}(\mathrm{alg}\ \mathrm{Cl}(V))$ denotes the algebra automorphisms of $\mathrm{Cl}(V)$.

Definition 9.19. *A Clifford automorphism of* $\mathrm{Cl}(V)$*,* $f \in \mathrm{Aut}(\mathrm{Cl}(V))$*, is an algebra automorphism of* alg $\mathrm{Cl}(V)$ *that also maps* V *to* V.

Theorem 9.20. $\mathrm{Aut}(\mathrm{Cl}(V)) = O(V)$.

Proof: Suppose $f \in \mathrm{Aut}(\mathrm{Cl}(V))$. Then $f \in O(V)$, since $\langle f(x), f(x) \rangle = -f(x)f(x) = -f(x^2) = f(\langle x, x \rangle) = \langle x, x \rangle f(1) = \langle x, x \rangle$, for all $x \in V$. Conversely, each $f \in O(V)$ has a unique extension to an algebra automorphism of $\mathrm{Cl}(V)$ by Proposition 9.14 and Remark 9.18 applied to $g = f^{-1}$. ∎

THE CLIFFORD INVOLUTIONS

The Clifford algebra $\mathrm{Cl}(V)$ comes equipped with a particularly distinguished automorphism and two anti-automorphisms.

Definition 9.21. *The isometry $x \mapsto -x$ on V extends to an automorphism of the Clifford algebra $\mathrm{Cl}(V)$, because of Theorem 9.20. This automorphism of $\mathrm{Cl}(V)$ will be denoted by $x \mapsto \tilde{x}$ and is referred to as the* canonical automorphism *of* $\mathrm{Cl}(V)$.

Note that $\tilde{\tilde{x}} = x$ because of Remark 9.18.
Define the *even part* of $\mathrm{Cl}(V)$ by

$$(9.22) \qquad \mathrm{Cl}^{\mathrm{even}}(V) = \{x \in \mathrm{Cl}(V) : \tilde{x} = x\},$$

and the *odd part* of $\mathrm{Cl}(V)$ by

$$(9.23) \qquad \mathrm{Cl}^{\mathrm{odd}}(V) = \{x \in \mathrm{Cl}(V) : \tilde{x} = -x\}.$$

Under the canonical vector space isomorphism $\Lambda V \cong \mathrm{Cl}(V)$,

$$(9.24) \qquad \mathrm{Cl}^{\mathrm{even}}(V) \cong \Lambda^{\mathrm{even}} V$$

and

$$(9.25) \qquad \mathrm{Cl}^{\mathrm{odd}}(V) \cong \Lambda^{\mathrm{odd}} V.$$

Defintion 9.26. *The anti-automorphism of $\otimes V$ defined by reversing the order in a simple product, i.e., sending $v_1 \otimes \cdots \otimes v_p$ to $v_p \otimes \cdots \otimes v_1$, maps $I(V)$ to $I(V)$, and hence determines an anti-automorphism of the Clifford algebra $\mathrm{Cl}(V) = \otimes V / I(V)$. This anti-automorphism obviously squares to the identity (i.e., is an involution). It will be denoted by $x \mapsto \check{x}$ and referred to as the* check involution.

Note that the composition $\vee \circ \sim \circ \vee$ is an automorphism of $\mathrm{Cl}(V)$ that equals minus the identity of V. Thus, by the uniqueness in Lemma 9.7, $\vee \circ \sim \circ \vee = \sim$. This proves that $\vee$ and $\sim$ commute. The second anti-automorphism is defined to be the composite $\wedge = \vee \circ \sim = \sim \circ \vee$ and is referred to as the *hat involution* $x \mapsto \hat{x}$.

We say that $u \in \mathrm{Cl}(V)$ is of *degree p* if $u \in \Lambda^p V \subset \Lambda V \cong \mathrm{Cl}(V)$.

Proposition 9.27. *If $u \in \mathrm{Cl}(V)$ is of degree p, then $\tilde{u} = \pm u$, $\hat{u} = \pm u$, and $\check{u} = \pm u$ with the plus or minus depending only on $p \bmod 4$ as indicated in the following table.*

	$p \bmod 4$	0	1	2	3
(9.28)	$\sim$	$+$	$-$	$+$	$-$
	$\wedge$	$+$	$-$	$-$	$+$
	$\vee$	$+$	$+$	$-$	$-$

Proof: It suffices to verify the last row of the table for check $\vee$. Also, we may assume that u is an axis p-plane. Since $e_i \cdot e_j = -e_j \cdot e_i$ for $i \neq j$, and $p - 1 + p - 2 + \cdots + 1 = \frac{1}{2}p(p - 1)$, the check of the Ith axis p-plane is

$$\left(e_{i_1} \cdot \ldots \cdot e_{i_p}\right)^{\vee} = e_{i_p} \cdot \ldots \cdot e_{i_1} = (-1)^{\frac{1}{2}p(p-1)} e_{i_1} \cdot \ldots \cdot e_{i_p}. \quad \blacksquare$$

THE CLIFFORD INNER PRODUCT

The canonical vector space isomorphism $\Lambda V \cong \mathrm{Cl}(V)$ provides the Clifford algebra $\mathrm{Cl}(V)$ with a natural inner product $\langle \, , \, \rangle$, namely, the natural inner product on ΛV. This will be referred to as the *Clifford inner product*. Thus, if $e_1, \ldots, e_n$ is an orthonormal basis for V, then $\{e_I : I \text{ increasing}\}$, with $e_I \equiv e_{i_1} \cdot \ldots \cdot e_{i_p}$, is an orthonormal basis for $\mathrm{Cl}(V)$. Note $e_{i_1} \cdot \ldots \cdot e_{i_p} = e_1 \wedge \cdots \wedge e_{i_p}$ are the same. The square norm on $\mathrm{Cl}(V)$ determined by the inner product will be denoted $\| \ \|$, as usual for ΛV. The inner product in $\mathrm{Cl}(V)$ can be computed using the hat involution:

Proposition 9.29.

$$\langle x, y \rangle = \langle 1, \hat{x} \cdot y \rangle = \langle 1, y \cdot \hat{x} \rangle, \quad \text{for all } x, y \in \mathrm{Cl}(V).$$

Proof: Let $\langle x, y \rangle' \equiv \langle 1, \hat{x} \cdot y \rangle$. The bilinear forms $\langle \, , \, \rangle$ and $\langle \, , \, \rangle'$ on $\mathrm{Cl}(V)$ are equal because $\{e_I : I \text{ increasing}\}$ is an orthonormal basis for $\langle \, , \, \rangle'$ as well as $\langle \, , \, \rangle$.

$$\langle e_I, e_I \rangle' = \langle 1, \hat{e}_{i_p} \cdot \ldots \cdot \hat{e}_{i_1} \cdot e_{i_1} \cdot \ldots \cdot e_{i_p} \rangle$$
$$= \|e_{i_1}\| \cdots \|e_{i_p}\| = \|e_{i_1} \wedge \cdots \wedge e_{i_p}\| = \langle e_I, e_I \rangle,$$

since $\hat{u} \cdot u = -u^2 = \|u\|$ for all $u \in V$. Similarly, $\langle e_I, e_J \rangle'$ can be seen to vanish if $I \neq J$, completing the proof. $\quad \blacksquare$

The adjoint of Clifford multiplication by a is just Clifford multiplication by $\hat{a}$.

Proposition 9.30. *Given* $a \in \mathrm{Cl}(V)$,

$$\langle a \cdot u, v \rangle = \langle u, \hat{a} \cdot v \rangle$$

and

$$\langle u \cdot a, v \rangle = \langle u, v \cdot \hat{a} \rangle$$

for all $u, v \in \mathrm{Cl}(V)$.

Proof: $\langle a \cdot u, v \rangle = \langle 1, (a \cdot u)^{\wedge} \cdot v \rangle = \langle 1, \hat{u} \cdot \hat{a} \cdot v \rangle = \langle u, \hat{a} \cdot v \rangle. \quad \blacksquare$

We shall say $u \in \mathrm{Cl}(V)$ is a *simple product* if $u = u_1 \cdot \ldots \cdot u_p$ with $u_1, \ldots, u_p \in V$.

Proposition 9.31. *If $u \in \text{Cl}(V)$ is a simple product $u = u_1 \cdot \ldots \cdot u_p$ then*

(9.32) $$\|u\| = u \cdot \hat{u} = \hat{u} \cdot u = \|u_1\| \cdots \|u_p\|.$$

Proof: $\hat{u} \cdot u = \hat{u}_p \cdot \ldots \cdot \hat{u}_1 \cdot u_1 \cdot \ldots \cdot u_p = \|u_1\| \cdots \|u_p\|$, since $\hat{x} \cdot x = -x \cdot x = \|x\|$ for all $x \in V$. In particular, $\hat{u} \cdot u$ is a scalar multiple of 1, so that $\|u\| = \langle 1, \hat{u} \cdot u \rangle = \hat{u} \cdot u$ by Proposition 9.29. ∎

Given $u \in \text{Cl}(V)$, let $(u)_k \in \Lambda^k V$ denote the orthogonal projection of u on $\Lambda^k V$.

Corollary 9.33 (Hadamard). *For all $u_1, \ldots, u_p \in V$,*

$$\|u_1\| \cdots \|u_p\| = \|u_1 \wedge \cdots \wedge u_p\| + \sum_{k=0}^{p-1} \|(u_1 \cdot \ldots \cdot u_p)_k\|.$$

Thus, if $V, \langle\,,\,\rangle$ is positive definite,

(9.34) $$|u_1 \wedge \cdots \wedge u_p| \le |u_1| \cdots |u_p|$$

with equality if and only if $u_1, \ldots, u_p$ are orthogonal.

Proof: Note that $u_1, \ldots, u_p$ are orthogonal if and only if $u_1 \cdot \ldots \cdot u_p = u_1 \wedge \cdots \wedge u_p$. ∎

The Clifford inner product can also be expressed in terms of a trace (see Theorem 9.65).

THE MAIN SYMMETRY

Suppose $e_1, \ldots, e_n$ is an orthonormal basis for $\mathbf{R}(r, s)$ ($n \equiv r + s$), and $E, E_1, \ldots, E_n$ is an orthonormal basis for either $\mathbf{R}(r + 1, s)$ (with $E^2 = -\|E\| = -1$) or $\mathbf{R}(s, r+1)$ (with $E^2 = -\|E\| = 1$). Define $\phi(e_j) \equiv EE_j, j = 1, \ldots, n$. Then $\phi(e_i)\phi(e_j) = EE_i EE_j = -E^2 E_i E_j = -\phi(e_j)\phi(e_i)$ if $i \ne j$. Also, $\phi(e_i)\phi(e_i) = -E^2 E_i^2 = E^2 \|E_i\|$. We wish to apply Lemma 9.7 (in the form discussed in Remark 9.10) and extend ϕ to an algebra homomorphism. The remaining condition to verify is

(9.35) $$\phi(e_i)\phi(e_i) = -\|e_i\| \quad \text{for } i = 1, \cdots, n.$$

There are two cases to consider. First, if $E^2 = -1$ and $\|E_j\| = \|e_j\|$ for all j, then (9.35) is valid and ϕ extends to an algebra homomorphism

(9.36) $$\phi : \text{Cl}(r, s) \to \text{Cl}(r + 1, s).$$

Second, if $E^2 = 1$ and $\|E_j\| = -\|e_j\|$ for all j, then (9.35) is valid and ϕ extends to an algebra homomorphism

(9.37) $\phi : \mathrm{Cl}(r, s) \to \mathrm{Cl}(s, r + 1)$.

Theorem 9.38. *The maps defined above provide algebra isomorphisms:*

(9.39) $\mathrm{Cl}(r, s) \cong \mathrm{Cl}(r + 1, s)^{\mathrm{even}}$

and

(9.40) $\mathrm{Cl}(r, s) \cong \mathrm{Cl}(s, r + 1)^{\mathrm{even}}$,

which preserve the hat involutions.

Proof: Certainly the image of ϕ is contained in the even part of the target Clifford algebra. Each product $E E_i$ belongs to the image, and hence $E_i E_j = \pm E E_i E E_j$ also belongs to the image. Thus, ϕ surjects onto the even part. By counting dimensions, ϕ must be an isomorphism. Note that

(9.41) ϕ preserves even degrees.
(9.42). ϕ adds one to odd degrees

Therefore, the hat involutions are preserved. ∎

    Combining these two isomorphisms yields the main symmetry for both $\mathrm{Cl}^{\mathrm{even}}$ and Cl.

Theorem 9.43.

(9.44) $\mathrm{Cl}(r, s)^{\mathrm{even}} \cong \mathrm{Cl}(s, r)^{\mathrm{even}}$

and

(9.45) $\mathrm{Cl}(r - 1, s) \cong \mathrm{Cl}(s - 1, r)$.

Moreover, these isomorphisms preserve the hat involutions.

Remark 9.46. In fact, the isomorphism $\mathrm{Cl}(r, s)^{\mathrm{even}} \cong \mathrm{Cl}(s, r)^{\mathrm{even}}$ preserves degree, while the isomorphism $\mathrm{Cl}(r - 1, s) \cong \mathrm{Cl}(s - 1, r)$ preserves $\Lambda^{2m} \mathbf{R}^n \oplus \Lambda^{2m-1} \mathbf{R}^n$.

THE CLIFFORD CENTER

The *center* of the Clifford algebra $\text{Cl}(V)$ is

(9.47) $\text{cen } \text{Cl}(V) \equiv \{a \in \text{Cl}(V) : a \cdot x = x \cdot a \text{ for all } x \in \text{Cl}(V)\}.$

Since $\text{Cl}(V)$ has a basis of simple products of vectors in V,

(9.47′) $\text{cen } \text{Cl}(V) = \{a \in \text{Cl}(V) : a \cdot x = x \cdot a \text{ for } x \in V\}.$

The *twisted center* of the Clifford algebra $\text{Cl}(V)$ is

(9.48) $\text{twcen } \text{Cl}(V) = \{a \in \text{Cl}(V) : a \cdot x = \tilde{x} \cdot a \text{ for all } x \in \text{Cl}(V)\}.$

Note that

(9.48′) $\text{twcen } \text{Cl}(V) = \{a \in \text{Cl}(V) : a \cdot x = -x \cdot a \text{ for all } x \in V\}.$

Of course, the real scalars $\mathbf{R} = \Lambda^0 V$ always belong to the center.

Lemma 9.49. *If $n = \dim V$ is even, then*

$$\text{cen } \text{Cl}(V) = \Lambda^0 V \quad \text{and} \quad \text{twcen } \text{Cl}(V) = \Lambda^n V.$$

If $n = \dim V$ is odd, then

$$\text{cen } \text{Cl}(V) = \Lambda^0 V \oplus \Lambda^n V \quad \text{and} \quad \text{twcen } \text{Cl}(V) = \{0\}.$$

Remark 9.50. Perhaps a more convenient way to remember Lemma 9.49 is in terms of a unit volume element $\lambda \in \Lambda^n V$. First, note that the center and the twisted center always belong to $\Lambda^0 V \oplus \Lambda^n V$. Second, note that

(9.51)   (n odd) λ commutes with all of the Clifford algebra $\text{Cl}(V)$.

(9.52)   (n even) λ commutes with $\text{Cl}^{\text{even}}(V)$ and anticommutes with $\text{Cl}^{\text{odd}}(V)$.

All the information in Lemma 9.49 is contained in this remark.

Proof: First note that for any non-null vector $e \in V$, we have the following

(9.53)   If a is even and $a \cdot e = e \cdot a$, then a does not involve e.

(9.54)   If a is even and $a \cdot e = -e \cdot a$, then a does involve e.

(9.55)   If a is odd and $a \cdot e = -e \cdot a$, then a does not involve e.

(9.56)   If a is odd and $a \cdot e = e \cdot a$, then a does involve e.

For example, to prove (9.53), express a as $a = x + e \cdot y$, where $x, y \in \mathrm{Cl}(V)$ do not involve e. Then $a \cdot e = x \cdot e + e \cdot y \cdot e = x \cdot e - e^2 \cdot y$, since y is odd; and $e \cdot a = e \cdot x + e^2 \cdot y = x \cdot e + e^2 \cdot y$, since x is even. Therefore, $y = 0$, so that $a = x$ does not involve e.

Now choose an orthonormal basis $e_1, \ldots, e_n$ for V. An element a belongs to the center if and only if a commutes with each vector e_j, if and only if both the even and the odd part of a commute with each vector e_j. Thus, it suffices to consider the even and odd parts of a separately.

Now if a is even and in the center, repeated application of (9.53) implies that $a \in \Lambda^0 V = \mathbf{R}$ does not involve any of the $e_1, \ldots, e_n$. The other cases are similarly handled. ∎

SELF DUALITY

A choice of orientation for $\mathbf{R}(r, s)$ is equivalent to a choice λ of one of the two unit volume elements for $\mathbf{R}(r, s)$. However, the conditions $\lambda^2 = 1$ and $\lambda^2 = -1$ are independent of the choice, λ or $-\lambda$.

Proposition 9.57. *Suppose λ is a unit volume element for $\mathbf{R}(r, s)$.*

$$(9.58) \qquad \lambda^2 = 1 \text{ if } r - s = 0, 3 \bmod 4,$$
$$(9.59) \qquad \lambda^2 = -1 \text{ if } r - s = 1, 2 \bmod 4.$$

Proof: First note that $\lambda \check{\lambda} = (-1)^r$ for $\lambda \equiv e_1 \cdots e_{r+s}$. Second, recall from Proposition 9.27, that

$$\check{\lambda} = \lambda \text{ if } n \equiv r + s = 0, 1 \bmod 4,$$
$$\check{\lambda} = -\lambda \text{ if } n \equiv r + s = 2, 3 \bmod 4,$$

Thus, if $n = 2p$ is even, then $\check{\lambda} = (-1)^p \lambda$, so that $\lambda^2 = (-1)^{r-p} = (-1)^{(r-s)/2}$; while if $n = 2p + 1$ is odd, then $\check{\lambda} = (-1)^p \lambda$, so that $\lambda^2 = (-1)^{r-p} = (-1)^{(r-s+1)/2}$. ∎

Definition 9.60. *Suppose $r - s = 0, 3 \bmod 4$, or equivalently $\lambda^2 = 1$. An element $a \in \mathrm{Cl}(r, s)$ is said to be self-dual if $\lambda a = a$ and anti-self-dual if $\lambda a = -a$. Let*

$$(9.61) \qquad \mathrm{Cl}(r, s)^{\pm} \equiv \{a \in \mathrm{Cl}(r, s) : \lambda a = \pm a\}$$

denote the self-dual and anti-self dual parts of the Clifford algebra $\text{Cl}(r, s)$.

Proposition 9.62. *Suppose $r - s = 3$ mod 4. Then*

$$(9.63) \qquad \text{Cl}(r, s) = \text{Cl}(r, s)^+ \oplus \text{Cl}(r, s)^-,$$

and both $\text{Cl}(r, s)^+$ and $\text{Cl}(r, s)^-$ are two-sided ideals in $\text{Cl}(r, s)$.

In fact, the only case in which $\text{Cl}(r, s)$ has a nontrivial proper two-sided ideal is this case $r - s = 3$ mod 4 (see Problem 11.4 and Lemma 8.6).

Remark 9.64. $\text{Cl}(r, s)^\pm$, with the multiplicative identity defined to be $\frac{1}{2}(1 \pm \lambda)$, is an algebra (assuming $r - s = 3$ mod 4).

TRACE

The Clifford inner product can be expressed in terms of the trace given a representation of the Clifford algebra.

Theorem 9.65. *Suppose $\rho : \text{Cl}(r, s) \to \text{End}_{\mathbf{R}}(V)$ is an $\mathbf{R}$-representation of the Clifford algebras $\text{Cl}(r, s)$ on a vector space V of real dimension N. Then*

$$(9.66) \qquad \langle a, b \rangle = \frac{1}{N} \, \text{trace}_{\mathbf{R}} \, \rho(\hat{a}b) \quad \text{for all } a, b \in \text{Cl}(r, s),$$

unless $r - s = 3$ mod 4 and $\text{trace}_{\mathbf{R}} \, \rho(\lambda) \neq 0$.

Proof: Since $\langle a, b \rangle = \langle 1, \hat{a}b \rangle$, it suffices to show that

$$(9.67) \qquad \langle 1, a \rangle = \frac{1}{N} \, \text{trace}_{\mathbf{R}} \, \rho(a) \quad \text{for all } a \in \text{Cl}(r, s).$$

Because $\text{trace}_{\mathbf{R}} \, \rho(a)$ is a linear functional on $\text{Cl}(r, s)$, there exists an element $c \in \text{Cl}(r, s)$ such that

$$(9.68) \qquad \frac{1}{N} \, \text{trace}_{\mathbf{R}} \, \rho(a) = \langle c, a \rangle \quad \text{for all } a \in \text{Cl}(r, s).$$

The property $\text{trace}_{\mathbf{R}} \, AB = \text{trace}_{\mathbf{R}} \, BA$ implies that $\langle c, ab \rangle = \langle c, ba \rangle$ for all $a, b \in \text{Cl}(r, s)$. Therefore, $\langle c\hat{a}, b \rangle = \langle \hat{a}c, b \rangle$, or $c\hat{a} = \hat{a}c$ for all a, so that $c \in \text{center } \text{Cl}(r, s)$. Consequently, if the dimension $n = r + s$ of $\mathbf{R}(r, s)$ is even, then $c \in \mathbf{R}$, so that

$$(9.69) \qquad (r + s \text{ even}) \qquad \frac{1}{N} \, \text{trace}_{\mathbf{R}} \, \rho(a) = c\langle 1, a \rangle.$$

Setting $a = 1$, yields (9.66). While if the dimension $n = r + s$ is odd, then $c = \alpha + \beta\lambda$ with $\alpha, \beta \in \mathbf{R}$. Thus,

$$(9.70) \qquad (r + s \text{ odd}) \qquad \frac{1}{N} \, \text{trace}_{\mathbf{R}} \, \rho(a) = \alpha\langle 1, a\rangle + \beta\langle\lambda, a\rangle.$$

Setting $a = 1$ yields $\alpha = 1$, while setting $a = \lambda$ yields

$$\beta = \frac{(-1)^s}{N} \, \text{trace}_{\mathbf{R}} \, \rho(\lambda).$$

Since $r + s$ is odd, $r - s$ is also odd and hence equals either 1 or 3 mod 4. If $r - s = 1 \bmod 4$, then $\lambda^2 = -1$ and hence $\rho(\lambda)^2 = -1$. In this case, $\rho(\lambda)$ is a complex structure on V and therefore $\text{trace}_{\mathbf{R}} \, \rho(\lambda) = 0$ is automatic. ∎

Remark 9.71. This proof shows that

$$(9.72) \qquad \frac{1}{N} \, \text{trace}_{\mathbf{R}} \, \rho(a) = \langle 1, a\rangle + \frac{(-1)^s}{N} \, (\text{trace}_{\mathbf{R}} \, \rho(\lambda))\langle\lambda, a\rangle$$

for all $a \in \text{Cl}(r, s)$, no matter what the signature r, s or the representation.

In particular, for any representation ρ of a Clifford algebra $\text{Cl}(r, s)$,

$$(9.73) \qquad\qquad \text{trace}_{\mathbf{R}} \, \rho(u) = 0 \quad \text{for each } u \in \mathbf{R}(r, s).$$

In fact, for any $u \in \text{Cl}(r, s) \cong \Lambda\mathbf{R}(r, s)$ which does not have a degree 0 or degree $n \equiv r + s$ part, (9.73) also holds.

THE COMPLEX CLIFFORD ALGEBRAS

Suppose V is a complex vector space with nondegeneate symmetric complex bilinear form $\langle\,,\,\rangle$. Then, proceeding exactly as in the real case, the *complex Clifford algebra* $\text{Cl}_{\mathbf{C}}(V)$ is defined by

$$(9.74) \qquad\qquad \text{Cl}_{\mathbf{C}}(V) \equiv \otimes V / I(V),$$

where $I(V)$ is the two-sided ideal in $\otimes V$ generated by

$$(9.75) \qquad\qquad x \otimes x + \langle x, x\rangle \quad \text{with } x \in V.$$

All of the results of this chapter carry over, with the field $\mathbf{R}$ replaced by the field $\mathbf{C}$, except for the following two items. First, the Hadamard Inequality (9.34) for the positive definite case must be deleted. Second, a unit volume element λ for V complex is only determined up to a multiple

$e^{i\theta}\lambda$, so that Proposition 9.57 loses any significance. However, the concepts of self-dual and anti-self-dual are now valid in all dimensions and independent of the choice of volume element. Some of the results that remain valid for $\mathrm{Cl_C}(V)$ have been reformulated as exercises.

Since all of the complex inner product spaces $\mathbf{C}(r, s)$ of the same dimension $n \equiv r + s$ are isometric, and the isometry has a unique extension to a complex Clifford algebra automorphism, there is (essentially) only one complex Clifford algebra for each dimension, denoted $\mathrm{Cl_C}(n)$.

Proposition 9.76. *Let $\mathcal{R}$ denote conjugation (or the reality operator) on $\mathbf{C}(r, s) \equiv \mathbf{R}(r, s) \otimes_\mathbf{R} \mathbf{C}$, the complexification of $\mathbf{R}(r, s)$. Since $\langle \mathcal{R}x, \mathcal{R}x \rangle = \langle x, x \rangle$ for all $x \in \mathbf{C}(r, s)$, $\mathcal{R}$ has a unique extension to a complex antilinear isomorphism of $\mathrm{Cl_C}(n)$, which is the conjugation for*

$$(9.77) \qquad \mathrm{Cl_C}(n) \cong \mathrm{Cl}(r, s) \otimes_\mathbf{R} \mathbf{C}.$$

Proof: The fundamental lemma for complex Clifford algebras does not apply to $\mathcal{R}$ since $\mathcal{R}$ is not complex linear and does not preserve $\langle\,,\,\rangle$. However, $\mathcal{R}$ has a unique extension to a complex antilinear algebra automorphism of $\otimes V$ (define $\mathcal{R}(i) = -i$). This extension $\mathcal{R}$ maps $x \otimes x + \langle x, x \rangle$ to $\mathcal{R}(x \otimes x + \langle x, x \rangle) = \mathcal{R}x \otimes \mathcal{R}x + \overline{\langle x, x \rangle} = \mathcal{R}x \otimes \mathcal{R}x + \langle \mathcal{R}x, \mathcal{R}x \rangle$, and hence maps $I(\mathbf{C}(r, s))$ to $I(\mathbf{C}(r, s))$. Therefore, $\mathcal{R}$ descends to the quotient yielding a complex antilinear algebra automorphism of $\mathrm{Cl_C}(n)$. Since $\mathcal{R}$ fixes $\mathbf{R}(r, s)$, $\mathcal{R}$ fixes the real subalgebra $\mathrm{Cl}(r, s)$ of $\mathrm{Cl_C}(n)$ generated by $\mathbf{R}(r, s)$. Since $\mathcal{R}$ is complex antilinear, $\mathcal{R}$ is -1 on $i\mathrm{Cl}(r, s) \subset \mathrm{Cl_C}(n)$. Now, counting dimensions,

$$\mathrm{Cl_C}(n) = \mathrm{Cl}(r, s) \oplus i\mathrm{Cl}(r, s),$$

and hence

$$\mathrm{Cl_C}(n) = \mathrm{Cl}(r, s) \otimes_\mathbf{R} \mathbf{C}. \quad \blacksquare$$

PROBLEMS

1. Use (9.6) to show directly that the natural map of V to $\mathrm{Cl}(V)$ is injective.

2. Under the canonical isomorphism of vector spaces $\Lambda V \cong \mathrm{Cl}(V)$, show that, for all vectors $x_1, x_2, x_3, x_4 \in V$

   (a) $x_1 \cdot x_2 = x_1 \wedge x_2 - \langle x_1, x_2 \rangle$,

   (b) $x_1 \cdot x_2 \cdot x_3 = x_1 \wedge x_2 \wedge x_3 - \langle x_2, x_3 \rangle x_1 + \langle x_1, x_3 \rangle x_2 - \langle x_1, x_2 \rangle x_3$,

(c) $x_1 \cdot x_2 \cdot x_3 \cdot x_4 = x_1 \wedge x_2 \wedge x_3 \wedge x_4$

$\qquad - \langle x_1, x_2 \rangle x_3 \wedge x_4 + \langle x_1, x_3 \rangle x_2 \wedge x_4 - \langle x_1, x_4 \rangle x_2 \wedge x_3$

$\qquad - \langle x_3, x_4 \rangle x_1 \wedge x_2 + \langle x_2, x_4 \rangle x_1 \wedge x_3 - \langle x_2, x_3 \rangle x_1 \wedge x_4$

$\qquad + \langle x_1, x_2 \rangle \langle x_3, x_4 \rangle - \langle x_1, x_3 \rangle \langle x_2, x_4 \rangle + \langle x_1, x_4 \rangle \langle x_2, x_3 \rangle.$

3. (a) Show that $\mathrm{Cl}(1,0)$ and $\mathbf{C}$ are isomorphic as algebras with inner product and that $\wedge$ (hat) on $\mathrm{Cl}(1,0)$ corresponds to bar (conjugation) on $\mathbf{C}$.

(b) Show that $\mathrm{Cl}(0,1)$ and $\mathbf{L}$ are isomorphic as algebras with inner product and that $\wedge$ (hat) on $\mathbf{C}(0,1)$ corresponds to bar (conjugation) on $\mathbf{L}$.

Note that $\vee$ (check) is the identity on $\mathrm{Cl}(1,0)$ and $\mathrm{Cl}(0,1)$ so that $\wedge = \sim$ on both $\mathrm{Cl}(1,0)$ and $\mathrm{Cl}(0,1)$.

4. (a) Show that as algebras with inner products

$$\mathrm{Cl}(2,0) \cong \mathbf{H},$$

and that the isomorphism can be chosen so that

$$\mathrm{Cl}^{\mathrm{even}}(2,0) \cong \mathbf{C} \subset \mathbf{H}, \quad \mathrm{Cl}^{\mathrm{odd}}(2,0) \cong j\mathbf{C} \subset \mathbf{H},$$

and the hat on $\mathrm{Cl}(2,0)$ corresponds to the bar (conjugation) on $\mathbf{H}$.

(b) Show that as algebras with inner products

$$\mathrm{Cl}(1,1) \cong M_2(\mathbf{R}),$$

and that the isomorphism can be chosen so that

$$\mathrm{Cl}^{\mathrm{even}}(1,1) \cong \mathbf{R} \oplus \mathbf{R} \subset M_2(\mathbf{R}) \text{ as diagonal } 2 \times 2 \text{ matrices,}$$

and the $\wedge$ (hat) on $\mathrm{Cl}(1,1)$ corresponds to the bar (conjugation) on the normed aglebra $M_2(\mathbf{R})$.

Alternatively, show that the isomorphism $\mathrm{Cl}(1,1) \cong M_2(\mathbf{R})$ can be chosen so that

$$\mathrm{Cl}^{\mathrm{even}}(1,1) \cong \mathbf{L} \equiv \left\{ \begin{pmatrix} a & b \\ b & a \end{pmatrix} : a, b \in \mathbf{R} \right\}$$

the space of $\mathbf{L}$-linear maps,

$$\mathrm{Cl}^{\mathrm{odd}}(1,1) \cong \left\{ \begin{pmatrix} a & b \\ -b & -a \end{pmatrix} : a, b \in \mathbf{R} \right\}$$

the space of $\mathbf{L}$-antilinear maps,

and, as before, the $\wedge$ (hat) on $\text{Cl}(1,1)$ corresponds to the bar (conjugation) on the normed algebra $M_2(\mathbf{R})$.

(c) Show that as algebras with inner product

$$\text{Cl}(0,2) \cong M_2(\mathbf{R})$$

and that the isomorphism can be chosen so that

$$\text{Cl}^{\text{even}}(0,2) \cong \mathbf{C} \cong \left\{ \begin{pmatrix} a & -b \\ b & a \end{pmatrix} : a, b \in \mathbf{R} \right\} \quad \text{(C-linear)}$$

$$\text{Cl}^{\text{odd}}(1,1) \cong \left\{ \begin{pmatrix} a & b \\ b & -a \end{pmatrix} : a, b \in \mathbf{R} \right\} \text{(L-antilinear)},$$

and the $\wedge$ (hat) on $\text{Cl}(0,2)$ corresponds to bar (conjugation) on the normed algebra $M_2(\mathbf{R})$.

5. Suppose $u \in V$ is nonnull. Given $x \in \Lambda^p V$ show that $u \, \llcorner \, x = 0$ if and only if $x = u \, \llcorner \, y$ for some $y \in \Lambda^{p+1} V$.

6. Give the proof of Remark 9.16.

7. Complete the proof of Lemma 9.49.

8. The *centralizer of A in B* is defined to be

$$\{b \in B : ab = ba \text{ for all } a \in A\}.$$

(a) Show that $a \in \text{Cl}(p,q)$ belongs to the centralizer of $\text{Cl}(p,q)^{\text{even}}$ in $\text{Cl}(p,q)$ if and only if $a \cdot x = x \cdot a$ for all elements $x \in G(2, \mathbf{R}(p,q)) \subset \Lambda^2 \mathbf{R}(p,q)$.

(b) Show that the centralizer of $\text{Cl}(p,q)^{\text{even}}$ in $\text{Cl}(p,q)$ is $\Lambda^0 \mathbf{R}(p,q) \oplus \Lambda^n \mathbf{R}(p,q) \equiv \text{span } \{1, \lambda\}$.

(c) Show that the center of $\text{Cl}(p,q)^{\text{even}}$ is $\Lambda^0 \mathbf{R}(p,q) = \mathbf{R}$ if $n = p+q$ is odd and $\Lambda^0 \mathbf{R}(p,q) \oplus \Lambda^n \mathbf{R}(p,q)$ if $n = p+q$ is even.

9. Give the proof of Proposition 9.62.

10. Suppose that $\rho : \text{Cl}(V) \to \text{End}_{\mathbf{R}}(W)$ is an injective $\mathbf{R}$-representation of $\text{Cl}(V)$ on W, where W, ε is an inner product space; and consider ρ as the identity so that $\text{Cl}(V) \subset \text{End}_{\mathbf{R}}(W)$ is a subalgebra of $\text{End}_{\mathbf{R}}(W)$.

(a) Assume that, given $a \in \text{Cl}(V)$,

$$\varepsilon(ax, y) = \varepsilon(x, \hat{a}y) \text{ for all } x, y \in W.$$

If V is not positive definite, show that there exists an anti-isometry of W, and hence that the signature of W is split (if W has a signature).

(b) Assume that, given $a \in \mathrm{Cl}(V)$,

$$\varepsilon(ax, y) = \varepsilon(x, \breve{a}y) \text{ for all } x, y \in W.$$

If V is not negative definite, show that there exists an anti-isometry of W, and hence that the signature of W is split (if W has a signature).

11. Suppose $\mathbf{R}(p, q)$ is oriented and let λ denote the (unique) unit volume element. The *Hodge star operator* $\star$ on $\Lambda\mathbf{R}(p, q)$ is defined by

$$a \wedge (\star b) = \langle a, b \rangle \lambda, \text{ for all } a, b \in \Lambda\mathbf{R}(p, q).$$

Show that

$$\star b = \hat{b}\lambda, \text{ for all } b \in \Lambda\mathbf{R}(p, q) \cong \mathrm{Cl}(p, q).$$

That is, $\star$ and (right) multiplication by λ agree on elements of degree 0 or 3 mod 4, while $\star$ equals (right) multiplication by $-\lambda$ on elements of degree 1 or 2 mod 4.

10. The Groups Spin and Pin

In this chapter, we construct several groups contained in the Clifford algebra $\text{Cl}(V)$. First, consider the multiplicative group $\text{Cl}^*(V)$ of invertible elements in the Clifford algebra $\text{Cl}(V)$. (Later we shall see that this group is just one or two copies of a general linear group over either $\mathbf{R}, \mathbf{C}$, or $\mathbf{H}$ depending on the signature.) Since

$$u^2 = -\|u\| \quad \text{for each vector } u \in V \subset \text{Cl}(V),$$

a nonnull vector $u \in V$ belongs to $\text{Cl}^*(V)$, with inverse $u^{-1} = -u/\|u\|$, while a null vector $u \in V$ has no inverse in $\text{Cl}(V)$.

Definition 10.1. *The Pin group is the subgroup of $\text{Cl}^*(V)$ generated by unit vectors in V. That is,*

(10.2) $\text{Pin} \equiv \{a \in \text{Cl}^*(V) : a = u_1 \cdots u_r \text{ with each } u_j \in V \text{ and } |u_j| = 1\}.$

Note that each $a \in \text{Pin}$ is either even or odd, since it is a simple product of vectors.

Defintion 10.3. *The spin group is defined by* $\text{Spin} \equiv \text{Pin} \cap \text{Cl}^{\text{even}}(V)$. *That is,*

(10.4) $\qquad \begin{aligned} \text{Spin} = \{&a \in \text{Cl}^*(V) : a = u_1 \cdots u_{2r} \text{ with each } u_j \in V \\ &\text{and } |u_j| = 1\}. \end{aligned}$

The alternate notation $\text{Pin}\,(V), \text{Spin}\,(V)$ is slightly more precise. If $V \equiv \mathbf{R}(p,q)$ is the standard model, then the notation used will be $\text{Pin}\,(p,q)$ and $\text{Spin}\,(p,q)$.

Clifford multiplication by a vector on the left is just exterior minus interior multiplication, therefore;

(10.5) $xy = -yx$ if $x, y \in V$ are orthogonal,

while

(10.6) $xy = yx$ if $x, y \in V$ are colinear.

This proves the next lemma.

Lemma 10.7. *If $u \in V$ is nonnull, then R_u, reflection along u, is given in terms of Clifford multiplication by*

(10.8) $R_u x = -uxu^{-1}$ *for all $x \in V$.*

Motivated by this lemma consider the *twisted adjoint representation* $\widetilde{\text{Ad}}$ of the group $\text{Cl}^*(V)$ on $\text{Cl}(V)$:

(10.9) $\widetilde{\text{Ad}}_a x \equiv \tilde{a} x a^{-1}$ for all $x \in \text{Cl}(V)$.

Note that $(\widetilde{\text{Ad}}_a \, \widetilde{\text{Ad}}_b)(x) = \tilde{a} \tilde{b} x b^{-1} a^{-1} = \widetilde{\text{Ad}}_{ab}(x)$, so that $\text{Cl}^*(V) \xrightarrow{\widetilde{\text{Ad}}} \text{GL}(\text{Cl}(V))$ is a group representation.

The *adjoint representation* Ad of $\text{Cl}^*(V)$ on $\text{Cl}(V)$ is defined by

(10.10) $\text{Ad}_a x \equiv a x a^{-1}$ for all $x \in \text{Cl}(V)$.

Theorem 10.11.

(a) *The sequence*

(10.12) $1 \to \mathbf{Z}_2 \to \text{Pin} \xrightarrow{\widetilde{\text{Ad}}} O(V) \to I.$

is exact.

(b) *The sequence*

(10.13) $1 \to \mathbf{Z}_2 \to \text{Spin} \xrightarrow{\widetilde{\text{Ad}}} SO(V) \to I.$

is exact.

Proof: If $u \in V$ is nonnull, then $\widetilde{\mathrm{Ad}}_u = R_u \in O(V)$. Therefore, if $a \in \mathrm{Pin}$, then $\widetilde{\mathrm{Ad}}_a \in O(V)$. The weak form of the Cartan–Dieudonné Theorem (Remark 4.25) states that each orthogonal transformation can be expressed as the product of reflections along nondegenerate lines, so that $\widetilde{\mathrm{Ad}}$ is surjective.

Now suppose $a \in \mathrm{Pin}$ and $\widetilde{\mathrm{Ad}}_a = Id$. That is, assume $\tilde{a}x = xa$ for all $x \in V$. First, suppose a is odd. Then $-ax = xa$ for all $x \in V$. Thus $a\tilde{x} = xa$ for all $x \in \mathrm{Cl}(V)$, so that a belongs to the twisted center. But (Lemma 9.49) the twisted center of $\mathrm{Cl}(V)$ cannot have an odd part. Second, suppose a is even. Then $ax = xa$ for all $x \in V$ and hence for all $x \in \mathrm{Cl}(V)$; that is, a belongs to the center of $\mathrm{Cl}(V)$. Therefore, by (Lemma 9.49), $a \in \Lambda^0 V \cong \mathbf{R}$.

Finally, since $a = u_1 \cdots u_r$ is a simple product of unit vectors, by Proposition 9.29,

$$(10.14) \qquad \|a\| = \|u_1\| \cdots \|u_r\| = \pm 1 \quad \text{for all } a \in \mathrm{Pin}.$$

This proves $a \in \mathbf{Z}_2 \equiv \{1, -1\} \subset \mathrm{Pin}$, completing the proof of part (a).

Part (b) follows easily from part (a) and the fact that $\det R_u = -1$ for each reflection R_u. ∎

Remark 10.15. Note that the full strength of the Cartan–Dieudonné Theorem 4.23 implies that

$$(10.16) \qquad \mathrm{Pin} = \{a \in \mathrm{Cl}^*(V) : a = u_1 \cdots u_r, r \le n\}.$$

THE GRASSMANNIANS AND REFLECTIONS

Recall that $G(r, V)$, the grassmannian of all unit, oriented, nondegenerate r-planes through the origin in V, consists of all simple vectors in $\Lambda^r V$ that are of unit length. That is,

$$u \in G(r, V) \text{ if } u = u_1 \wedge \cdots \wedge u_r,$$
$$\text{with } u_1, \ldots, u_r \in V \text{ and } |u| = 1 \ (\|u\| = \pm 1).$$

Of course, we may choose $u_1, \ldots, u_r$ orthonormal without changing u. The span of u, denoted span u, is just the span of $\{u_1, \ldots, u_r\}$.

Given $u \in G(r, V)$, *reflection along* u, denoted R_u, is defined by

$$(10.17) \qquad R_u x \equiv -x \quad \text{if } x \in \text{span } u,$$

and

$$(10.18) \qquad R_u x \equiv x \quad \text{if } x \in (\text{span } u)^\perp.$$

Lemma 10.7 can be generalized.

Lemma 10.19. *If $u \in G(r, V)$ is a unit nondegenerate r-plane in V, then*

$$R_u v = \widetilde{\text{Ad}}_u \ x \quad \text{for all } x \in V.$$

Remark 10.20. Therefore, each reflection $R_u \in O(V)$ along a subspace span u of V is replaced in the double cover $\text{Pin}(V)$ of $O(V)$ by either of the two elements $\pm u \in G(r, V) \subset \Lambda^r V \subset \text{Cl}(V)$ in the Clifford algebra.

Proof of Lemma 10.19. Each $u \in G(r, V)$ can be expressed as $u = u_1 \wedge \ldots \wedge u_r$, where $u_1, \cdots, u_r$ is chosen to be an oriented orthonormal basis for the nondegenerate subspace span $u \subset V$. Since $u_1, \ldots, u_r$ are othogonal, $R_u = R_{u_1} \circ \cdots \circ R_{u_r}$ and $u = u_1 \wedge \cdots \wedge u_r = u_1 \cdots u_r$, so that $\widetilde{\text{Ad}}_u = \widetilde{\text{Ad}}_{u_1} \circ \cdots \circ \widetilde{\text{Ad}}_{u_r}$. Finally note that $R_{u_j} = \widetilde{\text{Ad}}_{u_j}$ for each j by Lemma 10.7. ∎

By definition, the group Pin is generated by the element $G(1, V) \subset$ Pin. Although the definition of Spin appears to be "generative," it suffers from the defect that the generators $u \subset G(1, V)$ are not in Spin. This defect can be corrected in several ways. First, if e is any unit vector and S^{n-1} denotes the unit sphere in V, then $e \cdot S^{n-1}$ generates Spin, since $u \cdot v = e \cdot w \cdot e \cdot v$ for some $w = \pm e u e \in S^{n-1}$.

In addition, we have

Proposition 10.21. $(n \equiv \dim V \geq 3)$. *The group Spin is the subgroup of $\text{Cl}^*(V)$ generated by $G(2, V)$. That is,*

$$(10.22) \qquad \begin{aligned} \text{Spin} \ = \ \{ a \in \text{Cl}^*(V) : a = u_1 \cdots u_r \ \text{with each} \\ u_j \in G(2, V) \subset \Lambda V \cong \text{Cl}(V) \}. \end{aligned}$$

Moreover, r can be chosen even and $\leq n$.

This proposition is the "double cover" of a theorem in Chapter 4. However, the proof is so short that it is repeated here.

Proof: First, suppose $n = 3$. Each $a \in$ Spin can be expressed as $a = u_1 u_2$ with $u_1, u_2 \in V$ unit vectors. Let λ denote a unit volume form for V. Then $v_1 \equiv u_1 \lambda \in G(2, V)$ and $v_2 \equiv \lambda u_2 \in G(2, V)$. Thus, $a = u_1 u_2 = \pm v_1 v_2$, where $\lambda^2 = \pm 1$.

Second, suppose $n \geq 4$. Then, given a product $u_1 u_2$ of unit vectors $u_1, u_2 \in V$, there exists a unit vector v orthogonal to both u_1 and u_2 (see Problem 4.10(b)). Hence $u_1 u_2 = \pm u_1 v v u_2 = (\pm u_1 \wedge v) \cdot (v \wedge u_2)$. ∎

If n is odd, then Spin is generated by another sphere as well as by the grassmannian $G(2, V)$ and $e \cdot S^{n-1}$.

Proposition 10.23. $(n \geq 3)$ *Suppose* $n \equiv \dim V$ *is odd. The group* Spin *is generated by the sphere* $G(n-1, V)$ *in* $\Lambda^{n-1}V$.

See Chapter 14 for applications (for example, $\mathrm{Spin}(7) \subset \mathrm{Cl}^{\mathrm{even}}(8) \cong M_8(\mathbf{R})$ is generated by $\{R_u : u \in S^6 \subset \mathrm{Im}\, \mathbf{O}\}$).

Proof: Since n is odd, the unit volume element λ is in the center of $\mathrm{Cl}(V)$. Therefore, for u, v an orthonormal pair and $\lambda^2 = \pm 1$,

$$u \wedge v = uv = (\pm \lambda u)(\lambda v), \quad \text{with} \quad \pm \lambda u, \lambda v \in G(n-1, V). \quad \blacksquare$$

ADDITIONAL GROUPS FOR NONDEFINITE SIGNATURE

If the signature p, q is not definite (i.e., $p \geq 1$ and $q \geq 1$), then, in addition to the groups Spin and Pin, there are three other closely related groups (cf. Definition 4.48).

Note that the square norm (or Clifford norm) $\|a\| \equiv \langle a, a \rangle = \langle a\hat{a}, 1 \rangle = \langle \hat{a}a, 1 \rangle$ is multiplicative on Pin because of (10.14), i.e.,

$$\|ab\| = \|a\| \, \|b\| \quad \text{for all } a, b \in \mathrm{Pin}.$$

That is, the Clifford norm

(10.24) $$\|\cdot\| : \mathrm{Pin} \to \mathbf{Z}_2$$

is a group homomorphism. The twisted Clifford norm, defined by

$$\langle a, b \rangle' \equiv \langle a, \tilde{b} \rangle \quad \text{for } a, b \in \mathrm{Cl}(p, q),$$

provides another group homomorphism from Pin to $\mathbf{Z}_2$. The kernels of these group homomorphisms are subgroups of Pin.

Definition 10.25.

(10.26) $$\mathrm{Pin}^\wedge \equiv \{a \in \mathrm{Pin} : a\hat{a} = 1\}$$
(10.27) $$\mathrm{Pin}^\vee \equiv \{a \in \mathrm{Pin} : a\check{a} = 1\}.$$

Note that the intersection of any two of the three subgroups of Pin,

(10.28) $$\mathrm{Spin}, \quad \mathrm{Pin}^\wedge, \quad \text{and} \quad \mathrm{Pin}^\vee,$$

is the same as the intersection of all three subgroups. This common intersection is denoted

(10.29) $\text{Spin}^0 = \{a \in \text{Spin} \; : a\hat{a} = 1\}$

and called the *reduced spin group*.

Remark 10.30. If V is positive definite, then

(10.31) $\text{Pin}^\wedge = \text{Pin}, \quad \text{Pin}^\vee = \text{Spin}, \quad \text{and} \quad \text{Spin}^0 = \text{Spin};$

while if V is negative definite, then

(10.31') $\text{Pin}^\vee = \text{Pin}, \; \text{Pin}^\wedge = \text{Spin}, \; \text{and } \text{Spin}^0 = \text{Spin},$

so that the only groups are Spin and Pin.

Since $\|a\| = \|-a\|$, the Clifford norm $\| \cdot \|$ on Pin descends to the homomorphism

(10.32) $\| \; \|_S : O(V) \to \mathbf{Z}_2,$

called the *spinorial norm* (of an orthogonal transformation—see Definition 4.55). To prove this fact, assume that

$$A = R_{u_1} \circ \cdots \circ R_{u_r} \in O(V)$$

is expressed as the product of reflections. By definition, the spinorial norm of A is equal to the parity of number of reflections that are along timelike ($\|u_j\| = -1$) directions, which of course is also equal to the Clifford norm of $u_1 \cdots u_r$. Similarly, the twisted Clifford norm on Pin descends to the homomorphism $(\det A)\|A\|_S$ on $O(V)$.

The next result is immediate from Theorem 10.11 (cf. Lemma 4.54).

Proposition 10.33. $(p, q \geq 1)$ *The following sequences are exact.*

(10.34) $1 \to \mathbf{Z}_2 \to \text{Spin}^0 \xrightarrow{\widetilde{\text{Ad}}} SO^0 \to 1$

(10.35) $1 \to \mathbf{Z}_2 \to \text{Pin}^\wedge \xrightarrow{\widetilde{\text{Ad}}} O^\uparrow \to 1$

(10.36) $1 \to \mathbf{Z}_2 \to \text{Pin}^\vee \xrightarrow{\widetilde{\text{Ad}}} O^+ \to 1.$

There are certain distinguished classical groups containing Spin^0, Spin, $\text{Pin}^\vee$, or Pin, with equality occurring in low dimensions. We shall concentrate on the case Spin^0, leaving the other cases to the reader.

Definition 10.37. *The classical companion group, denoted* $\mathrm{Cp}(V)$, *is defined by*

$$(10.38) \qquad \mathrm{Cp}(V) \equiv \{a \in \mathrm{Cl}^{\mathrm{even}}(V) : a\hat{a} = 1\}.$$

The connected component of the identity element in $\mathrm{Cp}(V)$ *is a subgroup that is called the reduced classical companion group and is denoted by* $\mathrm{Cp}^0(V)$.

Note: $\mathrm{Spin}^0(V) \subset \mathrm{Cp}^0(V)$. By Problem 2(d), $\mathrm{Spin}^0(r,s)$ is connected (unless $p \leq 1$ and $q \leq 1$). Therefore $\mathrm{Spin}^0(V) \subset \mathrm{Cp}^0(V)$.

The identification of the groups $\mathrm{Cp}^0(r,s)$ as classical groups is a major part of Chapter 13.

THE CONFORMAL PIN (OR CLIFFORD) GROUP

Definition 10.39. *The conformal pin group* (*or Clifford group*) *consists of all simple products of nonnull vectors. That is,*

$$(10.40) \qquad \begin{aligned} \mathrm{Cpin} \equiv \{&a \in \mathrm{Cl}^*(V) : a = u_1 \cdots u_r \\ &\text{with each } u_j \in V \text{ nonnull}\}. \end{aligned}$$

The proof of Theorem 10.11 applies to the conformal Pin group yielding

Proposition 10.41. *The sequence*

$$(10.42) \qquad 1 \to \mathbf{R}^* \to \mathrm{Cpin} \xrightarrow{\widetilde{\mathrm{Ad}}} \mathrm{CO}(V) \to \mathrm{Id}$$

is exact.

The conformal Pin group has a very useful alternative description, as the subgroup of $\mathrm{Cl}^*(V)$ that maps V to V under the twisted adjoint representation.

Theorem 10.43.

$$(10.44) \qquad \mathrm{Cpin} = \{a \in \mathrm{Cl}^*(V) : \widetilde{\mathrm{Ad}}_a(V) \subset V\}.$$

The proof of this result depends on the next lemma.

Lemma 10.45. *The following are equivalent:*

(a) $a\hat{a} = \|a\|$,
(b) $\hat{a}a = \|a\|$,
(c) $a\hat{a} \in \mathbf{R}$,
(d) $\hat{a}a \in \mathbf{R}$.

Proof: For each $a \in \mathrm{Cl}(V)$,

$$\|a\| = \langle 1, a\hat{a} \rangle = \langle 1, \hat{a}a \rangle,$$

so the lemma follows easily. ∎

Corollary 10.46. *If $a \in \mathrm{Cl}(V)$ satisfies the equivalent conditions of Lemma 10.45, then*

$$\|ab\| = \|a\| \, \|b\| \quad \text{for all } b \in \mathrm{Cl}(V).$$

Proof: $\|ab\| = \langle 1, \widehat{ab}ab \rangle = \langle 1, \hat{b}\hat{a}ab \rangle = \|a\|\langle 1, \hat{b}b \rangle = \|a\| \, \|b\|$. ∎

Proof of Theorem 10.43: Let

$$\Gamma \equiv \{ a \in \mathrm{Cl}^*(V) : \tilde{a}xa^{-1} \in V \text{ if } x \in V \}$$

denote the group occurring on the right hand side of (10.44).

We wish to show that the subgroup Cpin of Γ is all of Γ. Because of the exact sequence (10.42),

$$1 \to \mathbf{R}^* \to \mathrm{Cpin} \xrightarrow{\widetilde{\mathrm{Ad}}} \mathrm{CO}(V) \to \mathrm{Id},$$

it suffices to prove that

(10.47) $$1 \to \mathbf{R}^* \to \Gamma \xrightarrow{\widetilde{\mathrm{Ad}}} \mathrm{CO}(V)$$

is exact.

Assume $a \in \Gamma$ and $\widetilde{\mathrm{Ad}}_a = \mathrm{Id}$. That is,

(10.48) $$\tilde{a}x = xa \quad \text{for all } x \in V.$$

Let a_+ denote the even part of a, and a_- denote the odd part of a. Then (10.48) is equivalent to

(10.48′) $$a_+x = xa_+ \quad \text{and} \quad a_-x = -xa_- \quad \text{for all } x \in V.$$

Thus, the even part of a is in the center of $\mathrm{Cl}(V)$, and the odd part of a is in the twisted center of $\mathrm{Cl}(V)$. However, referring back to Lemma 9.49, the twisted center of $\mathrm{Cl}(V)$ cannot have an odd part, and the even part of the center is always $\mathbf{R}^*$. Thus,

$$1 \to \mathbf{R}^* \to \Gamma \xrightarrow{\widetilde{\mathrm{Ad}}} \mathrm{GL}(V)$$

is exact.

Next, we show that

(10.49) $$\hat{a}a \in \mathbf{R} \quad \text{for all } a \in \Gamma$$

Because of (10.47), $\hat{a}a \in \mathbf{R}^*$ if $\hat{a}a \in \ker \widetilde{\mathrm{Ad}}$. The hypothesis $a \in \Gamma$ says that $\tilde{a}xa^{-1} \in V$ if $x \in V$. Therefore, the check anti-automorphism fixes $\tilde{a}xa^{-1}$ if $x \in V$. However, the check of $\tilde{a}xa^{-1}$ is $\check{a}^{-1}x\hat{a}$ so that $\tilde{a}xa^{-1} = \check{a}^{-1}x\hat{a}$ or $(\hat{a}a)x = x\hat{a}a$, proving that $\hat{a}a \in \ker \widetilde{\mathrm{Ad}} = \mathbf{R}^*$.

Now by Lemma 10.45, if $a \in \Gamma$, then $a^{-1} = \hat{a}/\|a\|$. Thus,

$$\|\tilde{a}xa^{-1}\| = \|a\|^{-2}\|\tilde{a}x\hat{a}\| = \|x\|,$$

because of Corollary 10.46 and the fact that $\tilde{}$ and $\hat{}$ are isometries. This proves that $\mathrm{Ad} : \Gamma \to \mathrm{CO}(V)$, thus completing the proof of Theorem 10.43. ∎

Sometimes it is useful to adopt the following as alternate (equivalent) definitions of the groups $\mathrm{Pin}, \mathrm{Spin}$, and Spin^0.

Corollary 10.50.

$$\mathrm{Pin} = \{a \in \mathrm{Cl}^*(V) : \widetilde{\mathrm{Ad}}_a(V) \subset V \text{ and } \|a\| = \pm 1\}$$
$$\mathrm{Spin} = \{a \in \mathrm{Cl}^*(V)^{\mathrm{even}} : \mathrm{Ad}_a(V) \subset V \text{ and } \|a\| = \pm 1\}$$
$$\mathrm{Spin}^0 = \{a \in \mathrm{Cl}^*(V)^{\mathrm{even}} : \mathrm{Ad}_a(V) \subset V \text{ and } \|a\| = 1\}.$$

DETERMINANTS

In several different situations it is useful to compute the determinant of an element of Spin or Pin acting on a vector space W.

The idea of the proof is the same in the two cases considered below. Suppose $\rho : G \to \mathrm{End}_{\mathbf{R}}(W)$ is a real representation of a group G on a vector space W. Further, suppose G is generated by elements u in a subset S with

(10.51) $$u^2 = \pm 1 \quad \text{for each } u \in S.$$

Note that this is true for $G \equiv \mathrm{Pin}$ with $S \equiv G(1, V)$; and true for $G \equiv \mathrm{Spin}$ with $S \equiv G(2, V)$ by Proposition 10.21 (if $n \geq 2$). Then,

$$(10.52) \qquad \qquad \rho(a)^2 = \pm 1 \quad \text{for all } a \in S.$$

Consequently,

$$(10.53) \qquad \qquad \det_{\mathbf{R}} \rho(a) = \pm 1 \quad \text{for all } a \in S,$$

and hence for all $a \in G$.

To prove (10.53) consider two cases.

Case 1:

$$(10.54) \qquad \begin{array}{l} \text{If } \rho(a)^2 = -1, \text{ then } \det_{\mathbf{R}} \rho(a) = 1, \\ \text{because } \rho(a) \text{ is a complex structure on } W. \end{array}$$

Case 2:
(10.55)

If $\rho(a)^2 = 1$, then $\rho(a)$ is reflection along W_- through W_+,

where $W_{\pm} \equiv \{x \in W : \rho(a)x = \pm x\}$ are the eigenspaces of $\rho(a)$.

Therefore,

$$(10.56) \qquad \qquad \det_{\mathbf{R}} \rho(a) = (-1)^{\dim W_-}.$$

PROBLEMS

1. Verify Remark 10.30.

2. Suppose $V, \langle \, , \, \rangle$ has signature p, q, and e_1 and e_2 are orthonormal in V.

   (a) Show that

   $$\cos \theta + \sin \theta e_1 e_2 = (-\cos \theta/2 \, e_1 + \sin \theta/2 \, e_2)(\cos \theta/2 \, e_1 - \sin \theta/2 \, e_2),$$
   $$\text{if } \|e_1\| = \|e_2\| = 1,$$

   and that

   $$\cos \theta + \sin \theta e_1 e_2 = (\cos \theta/2 \, e_1 - \sin \theta/2 \, e_2)(\cos \theta/2 \, e_1 + \sin \theta/2 \, e_2),$$
   $$\text{if } \|e_1\| = \|e_2\| = -1.$$

   (b) If either $p \geq 2$ or $q \geq 2$, find a subalgebra of $\mathrm{Cl}(V)$ that is isomorphic to $\mathbf{C}$ with $\mathrm{Spin}(V) \cap \mathbf{C} \cong S^1$ the unit circle in $\mathbf{C} \subset \mathrm{Cl}(V)$.

(c) If either $p \geq 2$ or $q \geq 2$, show that

$$\text{Spin}^0(V) \xrightarrow{\text{Ad}} \text{SO}^\uparrow(V)$$

is a nontrivial double cover, i.e., show that -1 can be connected to $+1$ in $\text{Spin}^0(V)$.

(d) If either $p \geq 2$ or $q \geq 2$, show that $\text{Spin}^0(p,q)$ is connected.

3. Give the proof of Proposition 10.41.

4. Compute, explicitly as (classical) subgroups of $\text{Cl}^*(V)$,
   (a) Pin and Spin if $p = 2$, $q = 0$, or $p = 0$, $q = 2$.
   (b) Pin, Pin$^\wedge$, Pin$^\vee$, Spin, and Spin^0 if $p = 1$, $q = 1$.

5. (a) Show that

$$1 \to \mathbf{Z}_2 \to \text{Pin} \xrightarrow{\text{Ad}} O(V) \to 1$$

is exact if n is even.

(b) Show that for n odd,

$$1 \to K \to \text{Pin} \xrightarrow{\text{Ad}} O(V) \to 1$$

is exact where $K \equiv \{\pm 1, \pm \lambda\}$ with λ a unit volume element. Note that

$$K = \mathbf{Z}_2 \times \mathbf{Z}_2 \quad \text{if } p - q = 3, 7 \bmod 8, \text{ while}$$
$$K = \mathbf{Z}_4 \quad \text{if } p - q = 1, 5 \bmod 8.$$

6. (a) Show that the natural inclusion of $\mathbf{R}(p,q)$ into $\mathbf{R}(p+1,q)$ induces an inclusion of algebras $\text{Cl}(p,q) \subset \text{Cl}(p+1,q)$.

(b) Prove that $\text{Pin}(p,q)$ is included in $\text{Pin}(p+1,q)$ by

$$\text{Pin}(p,q) \cong \left\{ a \in \text{Pin}(p+1,q) : \widetilde{\text{Ad}}_a(e_{p+1}) = e_{p+1} \right\},$$

where $e_1, \ldots, e_p, e_{p+1}; e_{p+2}, \ldots, e_{n+1}$ is the standard basis for $\mathbf{R}(p+1,q)$. Thus,

$$\begin{array}{ccc} \text{Pin}(p,q) & \subset & \text{Pin}(p+1,q) \\ \downarrow \widetilde{\text{Ad}} & & \downarrow \widetilde{\text{Ad}} \\ O(p,q) & \subset & O(p+1,q). \end{array}$$

7. Suppose $\rho : \text{Spin}(r,s) \to \text{End}_{\mathbf{C}}(W)$ is a group representation with dimension $n \equiv r + s \geq 3$. Show that

$$\det{}_{\mathbf{C}} \rho(a) \in \{\pm 1, \pm i\} \quad \text{for all } a \in \text{Spin}(r,s).$$

8. Define the complex Pin group by

$$\text{Pin}(n, \mathbf{C}) = \big\{ a \in \text{Cl}_{\mathbf{C}}^*(n) : a = u_1 \cdots u_r \text{ with}$$
$$u_j \in \mathbf{C}^n \text{ and } \|u_j\| = 1, j = 1, \ldots, r \big\},$$

and the complex Spin group by

$$\text{Spin}(n, \mathbf{C}) \equiv \text{Pin}(n, \mathbf{C}) \cap \text{Cl}^{\text{even}}(n).$$

Show that (see Problem 4.17)

(a) $1 \to \mathbf{Z}_2 \to \text{Pin}(n, \mathbf{C}) \xrightarrow{\widetilde{\text{Ad}}} O(n, \mathbf{C}) \to 1$ is exact,

(b) $1 \to \mathbf{Z}_2 \to \text{Spin}(n, \mathbf{C}) \xrightarrow{\text{Ad}} SO(n, \mathbf{C}) \to 1$ is exact,

(c) $\text{Pin}(n, \mathbf{C}) = \{ a \in \text{Cl}_{\mathbf{C}}^*(n) : \widetilde{\text{Ad}}_a(\mathbf{C}^n) \subset \mathbf{C}^n \text{ and } \|a\| = 1 \}$.

9. Suppose $A \in O(V)$ can be written as the product of reflections $A = R_{u_1} \circ \cdots \circ R_{u_r}$ with $u_1, \ldots, u_r$ linearly independent. Let $a \equiv u_1 \cdots u_r \in \text{Pin}(V)$. Let $(a)_k$ denote the degree k component of a in $\Lambda^k V \subset \Lambda V \cong \text{Cl}(V)$. Note $(a)_k = 0$ for $k > r$ and $(a)_r = u_1 \wedge \cdots \wedge u_r$.

Suppose $A = R_{v_1} \circ \cdots \circ R_{v_k}$ is any other representation of A as the product of reflections.

(a) Show that $k < r$ implies $u_1, \ldots, u_r$ are linearly dependent, so that $k \geq r$ (i.e., r is the Cartan–Dieudonné rank of A).

(b) Show that if $k = r$, then $u_1 \wedge \cdots \wedge u_r = \pm v_1 \wedge \cdots \wedge v_r$. (Thus, there is a well-defined vector space $\text{span}\{u_1, \ldots, u_r\} = \text{span}\{v_1, \ldots, v_r\}$ associated with A.)

10. Show that, for $n \geq 3$, $\text{Spin}^0(p, q)$ is generated by $G^+(2, V) \cup G^-(2, V)$ (the definite 2-planes).

11. Show that the centralizer of $\text{Spin}^0(p, q)$ in $\text{Cl}(p, q)$ is $\Lambda^0 \mathbf{R}(p, q) \oplus \Lambda^n \mathbf{R}(p, q)$, and hence the center of $\text{Spin}^0(p, q)$ is equal to

(a) $\{\pm 1\} = \mathbf{Z}_2$ if n is odd or q is odd.

(b) $\{\pm 1, \pm \lambda\} = \mathbf{Z}_2 \oplus \mathbf{Z}_2$ if $p - q = 0 \mod 4$ and n, q are even.

(c) $\{\pm 1, \pm \lambda\} = \mathbf{Z}_4$ if $p - q = 2 \mod 4$ and n, q are even.

11. The Clifford Algebras $Cl(r,s)$ as Algebras

Recall the algebra isomorphisms

$$(11.1) \qquad Cl(1,0) \cong \mathbf{C}, \qquad Cl(0,1) \cong \mathbf{L} \cong \mathbf{R} \oplus \mathbf{R}$$

$$(11.2) \qquad Cl(2,0) \cong \mathbf{H}, \quad Cl(1,1) \cong M_2(\mathbf{R}), \quad Cl(0,2) \cong M_2(\mathbf{R})$$

established in Problems 9.3 and 9.4. Each of the Clifford algebras $Cl(r,s)$ is isomorphic, as an associative algebra with unit, to a matrix algebra. Some of the important extra structure of $Cl(r,s)$, that is hidden in the corresponding matrix algebra will be exposed in terms of "pinors" in Chapter 13.

Theorem 11.3. *The Clifford algebras $Cl(r,s)$ of signature r,s are isomorphic, as associative algebras with unit, to the matrix algebras listed below.*

$r - s$ *mod* 8 :

$0, 6$	$Cl(r,s) \cong M_N(\mathbf{R})$
$2, 4$	$Cl(r,s) \cong M_N(\mathbf{H})$
$1, 5$	$Cl(r,s) \cong M_N(\mathbf{C})$
3	$Cl(r,s) \cong M_N(\mathbf{H}) \oplus M_N(\mathbf{H})$
7	$Cl(r,s) \cong M_N(\mathbf{R}) \oplus M_N(\mathbf{R})$

Remark 11.4. The integer N is, of course, trivial to compute in terms of the dimension $n \equiv r + s$, since the Clifford algebra $\mathrm{Cl}(r, s)$ has dimension 2^n as a real vector space.

Before giving the proof of Theorem 11.3, we examine some reformulations and corollaries.

Some readers may find the following tabular form of Theorem 11.3 convenient for reference.

Table 11.5. Matrix Algebras Isomorphic to $\mathrm{Cl}(r, s)$

$M_{16}(\mathbf{R})$	$M_{16}(\mathbf{C})$	$M_{16}(\mathbf{H})$	$M_{16}(\mathbf{H})$ $\oplus$ $M_{16}(\mathbf{H})$	$M_{32}(\mathbf{H})$	$M_{64}(\mathbf{C})$	$M_{128}(\mathbf{R})$	$M_{128}(\mathbf{R})$ $\oplus$ $M_{128}(\mathbf{R})$	$M_{256}(\mathbf{R})$
$M_8(\mathbf{C})$	$M_8(\mathbf{H})$	$M_8(\mathbf{H})$ $\oplus$ $M_8(\mathbf{H})$	$M_{16}(\mathbf{H})$	$M_{32}(\mathbf{C})$	$M_{64}(\mathbf{R})$	$M_{64}(\mathbf{R})$ $\oplus$ $M_{64}(\mathbf{R})$	$M_{128}(\mathbf{R})$	$M_{128}(\mathbf{C})$
$M_4(\mathbf{H})$	$M_4(\mathbf{H})$ $\oplus$ $M_4(\mathbf{H})$	$M_8(\mathbf{H})$	$M_{16}(\mathbf{C})$	$M_{32}(\mathbf{R})$	$M_{32}(\mathbf{R})$ $\oplus$ $M_{32}(\mathbf{R})$	$M_{64}(\mathbf{R})$	$M_{64}(\mathbf{C})$	$M_{64}(\mathbf{H})$
$M_2(\mathbf{H})$ $\oplus$ $M_2(\mathbf{H})$	$M_4(\mathbf{H})$	$M_8(\mathbf{C})$	$M_{16}(\mathbf{R})$	$M_{16}(\mathbf{R})$ $\oplus$ $M_{16}(\mathbf{R})$	$M_{32}(\mathbf{R})$	$M_{32}(\mathbf{C})$	$M_{32}(\mathbf{H})$	$M_{32}(\mathbf{H})$ $\oplus$ $M_{32}(\mathbf{H})$
$M_2(\mathbf{H})$	$M_4(\mathbf{C})$	$M_8(\mathbf{R})$	$M_8(\mathbf{R})$ $\oplus$ $M_8(\mathbf{R})$	$M_{16}(\mathbf{R})$	$M_{16}(\mathbf{C})$	$M_{16}(\mathbf{H})$	$M_{16}(\mathbf{H})$ $\oplus$ $M_{16}(\mathbf{H})$	$M_{32}(\mathbf{H})$
$M_2(\mathbf{C})$	$M_4(\mathbf{R})$	$M_4(\mathbf{R})$ $\oplus$ $M_4(\mathbf{R})$	$M_8(\mathbf{R})$	$M_8(\mathbf{C})$	$M_8(\mathbf{H})$	$M_8(\mathbf{H})$ $\oplus$ $M_8(\mathbf{H})$	$M_{16}(\mathbf{H})$	$M_{32}(\mathbf{C})$
$M_2(\mathbf{R})$	$M_2(\mathbf{R})$ $\oplus$ $M_2(\mathbf{R})$	$M_4(\mathbf{R})$	$M_4(\mathbf{C})$	$M_4(\mathbf{H})$	$M_4(\mathbf{H})$ $\oplus$ $M_4(\mathbf{H})$	$M_8(\mathbf{H})$	$M_{16}(\mathbf{C})$	$M_{32}(\mathbf{R})$
$\mathbf{R} \oplus \mathbf{R}$	$M_2(\mathbf{R})$	$M_2(\mathbf{C})$	$M_2(\mathbf{H})$	$M_2(\mathbf{H})$ $\oplus$ $M_2(\mathbf{H})$	$M_4(\mathbf{H})$	$M_8(\mathbf{C})$	$M_{16}(\mathbf{R})$	$M_{16}(\mathbf{R})$ $\oplus$ $M_{16}(\mathbf{R})$
$\mathbf{R}$	$\mathbf{C}$	$\mathbf{H}$	$\mathbf{H} \oplus \mathbf{H}$	$M_2(\mathbf{H})$	$M_4(\mathbf{C})$	$M_8(\mathbf{R})$	$M_8(\mathbf{R})$ $\oplus$ $M_8(\mathbf{R})$	$M_{16}(\mathbf{R})$

s ↑

r →

The even part of the Clifford algebra, Cl(r, s)^{even} may be read of from the entry to the left of Cl(r, s) in this table since (see Theorem 9.38)

$$(11.6) \qquad \text{Cl}(r, s)^{\text{even}} \cong \text{Cl}(r - 1, s), \quad r \geq 1.$$

For an entry

$$(11.7) \qquad \text{Cl}(0, s)^{\text{even}} \cong \text{Cl}(s - 1, 0), \quad s \geq 1.$$

Note that (11.7) is equivalent to extending the table one column to the left, in the obvious manner, so that (11.6) can be applied in the case $r = 0$.

Corollary 11.8. *The even Clifford algebras* Cl(r, s)^{even} *are isomorphic, as associative algebras with unit, to the matrix algebras described below.*

$r - 2 \bmod 8 :$

0	Cl(r, s)$^{\text{even}} \cong M_N(\mathbf{R}) \oplus M_N(\mathbf{R})$
1, 7	Cl(r, s)$^{\text{even}} \cong M_N(\mathbf{R})$
3, 5	Cl(r, s)$^{\text{even}} \cong M_N(\mathbf{H})$
2, 6	Cl(r, s)$^{\text{even}} \cong M_N(\mathbf{C})$
4	Cl(r, s)$^{\text{even}} \cong M_N(\mathbf{H}) \oplus M_N(\mathbf{H})$.

Given a choice of unit volume element λ for $\mathbf{R}(r, s)$, recall that an element $a \in$ Cl(r, s) is said to be *self-dual* if $\lambda a = a$ and *anti-self-dual* if $\lambda a = -a$. Also recall that Cl$^+(r, s)$ denotes the space of self-dual Clifford elements and Cl$^-(r, s)$ denotes the space of anti-self-dual Clifford elements. If Cl(r, s) or Cl(r, s)^{even} is isomorphic to the sum $M_N(F) \oplus M_N(F)$ of two simple matrix algebras, the summands can be easily identified in Clifford terms as the self-dual and anti-self-dual parts, as follows.

Lemma 11.9.

(a) *In the following cases,*

$r - s \bmod 8 :$

3	Cl(r, s)	$\cong M_N(\mathbf{H}) \oplus M_N(\mathbf{H})$
7	Cl(r, s)	$\cong M_N(\mathbf{R}) \oplus M_N(\mathbf{R})$
4	Cl(r, s)^{even}	$\cong M_N(\mathbf{H}) \oplus M_N(\mathbf{H})$
0	Cl(r, s)^{even}	$\cong M_N(\mathbf{R}) \oplus M_N(\mathbf{R})$

a unit volume element λ for $\mathbf{R}(r, s)$ must correspond to either $(1, -1)$ or $(-1, 1)$.

Selecting $\lambda \equiv (1, -1)$ yields

$r - s \bmod 8 :$

3	$\mathrm{Cl}(r,s)^+ \cong M_N(\mathbf{H}) \oplus \{0\}$
	$\mathrm{Cl}(r,s)^- \cong \{0\} \oplus M_N(\mathbf{H})$
7	$\mathrm{Cl}(r,s)^+ \cong M_N(\mathbf{R}) \oplus \{0\}$
	$\mathrm{Cl}(r,s)^- \cong \{0\} \oplus M_N(\mathbf{R})$
4	$\mathrm{Cl}(r,s)^{\mathrm{even}}_+ \cong M_N(\mathbf{H}) \oplus \{0\}$
	$\mathrm{Cl}(r,s)^{\mathrm{even}}_- \cong \{0\} \oplus M_N(\mathbf{H})$
0	$\mathrm{Cl}(r,s)^{\mathrm{even}}_+ \cong M_N(\mathbf{R}) \oplus \{0\}$
	$\mathrm{Cl}(r,s)^{\mathrm{even}}_- \cong \{0\} \oplus M_N(\mathbf{R})$.

(b) *In the cases where* $\mathrm{Cl}(r,s) \cong M_N(\mathbf{C})$(i.e., $r - s = 1, 5 \bmod 8$) *or the cases where* $\mathrm{Cl}(r,s)^{\mathrm{even}} \cong M_N(\mathbf{C})$(i.e., $r - s = 2, 6 \bmod 8$), *a unit volume element* λ *for* $\mathbf{R}(r,s)$ *must correspond to* $\pm i \in M_N(\mathbf{C})$. *In these cases, the standard choice is* $\lambda \equiv i$.

Proof: (a) Suppose $A \cong M$ is one of the isomorphisms listed in this lemma. Of course, the identity $1 \in A$ must correspond to the identity $1 \in M$, and $-1 \in A$ must correspond to $-1 \in M$. The remaining elements in the center of M that square to 1 are $(1, -1)$ and $(-1, 1)$. The remaining elements in the center of A that square to 1 are $\pm\lambda$ since $A \cong M$ and M has center $\mathbf{R} \oplus \mathbf{R}$.

(b) The elements of the center of $M_N(\mathbf{C})$ that square to -1 are $\pm i$. Hence $\pm i$ must correspond to $\pm\lambda \in A$. ∎

THE PINOR REPRESENTATIONS

The irreducible representations of the Clifford algebra $\mathrm{Cl}(r,s)$ are called the *pinor representations*, while the irreducible representations of the even parts $\mathrm{Cl}(r,s)^{\mathrm{even}}$ are called the *spinor representations*.

Definition 11.10. *The pinor representation of* $\mathrm{Cl}(r,s)$ *and the space of pinors* $\mathbf{P}(r,s)$ *are defined as follows. (If the dimension $n \equiv r + s$ is odd, assume a choice of unit volume element for* $\mathbf{R}(r,s)$*—an orientation—has been made.)*

For r − s = 0, 6 mod 8, *an irreducible representation*

$$\mathrm{Cl}(r,s) \cong \mathrm{End}_{\mathbf{R}}(P),$$

is called the pinor representation, and the real vector space $\mathbf{P}$ *is called the space of pinors.*

For r − s = 2, 4 mod 8, an *irreducible* **H**-*representation*

$$\mathrm{Cl}(r, s) \cong \mathrm{End}_{\mathbf{H}}(\mathbf{P}),$$

is called the pinor representation, and the right **H**-*space* **P** *is called the space of pinors.*

For r − s = 1, 5 mod 8, an *irreducible* **C**-*representation*

$$\mathrm{Cl}(r, s) \cong \mathrm{End}_{\mathbf{C}}(\mathbf{P}),$$

with the unit volume element in Cl(r, s) *corresponding to the complex structure on* **P**, *is called the pinor representation, and the complex vector space* **P** *is called the space of pinors.*

For r − s = 3 mod 8, an *irreducible* **H**-*representation*

$$\mathrm{Cl}(r, s)^{+} \cong \mathrm{End}_{\mathbf{H}}(\mathbf{P}_{+})$$

is called the positive pinor representation, while an irreducible **H**-*representation* Cl(r, s)⁻ ≅ End_H(**P**₋) *is called the negative pinor representation. The right* **H**-*space* **P** ≡ **P**₊ ⊕ **P**₋ *is called the space of pinors. Note that*

$$\mathrm{Cl}(r, s) \cong \mathrm{End}_{\mathbf{H}}(\mathbf{P}_{+}) \oplus \mathrm{End}_{\mathbf{H}}(\mathbf{P}_{-}),$$

and that the unit volume element is

$$\begin{pmatrix} 1 & 0 \\ 0 & -1 \end{pmatrix} \in \mathrm{End}_{\mathbf{H}}(\mathbf{P}_{+}) \oplus \mathrm{End}_{\mathbf{H}}(\mathbf{P}_{-}).$$

For r − s = 7 mod 8, an *irreducible representation*

$$\mathrm{Cl}(r, s)^{+} \cong \mathrm{End}_{\mathbf{R}}(\mathbf{P}_{+})$$

is called the positive pinor representation, while an irreducible representation

$$\mathrm{Cl}(r, s)^{-} \cong \mathrm{End}_{\mathbf{R}}(\mathbf{P}_{-})$$

is called the negative pinor representation. The real vector space **P** ≡ **P**₊ ⊕ **P**₋ *is called the space of pinors. Note that*

$$\mathrm{Cl}(r, s) \cong \mathrm{End}(\mathbf{P}_{+}) \oplus \mathrm{End}(\mathbf{P}_{-})$$

and that the unit volume element is

$$\begin{pmatrix} 1 & 0 \\ 0 & -1 \end{pmatrix} \in \mathrm{End}_{\mathbf{R}}(\mathbf{P}_{+}) \oplus \mathrm{End}_{\mathbf{R}}(\mathbf{P}_{-}).$$

Now Theorem 11.3 may be rewritten:

Theorem 11.3 (Pinor Version). *The Clifford algebras* $\mathrm{Cl}(r,s)$ *are isomorphic, as an associative algebras with unit, to the endomorphism algebras listed below.*

$r - s \bmod 8,$

$0, 6$	$\mathrm{Cl}(r,s) \cong \mathrm{End}_{\mathbf{R}}(\mathbf{P})$
$2, 4$	$\mathrm{Cl}(r,s) \cong \mathrm{End}_{\mathbf{H}}(\mathbf{P})$
$1, 5$	$\mathrm{Cl}(r,s) \cong \mathrm{End}_{\mathbf{C}}(\mathbf{P})$
3	$\mathrm{Cl}(r,s) \cong \mathrm{End}_{\mathbf{H}}(\mathbf{P}_+) \oplus \mathrm{End}_{\mathbf{H}}(\mathbf{P}_-)$
7	$\mathrm{Cl}(r,s) \cong \mathrm{End}_{\mathbf{R}}(\mathbf{P}_+) \oplus \mathrm{End}_{\mathbf{R}}(\mathbf{P}_-)$

Remark 11.11. If $r-s = 1, 5 \bmod 8$, the isomorphism $\mathrm{Cl}(r,s) \cong \mathrm{End}_{\mathbf{C}}(\mathbf{P})$ is required to map the unit volume element in $\mathrm{Cl}(r,s)$ to $i \in \mathrm{End}_{\mathbf{C}}(\mathbf{P})$. That is, this is the standard irreducible $\mathbf{C}$-representation of $\mathrm{Cl}(r,s)$. The conjugate $\mathbf{C}$-representation of $\mathrm{Cl}(r,s)$ defines the *conjugate pinor space* $\overline{\mathbf{P}}$:

$$(11.12) \qquad\qquad \mathrm{Cl}(r,s) \cong \mathrm{End}_{\mathbf{C}}(\overline{\mathbf{P}}),$$

where the unit volume element in $\mathrm{Cl}(r,s)$ maps to $-i \in \mathrm{End}_{\mathbf{C}}(\overline{\mathbf{P}})$. In all odd dimensions, $\mathrm{Cl}(r,s)$ has exactly two irreducible representations (with the appropriate notion of equivalence of representations) that define two spaces of pinors, $\mathbf{P}, \overline{\mathbf{P}}$ or $\mathbf{P}_+, \mathbf{P}_-$ as the case may be.

The question of how unique or canonical the space $\mathbf{P}$ of pinors is may be answered by the uniqueness results of Chapter 8 for intertwining operators, which are summarized in the next remark.

Remark 11.13. Suppose that A is one of the real algebras $M_N(F)$ with $F \equiv \mathbf{R}$ or $F = \mathbf{H}$, and that $\rho_1 : A \to \mathrm{End}_F(\mathbf{P}_1)$, $\rho_2 : A \to \mathrm{End}_F(\mathbf{P}_2)$ are two irreducible representations of A. Then there exists an F-linear map $f : \mathbf{P}_1 \to \mathbf{P}_2$ (the *intertwining operator*) with

$$\rho_2(a) = f \circ \rho_1(a) \circ f^{-1} \quad \text{for all } a \in A.$$

Moreover, f is unique up to a *real* scalar multiple $c \in \mathbf{R}^*$. If it were not for the flexibility of this scale $c \in \mathbf{R}^*$, the concept of pinor would be canonical.

Next, consider the algebra $A \equiv M_N(\mathbf{C})$. First recall that, for any $\mathbf{C}$-representation,

$$\rho : A \to \mathrm{End}_{\mathbf{C}}(W)$$

sends $i \in A$ to either $i \in \mathrm{End}_{\mathbf{C}}(W)$ or $-i \in \mathrm{End}_{\mathbf{C}}(W)$, and that (up to $\mathbf{C}$-equivalence) there are exactly two irreducible $\mathbf{C}$-representations:

$$\rho : A \cong \mathrm{End}_{\mathbf{C}}(W) \quad \text{where } \rho(i) = i \ (standard), \text{ and}$$
$$\overline{\rho} : A \cong \mathrm{End}_{\mathbf{C}}(W) \quad \text{where } \rho(i) = -i \ (conjugate).$$

Suppose $\rho_1 : A \cong \mathrm{End}_{\mathbf{C}}(\mathbf{P}_1)$ and $\rho_2 : A \cong \mathrm{End}_{\mathbf{C}}(\mathbf{P}_2)$ are two equivalent **C**-representations of A. Then by definition there exists a **C**-linear map $f : \mathbf{P}_1 \to \mathbf{P}_2$ (*the intertwining operator*) with

$$\rho_2(a) = f \circ \rho_1(a) \circ f^{-1}, \quad \text{for all } a \in A.$$

Moreover, f is unique up to a complex scalar multiple $c \in \mathbf{C}^*$. Again, if it were not for the flexibility of this scale $c \in \mathbf{C}^*$, the map f would be unique, making the concept of the pinor space $\mathbf{P}$ canonical.

In summary, the pinor space $\mathbf{P}$ is "projectively canonical."

THE SPINOR REPRESENTATIONS

The definition of the space of spinors is analogous to Definition 11.10 for pinors, except that now a choice of volume element is required for even dimensions.

Definition 11.14. *The spinor representation of* $\mathrm{Cl}(r,s)^{\mathrm{even}}$ *and the space of spinors* $\mathbf{S}(r,s)$ *is defined as follows. (If the dimension* $n \equiv r + s$ *is even, assume that a choice of unit volume element for* $\mathbf{R}(r,s)$—*an orientation*— *has been made.)*

For r − s = 1, 7 mod 8, *an irreducible representation*

$$\rho : \mathrm{Cl}(r,s)^{\mathrm{even}} \cong \mathrm{End}_{\mathbf{R}}(\mathbf{S})$$

is called the spinor representation, and the real vector space $\mathbf{S}$ *is called the space of spinors.*

For r − s = 3, 5 mod 8, *an irreducible* **H**-*representation*

$$\rho : \mathrm{Cl}(r,s)^{\mathrm{even}} \cong \mathrm{End}_{\mathbf{H}}(\mathbf{S})$$

is called the spinor representation, and the right **H**-*vector space* $\mathbf{S}$ *is called the space of spinors.*

For r − s = 2, 6 mod 8, *an irreducible* **C**-*representation*

$$\rho : \mathrm{Cl}(r,s)^{\mathrm{even}} \cong \mathrm{End}_{\mathbf{C}}(\mathbf{S}),$$

with the unit volume element λ *equal to the complex structure* i *on* $\mathbf{S}$, *is called the spinor representation, and the complex vector space* $\mathbf{S}$ *is called the space of spinors. An irreducible* **C**-*representation*

$$\overline{\rho} : \mathrm{Cl}(r,s)^{\mathrm{even}} \cong \mathrm{End}_{\mathbf{C}}(\overline{\mathbf{S}}),$$

with the unit volume element λ equal to $-i$ where i is the complex structure on $\overline{S}$, is called the conjugate spinor representation, and $\overline{S}$ is called the space of conjugate spinors.

For $r - s = 0$ mod 8, *an irreducible representation*

$$\rho_+ : \mathrm{Cl}(r,s)_+^{\mathrm{even}} \cong \mathrm{End}_{\mathbf{R}}(S_+)$$

is called the positive spinor representation, while an irreducible representation

$$\rho_- : \mathrm{Cl}(r,s)_-^{\mathrm{even}} \cong \mathrm{End}_{\mathbf{R}}(S_-)$$

is called the negative spinor representation. The real vector space $S \equiv S_+ \oplus S_-$ is called the space of spinors.

For $r - s = 4$ mod 8, *an irreducible $\mathbf{H}$-representation*

$$\rho_+ : \mathrm{Cl}(r,s)_+^{\mathrm{even}} \cong \mathrm{End}_{\mathbf{H}}(S_+)$$

is called the positive spinor representation, while an irreducible $\mathbf{H}$-representation

$$\rho_- : \mathrm{Cl}(r,s)_-^{\mathrm{even}} \cong \mathrm{End}_{\mathbf{H}}(S_-)$$

is called the negative spinor representation. The right $\mathbf{H}$-vector space $S \equiv S_+ \oplus S_-$ is called the space of spinors.

The spinor representation(s) of $\mathrm{Cl}(r,s)^{\mathrm{even}}$ restrict to representations of the group $\mathrm{Spin}(r,s) \subset \mathrm{Cl}(r,s)^{\mathrm{even}}$. It is convenient to use the same symbol for these representations of $\mathrm{Spin}(r,s)$ (i.e., ρ or $\rho, \overline{\rho}$ or ρ_+, ρ_-, as the case may be). Now Corollary 11.8 may be rewritten:

Corollary 11.15. (The spin representations).

$r - s$ mod 8 :

0	$\rho_+ \oplus \rho_- : \mathrm{Spin}(r,s) \subset \mathrm{Cl}(r,s)^{\mathrm{even}}$
	$\cong \mathrm{End}_{\mathbf{R}}(S_+) \oplus \mathrm{End}_{\mathbf{R}}(S_-)$
	$\rho_\pm : \mathrm{Cl}(r,s)_\pm^{\mathrm{even}} \cong \mathrm{End}_{\mathbf{R}}(S_\pm)$
4	$\rho_+ \oplus \rho_- : \mathrm{Spin}(r,s) \subset \mathrm{Cl}(r,s)^{\mathrm{even}}$
	$\cong \mathrm{End}_{\mathbf{H}}(S_+) \oplus \mathrm{End}_{\mathbf{H}}(S_-)$
	$\rho_\pm : \mathrm{Cl}(r,s)_\pm^{\mathrm{even}} \cong \mathrm{End}_{\mathbf{H}}(S_\pm)$
2, 6	$\rho \quad : \mathrm{Spin}(r,s) \subset \mathrm{Cl}(r,s)^{\mathrm{even}} \cong \mathrm{End}_{\mathbf{C}}(S)$
	$\overline{\rho} \quad : \mathrm{Spin}(r,s) \subset \mathrm{Cl}(r,s)^{\mathrm{even}} \cong \mathrm{End}_{\mathbf{C}}(\overline{S})$
1, 7	$\rho \quad : \mathrm{Spin}(r,s) \subset \mathrm{Cl}(r,s)^{\mathrm{even}} \cong \mathrm{End}_{\mathbf{R}}(S)$
3, 5	$\rho \quad : \mathrm{Spin}(r,s) \subset \mathrm{Cl}(r,s)^{\mathrm{even}} \cong \mathrm{End}_{\mathbf{H}}(S)$

THE FIRST PROOF

We shall give two proofs of the important Theorem 11.3. The first proof is by induction on the dimension $n \equiv r + s$ and is based on the following three lemmas.

Lemma 11.16. $\mathrm{Cl}(1,0) \cong \mathbf{C}, \mathrm{Cl}(0,1) \cong \mathbf{R} \oplus \mathbf{R}, \mathrm{Cl}(2,0) \cong \mathbf{H}, \mathrm{Cl}(1,1) \cong M_2(\mathbf{R}), \mathrm{Cl}(0,2) \cong M_2(\mathbf{R})$.

Proof: A set of "γ matrices" for each one of these five Clifford algebras $\mathrm{Cl}(r, s)$ is listed below:

For $\mathrm{Cl}(1,0)$, $\quad e_1 \equiv i \in \mathbf{C}$.

For $\mathrm{Cl}(0,1)$, $\quad e_1 \equiv \tau \in \mathbf{L}$.

For $\mathrm{Cl}(2,0)$, $\quad e_1 \equiv j, \; e_2 \equiv k \in \mathbf{H}$.

For $\mathrm{Cl}(1,1)$, $\quad e_1 \equiv \begin{pmatrix} 0 & -1 \\ 1 & 0 \end{pmatrix}, \; e_2 \equiv \begin{pmatrix} 0 & 1 \\ 1 & 0 \end{pmatrix} \in M_2(\mathbf{R})$.

For $\mathrm{Cl}(0,2)$, $\quad e_1 \equiv \begin{pmatrix} 0 & 1 \\ 1 & 0 \end{pmatrix}, \; e_2 \equiv \begin{pmatrix} 1 & 0 \\ 0 & -1 \end{pmatrix} \in M_2(\mathbf{R})$. ∎

Lemma 11.17.

(11.18) $\mathrm{Cl}(r+1, s+1) \cong \mathrm{Cl}(r, s) \otimes \mathrm{Cl}(1,1) \cong \mathrm{Cl}(r, s) \otimes_{\mathbf{R}} M_2(\mathbf{R})$.

(11.19) $\quad \mathrm{Cl}(s, r+2) \cong \mathrm{Cl}(r, s) \otimes \mathrm{Cl}(0,2) \cong \mathrm{Cl}(r, s) \otimes_{\mathbf{R}} M_2(\mathbf{R})$.

(11.20) $\quad \mathrm{Cl}(s+2, r) \cong \mathrm{Cl}(r, s) \otimes \mathrm{Cl}(2,0) \cong \mathrm{Cl}(r, s) \otimes_{\mathbf{R}} \mathbf{H}$.

Proof: Suppose $e_1, \ldots, e_n$ is an orthonormal basis for $\mathbf{R}(r, s) \subset \mathrm{Cl}(r, s)$. Let u_1, u_2 denote an orthonormal basis for $V \subset \mathrm{Cl}(V)$, where V is a two-dimensional inner product space, and let $\eta \equiv u_1 u_2$ denote the unit volume element.

Define $W \subset \mathrm{Cl}(r, s) \otimes_{\mathbf{R}} \mathrm{Cl}(V)$ to be the inner product space with orthonormal basis

(11.21) $\qquad e_1 \otimes \eta, \; \ldots, \; e_n \otimes \eta, \; 1 \otimes u_1, \; 1 \otimes u_2$,

and signature given by

(11.22) $\quad \begin{aligned} \|e_j \otimes \eta\| &= \eta^2 \|e_j\|, \quad j = 1, \ldots, n, \\ \|1 \otimes u_j\| &= \quad \|u_j\|, \quad j = 1, 2. \end{aligned}$

Note that any two of the basis vectors in (11.21) anticommute and that

(11.23) $\quad \begin{aligned} (e_j \otimes \eta)^2 &= e_j^2 \otimes \eta^2 = -\eta^2 \|e_j\| = -\|e_j \otimes \eta\|, \quad j = 1, \cdots, n \\ (1 \otimes u_j)^2 &= 1 \otimes u_j^2 = - \quad \|u_j\| = -\|1 \otimes u_j\|, \quad j = 1, 2. \end{aligned}$

Therefore, by the Fundamental Lemma of Clifford algebras the natural inclusion $\phi : W \to \mathrm{Cl}(r, s) \otimes \mathrm{Cl}(V)$ extends to an algebra homomorphism

$$\phi : \mathrm{Cl}(W) \to \mathrm{Cl}(r, s) \otimes \mathrm{Cl}(V).$$

This map ϕ must be surjective, since the basis for W generates the algebra $\mathrm{Cl}(r, s) \otimes \mathrm{Cl}(V)$. Therefore, by a dimension count, ϕ must be injective and hence provides the desired isomorphism

$$\mathrm{Cl}(W) \cong \mathrm{Cl}(r, s) \otimes \mathrm{Cl}(V).$$

The signature of W follows from (11.22), reversing that of $\mathbf{R}(r, s)$, when $V \equiv \mathbf{R}(2, 0)$ or $V \equiv \mathbf{R}(0, 2)$, since in both these cases $\eta^2 = -1$. ∎

Lemma 11.24.

(11.25) $M_k(F) \otimes_{\mathbf{R}} M_m(\mathbf{R}) \cong M_{km}(F)$ for $F = \mathbf{R}, \mathbf{C},$ or $\mathbf{H}$.

(11.26) $\mathbf{H} \otimes_{\mathbf{R}} \mathbf{H} \cong M_4(\mathbf{R}).$

(11.27) $\mathbf{C} \otimes_{\mathbf{R}} \mathbf{H} \cong M_2(\mathbf{C}).$

(11.28) $\mathbf{C} \otimes_{\mathbf{R}} \mathbf{C} \cong \mathbf{C} \oplus \mathbf{C}.$

Proof: The isomorphism (11.25) is a consequence of these two special cases:

(11.25a) $M_k(\mathbf{R}) \otimes M_m(\mathbf{R}) \cong M_{km}(\mathbf{R})$

and

(11.25b) $F \otimes M_m(\mathbf{R}) \cong M_m(F)$ for $F \equiv \mathbf{C}$ or $\mathbf{H}$.

The algebra homomorphism $\phi : \mathbf{H} \otimes_{\mathbf{R}} \mathbf{H} \to M_4(\mathbf{R})$, defined by

(11.29) $\phi(p \otimes q)(x) \equiv px\overline{q}$ for all $x \in \mathbf{R}^4 \cong \mathbf{H},$

is an isomorphism since ϕ is an isometry. (See Problem 1.)

    Let $\phi(i \otimes 1)$ (left multiplication by i) determine the complex structure on $\mathbf{H} \cong \mathbf{R}^4 \cong \mathbf{C}^2$. Then the subalgebra $\mathbf{C} \otimes_{\mathbf{R}} \mathbf{H}$ of $\mathbf{H} \otimes_{\mathbf{R}} \mathbf{H}$ corresponds to the subalgebra $M_2(\mathbf{C})$ of $M_4(\mathbf{R})$ under the map ϕ. This proves (11.27).

    Since $\mathbf{C}^2 \cong \mathbf{C} \oplus \mathbf{C}j \cong \mathbf{H}$, via the complex structure $\phi(i \otimes 1)$ (left multiplication by i), the matrix $\phi(1 \otimes i)$ (right multiplication by i) equals the 2×2 complex matrix $\begin{pmatrix} i & \\ & -i \end{pmatrix}$. (Note $(z + wj)i = iz - i(wj)$. Thus, $\mathbf{C} \otimes \mathbf{C} \cong \mathbf{C} \oplus \mathbf{C}$, contained in $M_2(\mathbf{C})$ as the diagonal 2×2 matrices. ∎

The reader is invited to give the proof of Theorem 11.3, based on these three lemmas, in Problem 2. A second proof of Theorem 11.3 will be given in the course of developing additional information about the Clifford algebras in the next chapter.

THE SPINOR STRUCTURE MAP ON $\mathbf{P}(r, s)$

Definition 11.30. *An invertible real linear map s on* $\mathbf{P}(r, s)$ *is called a spinor structure map if* $s^2 = \pm 1$ *and*

$$(11.31) \qquad \tilde{a} = sas^{-1} \quad \text{for all } a \in \text{Cl}(r, s) \subset \text{End}_{\mathbf{R}}(\mathbf{P}).$$

Thus, a spinor structure map on the space $\mathbf{P}(r, s)$ of pinors is sufficient extra structure to determine $\text{Cl}(r, s)^{\text{even}}$ as a subalgebra of $\text{End}_F(\mathbf{P}(r, s))$. Recall that, if the dimension $n \equiv r + s$ is even, then

$$(11.32) \qquad \tilde{a} = \lambda a \lambda^{-1} \quad \text{for all } a \in \text{Cl}(r, s),$$

while if the dimension $n \equiv r + s$ is odd, then

$$(11.33) \qquad \lambda a = a\lambda \quad \text{for all } a \in \text{Cl}(r, s);$$

where λ is a unit volume element.

EVEN DIMENSIONS

If the dimension $n \equiv r + s$ is even, then the spinor structure map s is defined to be a choice λ of the two unit volume elements. The next theorem describes how $s \equiv \lambda$ determines the spinor space $\mathbf{S}(r, s)$ from the pinor space $\mathbf{P}(r, s)$. This theorem is stated so that each of the four cases $r - s = 0, 2, 4, 6 \mod 8$ can be referred to independently of the other three cases. Consequently, there is a certain amount of repetition.

Theorem 11.34 (The Spinor and Pinor Representations in Even Dimensions). *Define the pinor structure map s to be a choice of unit volume element* λ.

For r − s = 0 mod 8: *Consider the pinor representation*

$$(11.35) \qquad \text{Cl}(r, s) \cong \text{End}_{\mathbf{R}}(\mathbf{P}).$$

Define $\mathbf{S_+}$ and $\mathbf{S_-}$ to be the eigenspaces of λ:

(11.36) $\mathbf{S_\pm} \equiv \{x \in \mathbf{P} : \lambda x = \pm x\}.$

The decomposition

(11.37) $\mathbf{P} = \mathbf{S_+} \oplus \mathbf{S_-}$

determines a 2×2 blocking

(11.38) $A = \begin{pmatrix} a & b \\ c & d \end{pmatrix}$ for each $A \in \operatorname{End}_{\mathbf{R}}(\mathbf{P}),$

and isomorphisms:

(11.39) $\operatorname{Cl}(r,s)^{\text{even}} \cong \left\{ \begin{pmatrix} a & 0 \\ 0 & b \end{pmatrix} : a \in \operatorname{End}_{\mathbf{R}}(\mathbf{S_+}),\ b \in \operatorname{End}_{\mathbf{R}}(\mathbf{S_-}) \right\},$

(11.40) $\operatorname{Cl}(r,s)^{\text{even}}_+ \cong \left\{ \begin{pmatrix} a & 0 \\ 0 & 0 \end{pmatrix} : a \in \operatorname{End}_{\mathbf{R}}(\mathbf{S_+}) \right\} \cong \operatorname{End}_{\mathbf{R}}(\mathbf{S_+}),$

(11.41) $\operatorname{Cl}(r,s)^{\text{even}}_- \cong \left\{ \begin{pmatrix} 0 & 0 \\ 0 & b \end{pmatrix} : b \in \operatorname{End}_{\mathbf{R}}(\mathbf{S_-}) \right\} \cong \operatorname{End}_{\mathbf{R}}(\mathbf{S_-})$

that are the spinor representations. Note that

(11.42) $\lambda = \begin{pmatrix} 1 & 0 \\ 0 & -1 \end{pmatrix}$

and that

(11.43) $\dim_{\mathbf{R}} \mathbf{S_+} = \dim_{\mathbf{R}} \mathbf{S_-}.$

For $r - s = 4 \bmod 8$: Consider the pinor representation

(11.44) $\operatorname{Cl}(r,s) \cong \operatorname{End}_{\mathbf{H}}(\mathbf{P}).$

Define $\mathbf{S_+}$ and $\mathbf{S_-}$ to be the eigenspaces of λ:

(11.45) $\mathbf{S_\pm} \equiv \{x \in \mathbf{P} : \lambda x = \pm x\}.$

The decomposition

$$(11.46) \qquad \mathbf{P} = \mathbf{S}_+ \oplus \mathbf{S}_-$$

determines a 2×2 blocking

$$(11.47) \qquad A = \begin{pmatrix} a & b \\ c & d \end{pmatrix} \quad \text{for each } A \in \text{End}_{\mathbf{H}}(\mathbf{P}),$$

and isomorphisms

$$(11.48) \qquad \text{Cl}(r,s)^{\text{even}} \cong \left\{ \begin{pmatrix} a & 0 \\ 0 & b \end{pmatrix} : a \in \text{End}_{\mathbf{H}}(S_+), \ b \in \text{End}_{\mathbf{R}}(S_-) \right\}$$
$$\cong \text{End}_{\mathbf{H}}(\mathbf{S}_+) \oplus \text{End}_{\mathbf{H}}(\mathbf{S}_-),$$

$$(11.49) \qquad \text{Cl}(r,s)^{\text{even}}_+ \cong \left\{ \begin{pmatrix} a & 0 \\ 0 & 0 \end{pmatrix} : a \in \text{End}_{\mathbf{H}}(S_+) \right\} \cong \text{End}_{\mathbf{H}}(\mathbf{S}_+),$$

$$(11.50) \qquad \text{Cl}(r,s)^{\text{even}}_- \cong \left\{ \begin{pmatrix} 0 & 0 \\ 0 & b \end{pmatrix} : b \in \text{End}_{\mathbf{H}}(S_-) \right\} \cong \text{End}_{\mathbf{H}}(\mathbf{S}_-)$$

that are the spinor representations. Note that

$$(11.51) \qquad \lambda \cong \begin{pmatrix} 1 & 0 \\ 0 & -1 \end{pmatrix}$$

and that

$$(11.52) \qquad \dim_{\mathbf{H}} \mathbf{S}_+ = \dim_{\mathbf{H}} \mathbf{S}_-.$$

For $\mathbf{r} - \mathbf{s} = \mathbf{2}$ mod 8: *Consider the pinor representation*

$$(11.53) \qquad \text{Cl}(r,s) \cong \text{End}_{\mathbf{H}}(\mathbf{P}),$$

with the right $\mathbf{H}$-structure on $\mathbf{P}$ given by I, J, K. Utilizing the complex structure I on $\mathbf{P}$, define $\mathbf{S}$ and $\overline{\mathbf{S}}$ to be the eigenspaces of λ:

$$(11.54) \qquad \mathbf{S} \equiv \{ x \in \mathbf{P} : \lambda x = xI \}, \quad \overline{\mathbf{S}} \equiv \{ x \in \mathbf{P} : \lambda x = -xI \}.$$

Then $\text{Cl}(r,s)^{\text{even}} \subset \text{End}_{\mathbf{H}}(\mathbf{P})$ maps $\mathbf{S}$ to $\mathbf{S}$ and $\overline{\mathbf{S}}$ to $\overline{\mathbf{S}}$, and the induced isomorphisms

$$(11.55) \qquad \rho : \text{Cl}(r,s)^{\text{even}} \cong \text{End}_{\mathbf{C}}(\mathbf{S}), \quad \overline{\rho} : \text{Cl}(r,s)^{\text{even}} \cong \text{End}_{\mathbf{C}}(\overline{\mathbf{S}})$$

are the spinor representations. As **R**-*representations,* ρ *and* $\overline{\rho}$ *are equivalent via the intertwining operator*

(11.56) $J : \mathbf{S} \to \overline{\mathbf{S}}$:

For r − s = 6 mod 8: *Consider the pinor representation*

(11.57) $\mathrm{Cl}(r, s) \cong \mathrm{End}_{\mathbf{R}}(\mathbf{P})$.

Define $\mathbf{S}$ *and* $\overline{\mathbf{S}}$ *to be the eigenspaces of* λ:

(11.58) $\mathbf{S} \equiv \{s \in \mathbf{P} \otimes_{\mathbf{R}} \mathbf{C} : \lambda x = ix\}$, $\overline{\mathbf{S}} \equiv \{s \in \mathbf{P} \otimes_{\mathbf{R}} \mathbf{C} : \lambda x = -ix\}$.

Then $\mathrm{Cl}(r, s)^{\mathrm{even}} \subset \mathrm{End}_{\mathbf{R}}(\mathbf{P}) \subset \mathrm{End}_{\mathbf{C}}(\mathbf{P} \otimes_{\mathbf{R}} \mathbf{C})$ *maps* $\mathbf{S}$ *to* $\mathbf{S}$ *and* $\overline{\mathbf{S}}$ *to* $\overline{\mathbf{S}}$, *and the induced isomorophisms*

(11.59) $\rho : \mathrm{Cl}(r, s)^{\mathrm{even}} \cong \mathrm{End}_{\mathbf{C}}(\mathbf{S})$, $\overline{\rho} : \mathrm{Cl}(r, s)^{\mathrm{even}} \cong \mathrm{End}_{\mathbf{C}}(\overline{\mathbf{S}})$

are the spinor representations. As **R**-*representations,* ρ *and* $\overline{\rho}$ *are equivalent via the intertwining operator*

(11.60) $C : \mathbf{S} \to \overline{\mathbf{S}}$,

the natural conjugation on $\mathbf{P} \otimes_{\mathbf{R}} \mathbf{C}$.

Proof: The cases $r - s = 0, 4, 6 \bmod 8$ are exercises (see Problem 6).

Suppose $r - s = 2 \bmod 8$. Since $A \in \mathrm{Cl}(r, s)^{\mathrm{even}}$ commutes with λ, A preserves the eigenspaces $\mathbf{S}$ and $\overline{\mathbf{S}}$ of λ. Let $a \equiv \rho(A)$ denote $A \in \mathrm{Cl}(r, s)^{\mathrm{even}}$ restricted to $\mathbf{S}$, and $\overline{a} \equiv \overline{\rho}(A)$ denote $A \in \mathrm{Cl}(r, s)^{\mathrm{even}}$ restricted to $\overline{\mathbf{S}}$. Note that $J : \mathbf{S} \to \overline{\mathbf{S}}$ is an anticomplex linear isomorphism. Since A commutes with J,

(11.61) $\overline{a}(xJ) = (ax)J$ for all $x \in \mathbf{S}$.

The theorem now follows easily. Note that $\rho(\lambda) = I$ and $\overline{\rho}(\lambda) = -I$. ∎

ODD DIMENSIONS

Now we turn to the odd dimensional cases. Several questions arise. Does a spinor structure map exist? How (non-) unique is it? Finally, how does s determine the spinor space $\mathbf{S}(r, s)$? These questions are most easily answered by turning to the inverse question first: How to construct the pinor representation from the spinor representation? The unit volume element λ provides the key to the answer to this question, since λ commutes with $\mathrm{Cl}(r, s)^{\mathrm{even}}$.

Lemma 11.61. *If* $r - s = 1, 5$ mod 8, *then*

(11.63) $\mathrm{Cl}(r, s) \cong \mathrm{Cl}(r, s)^{\mathrm{even}} \otimes_{\mathbf{R}} \mathbf{C}$,   *with* $\lambda = i$.

If $r - s = 3, 7$ mod 8, *then*

(11.64) $\mathrm{Cl}(r, s) \cong \mathrm{Cl}(r, s)^{\mathrm{even}} \otimes_{\mathbf{R}} \mathbf{L}$,   *with* $\lambda = \tau$,

or equivalently,

(11.64′) $\mathrm{Cl}(r, s) \cong \mathrm{Cl}(r, s)^{\mathrm{even}} \otimes (\mathbf{R} \oplus \mathbf{R})$,   *with* $\lambda = (1, -1)$.

These algebra isomorphisms are the identity on $\mathrm{Cl}(r, s)^{\mathrm{even}}$.

Proof: Since $\mathrm{Cl}(r, s) \cong \mathrm{Cl}(r, s)^{\mathrm{even}} \oplus \lambda \, \mathrm{Cl}(r, s)^{\mathrm{even}}$ and λ commutes with $\mathrm{Cl}(r, s)^{\mathrm{even}}$,

$$\mathrm{Cl}(r, s) \cong \mathrm{Cl}(r, s)^{\mathrm{even}} \otimes_{\mathbf{R}} A,$$

where A is the two-dimensional algebra $\mathrm{span}_{\mathbf{R}}\{1, \lambda\}$.

If $r - s = 1, 5$ mod 8, then $\lambda^2 = -1$ and $A \cong \mathbf{C}$ with $\lambda = i$; while if $r - s = 3, 7$ mod 8, then $\lambda^2 = 1$ and $A \cong \mathbf{L} \cong \mathbf{R} \oplus \mathbf{R}$ with $\lambda = \tau = (1, -1)$. ∎

This lemma provides part of the second proof of Theorem 11.3 that is given in the next chapter.

Corollary 11.65. The even dimensional cases of Theorem 11.3 imply the odd dimensional cases of Theorem 11.3.

Proof: Of course, $M_N(F) \otimes_{\mathbf{R}} (\mathbf{R} \oplus \mathbf{R}) \cong M_N(F) \oplus M_N(F)$ and $M_N(\mathbf{R}) \otimes_{\mathbf{R}} \mathbf{C} \cong M_N(\mathbf{C})$. Exactly as in Lemma 11.24, $M_N(\mathbf{H}) \otimes_{\mathbf{R}} \mathbf{C} \cong M_{2N}(\mathbf{C})$. ∎

Theorem 11.66 (The Spinor and Pinor Representations in Odd Dimensions).

For $r - s = 1$ mod 8: *Given the spinor representation*

(11.67) $\mathrm{Cl}(r, s)^{\mathrm{even}} \cong \mathrm{End}_{\mathbf{R}}(\mathbf{S})$,

define

(11.68) $\mathbf{P} \equiv \mathbf{S} \otimes_{\mathbf{R}} \mathbf{C}$.

Since $\mathrm{Cl}(r, s) = \mathrm{Cl}(r, s)^{\mathrm{even}} \otimes_{\mathbf{R}} \mathbf{C}$, *with* $\lambda = i$, *there is an induced isomorphism*

$$\mathrm{Cl}(r, s) \cong \mathrm{End}_{\mathbf{C}}(\mathbf{P})$$

that is the pinor representation. The natural conjugation s on $\mathbf{P} \equiv \mathbf{S} \otimes_{\mathbf{R}} \mathbf{C}$
is a spinor structure map. Note

$$\mathbf{S} \in \{x \equiv \mathbf{P} : sx = x\}.$$

All other spinor structure maps are of the form $e^{i\theta}s, \theta \in \mathbf{R}$.

For $r - s = 3$ mod 8: *Given the spinor representation*

(11.69) $\qquad\qquad\qquad \mathrm{Cl}(r, s)^{\mathrm{even}} \cong \mathrm{End}_{\mathbf{H}}(\mathbf{S}),$

define $\mathbf{P}_{\pm} \equiv \mathbf{S}$ *so that* $\mathbf{P} \equiv \mathbf{S} \oplus \mathbf{S}$. *Since* $\mathrm{Cl}(r, s) \cong \mathrm{Cl}(r, s)^{\mathrm{even}} \otimes (\mathbf{R} \oplus \mathbf{R})$,
with $\lambda = (1, -1)$, *there is an induced isomorphism*

$$\mathrm{Cl}(r, s) \cong \mathrm{End}_{\mathbf{H}}(\mathbf{S}) \oplus \mathrm{End}_{\mathbf{H}}(\mathbf{S})$$

that is the pinor representation. The natural reflection

$$s \equiv \begin{pmatrix} 0 & 1 \\ 1 & 0 \end{pmatrix}$$

that maps $(x, y) \in \mathbf{P}$ *to* $(y, x) \in \mathbf{P}$ *is a spinor structure map. All other
spinor structure maps are of the form*

$$\pm e^{\theta \tau} s = \pm \begin{pmatrix} 0 & e^{\theta \tau} \\ e^{-\theta \tau} & 0 \end{pmatrix}$$

on $\mathbf{P} \equiv \mathbf{S} \oplus \mathbf{S}$.

For $r - s = 5$ mod 8: *Given the spinor representation*

(11.70) $\qquad\qquad\qquad \mathrm{Cl}(r, s)^{\mathrm{even}} \cong \mathrm{End}_{\mathbf{H}}(\mathbf{S}),$

define $\mathbf{P}$ *to be* $\mathbf{S}$ *with the complex structure* I. *Since*

$$\mathrm{Cl}(r, s) \cong \mathrm{Cl}(r, s)^{\mathrm{even}} \otimes \mathbf{C},$$

with $\lambda = I$, *there is an induced isomorphism*

(11.71) $\qquad\qquad\qquad \mathrm{Cl}(r, s) \cong \mathrm{End}_{\mathbf{C}}(\mathbf{P})$

*that is the pinor representation. Given the isomorphism (11.70), the iso-
morphism (11.71) induced by (11.70) is given explicitaly by*

(11.72) $\qquad\qquad (a \otimes z)(x) = axz \quad \text{for all } x \in \mathbf{P},$

where $a \in \mathrm{Cl}(r,s)^{\text{even}} \cong \mathrm{End}_{\mathbf{H}}(\mathbf{S})$ and $z \in \mathbf{C} = \mathrm{span}_{\mathbf{R}}\{1, I\}$. *The map J (right multiplication) is a spinor structure map and all other spinor structure maps are of the form* $e^{I\theta}J$.

For r − s = 7 mod 8: *Given the spinor representation*

(11.73) $$\mathrm{Cl}(r, s)^{\text{even}} \cong \mathrm{End}_{\mathbf{R}}(\mathbf{S}),$$

define $\mathbf{P}_{\pm} \equiv \mathbf{S}$ *so that* $\mathbf{P} \equiv \mathbf{S} \oplus \mathbf{S}$. *Since* $\mathrm{Cl}(r, s) \cong \mathrm{Cl}(r, s)^{\text{even}} \otimes (\mathbf{R} \oplus \mathbf{R})$, *with* $\lambda = (1, -1)$, *there is an induced isomorphism*

$$\mathrm{Cl}(r, s) \cong \mathrm{End}_{\mathbf{R}}(\mathbf{S}) \oplus \mathrm{End}_{\mathbf{R}}(\mathbf{S})$$

that is the pinor representation. The natural reflection

$$s \equiv \begin{pmatrix} 0 & 1 \\ 1 & 0 \end{pmatrix}$$

that maps $(x, y) \in \mathbf{P}$ *to* $(y, x) \in \mathbf{P}$ *is a spinor structure map. All other spinor structure maps are of the form*

$$\pm e^{\theta\tau}s = \pm \begin{pmatrix} 0 & e^{\theta\tau} \\ e^{-\theta\tau} & 0 \end{pmatrix}$$

on $\mathbf{P} \cong \mathbf{S} \oplus \mathbf{S}$.

Proof: Except for the uniqueness results for a spinor structure map s, all other parts of the theorem are immediate consequences of Problem 1(c) and Problem 7.

Note that (n odd) a spinor structure map s for $\mathrm{Cl}(r, s) \subset \mathrm{End}_{\mathbf{R}}(\mathbf{P})$ must lie outside of cen $\mathrm{Cl}(r, s) = \mathrm{span}_{\mathbf{R}}\{1, \lambda\}$ but still lie inside of the centralizer of $\mathrm{Cl}(r, s)^{\text{even}}$ in $\mathrm{End}_{\mathbf{R}}(\mathbf{P})$. This leaves very little choice for s, and the precise range of possibilities for a spinor structure map s can be deduced from the next lemma (Problem 8). ∎

Lemma 11.74.

$r - s$ mod 8	cen $\mathrm{Cl}(r, s) \equiv \mathrm{span}_{\mathbf{R}}\{1, \lambda\}$	centralizer of $\mathrm{Cl}(r, s)^{\text{even}}$ in $\mathrm{End}_{\mathbf{R}}(\mathbf{P}(r, s))$
1	$\mathbf{C}$ with $\lambda \equiv i$	$M_2(\mathbf{R})$
3	$\mathbf{R} \oplus \mathbf{R}$ with $\lambda \equiv \tau \equiv (1, -1)$	$M_2(\mathbf{R})$
5	$\mathbf{C}$ with $\lambda \equiv I$	$\mathbf{H}$
7	$\mathbf{R} \oplus \mathbf{R}$ with $\lambda \equiv \tau \equiv (1, -1)$	$M_2(\mathbf{R})$.

Proof: In the case $r - s = 1$ mod 8, the lemma is just the statement that

(11.75)   the centralizer of $\mathrm{End}_{\mathbf{R}}(\mathbf{S})$ in $\mathrm{End}_{\mathbf{R}}(\mathbf{S} \otimes_{\mathbf{R}} \mathbf{C})$ is $M_2(\mathbf{R})$.

In fact, $1, i, s$, and is form a vector basis for this centralizer, where s denotes the natural conjugation on $\mathbf{S} \otimes_{\mathbf{R}} \mathbf{C}$.

In the case $r - s = 5$ mod 8, the lemma follows from the fact that

(11.76)   the centralizer of $\mathrm{End}_{\mathbf{H}}(\mathbf{S})$ in $\mathrm{End}_{\mathbf{R}}(\mathbf{S})$ is $\mathbf{H}$, the field of
               right scalar multiplications,

which is part (b) of Lemma 8.25.

The cases $r - s = 3$ mod 8 and $r - s = 7$ mod 8 are similar. The centralizer of

$$\mathrm{End}_F(\mathbf{S}) \cong \left\{ \begin{pmatrix} a & 0 \\ 0 & a \end{pmatrix} : a \in \mathrm{End}_F(\mathbf{S}) \right\}$$

in $\mathrm{End}_{\mathbf{R}}(\mathbf{S} \oplus \mathbf{S})$ is $M_2(\mathbf{R})$ for both $F \equiv \mathbf{R}$ and $F \equiv \mathbf{H}$ since $\mathbf{H}$ has center $\mathbf{R}$. ■

PROBLEMS

1. Let $\mathbf{H} \otimes_{\mathbf{R}} \mathbf{H}$ have the inner product defined by $\langle a \otimes b, c \otimes d \rangle \equiv \langle a, c \rangle \langle b, d \rangle$. Let $M_4(\mathbf{R})$ have the natural positive definite inner product defined by

$$\langle A, B \rangle = \frac{1}{4} \text{ trace } AB^t.$$

   (a) Prove that the map $\phi : \mathbf{H} \otimes_{\mathbf{R}} \mathbf{H} \to M_4(\mathbf{R})$ defined by (11.29) preserves the inner products.

   (b) Prove that ϕ is an isomorphism.

   (c) Suppose $V_{\mathbf{H}}$ is a right $\mathbf{H}$-space with I, J, K a standard basis for the scalars. Let $V_{\mathbf{C}}$ denote $V_{\mathbf{H}}$ with the complex structure I. Show that $(a \otimes z)(x) \equiv axz$ for $a \in \mathrm{End}_{\mathbf{H}}(V_{\mathbf{H}})$, $z \in \mathbf{C} \cong \text{span } \{1, I\}$, $x \in V$, defines an isomorphism

$$\mathrm{End}_{\mathbf{H}}(V_{\mathbf{H}}) \otimes_{\mathbf{R}} \mathbf{C} \cong \mathrm{End}_{\mathbf{C}}(V_{\mathbf{C}}).$$

2. Using Lemma 11.16, Lemma 11.17, and Lemma 11.24, give the proof of Theorem 11.3.

3. Prove that Table 11.5 is symmetric about the lines $q = p + k$ with $k = 3$ mod 4, and that Cl$(p, q) \cong$ Cl$(p - 4, q + 4)$ as algebras.

4. Equip $\mathbf{C}^n$ with the standard $\mathbf{C}$-symmetric inner product

$$\langle z, w \rangle \equiv z_1 w_1 + \cdots + z_n w_n.$$

(a) Define the associated complex Clifford algebra, denoted $\text{Cl}_{\mathbf{C}}(n)$.

(b) Prove that $\text{Cl}_{\mathbf{C}}(1) \cong \mathbf{C} \oplus \mathbf{C}$ and $\text{Cl}_{\mathbf{C}}(2) \cong M_2(\mathbf{C})$ as complex associative algebras with unit.

(c) Prove that, as complex associative algebras with unit,

$$\text{Cl}_{\mathbf{C}}(2p) \cong M_N(\mathbf{C}) \quad \text{with } N = 2^p,$$

and

$$\text{Cl}_{\mathbf{C}}(2p + 1) \cong M_N(\mathbf{C}) \oplus M_N(\mathbf{C}) \quad \text{with } N \equiv 2^p,$$

by showing

$$\text{Cl}_{\mathbf{C}}(n + 2) \cong \text{Cl}_{\mathbf{C}}(n) \otimes \text{Cl}_{\mathbf{C}}(2).$$

(d) Prove that Cl$(p, q) \otimes_{\mathbf{R}} \mathbf{C} \cong \text{Cl}_{\mathbf{C}}(n)$ for all signatures in a given dimension $n \equiv p + q$.

5. Prove that Table 11.5 has the following periodicity isomorphisms:

$$\text{Cl}(p + 1, q + 1) \cong C(p, q) \otimes M_2(\mathbf{R}),$$
$$\text{Cl}(p + 8, q) \cong C(p, q) \otimes M_{16}(\mathbf{R}),$$
$$\text{Cl}(p, q + 8) \cong C(p, q) \otimes M_{16}(\mathbf{R}).$$

6. Give the proof of Theorem 11.34 for the case $r - s = 0$ mod 8. Moreover, verify that the isomorphism Cl$(r, s) \cong \text{End}_{\mathbf{R}}(\mathbf{P})$ determines the following isomorphisms:

$$\text{Cl}(r, s)^{\text{even}} \cong \left\{ \begin{pmatrix} a & 0 \\ 0 & b \end{pmatrix} : a \in \text{End}_{\mathbf{R}}(\mathbf{S}_+), \ b \in \text{End}_{\mathbf{R}}(\mathbf{S}_-) \right\},$$

$$\text{Cl}(r, s)^{\text{odd}} \cong \left\{ \begin{pmatrix} 0 & a \\ b & 0 \end{pmatrix} : a \in \text{Hom}_{\mathbf{R}}(\mathbf{S}_-, \mathbf{S}_+), b \in \text{Hom}_{\mathbf{R}}(\mathbf{S}_+, \ \mathbf{S}_-) \right\},$$

$$\text{Cl}(r, s)_+ \cong \left\{ \begin{pmatrix} a & 0 \\ b & 0 \end{pmatrix} : a \in \text{End}_{\mathbf{R}}(\mathbf{S}_+), \ b \in \text{Hom}_{\mathbf{R}}(\mathbf{S}_+, \mathbf{S}_-) \right\}.$$

$$\text{Cl}(r, s)_- \cong \left\{ \begin{pmatrix} 0 & a \\ 0 & b \end{pmatrix} : a \in \text{Hom}_{\mathbf{R}}(\mathbf{S}_+, \mathbf{S}_-), \ b \in \text{End}_{\mathbf{R}}(\mathbf{S}_-) \right\},$$

$$\text{Cl}(r, s)_+^{\text{even}} \cong \left\{ \begin{pmatrix} a & 0 \\ 0 & 0 \end{pmatrix} : a \in \text{End}_{\mathbf{R}}(\mathbf{S}_+) \right\},$$

$$\text{Cl}(r, s)_-^{\text{even}} \cong \left\{ \begin{pmatrix} 0 & 0 \\ 0 & b \end{pmatrix} : b \in \text{End}_{\mathbf{R}}(\mathbf{S}_-) \right\}.$$

7. (a) Given an algebra isomorphism $A \cong \text{End}_{\mathbf{R}}(\mathbf{S})$, show that there is an induced isomorphism $A \otimes_{\mathbf{R}} \mathbf{C} \cong \text{End}_{\mathbf{C}}(\mathbf{S} \otimes_{\mathbf{R}} \mathbf{C})$. If s denotes the natural conjugation on $\mathbf{S} \otimes_{\mathbf{R}} \mathbf{C}$, then show that for $a \in A$, $z \in \mathbf{C}$,

$$s(a \otimes z)s = a \otimes \bar{z}.$$

(b) Given an algebra isomorphism $A \cong \text{End}_{\mathbf{R}}(\mathbf{S})$, show that

$$A \otimes (\mathbf{R} \oplus \mathbf{R}) \cong A \oplus A \cong \left\{ \begin{pmatrix} a & 0 \\ 0 & b \end{pmatrix} : a, b \in A \right\} \subset \text{End}_{\mathbf{H}}(\mathbf{S} \oplus \mathbf{S}),$$

and that

$$s\begin{pmatrix} a & 0 \\ 0 & b \end{pmatrix} s = \begin{pmatrix} b & 0 \\ 0 & a \end{pmatrix}, \quad \text{with } s \equiv \begin{pmatrix} 0 & 1 \\ 1 & 0 \end{pmatrix} \in \text{End}_{\mathbf{H}}(\mathbf{S} \oplus \mathbf{H}).$$

8. Using Lemma 11.74, verify that the only possible spinor structure maps (for odd dimensions) are those described in Theorem 11.66.

9. (a) Suppose V is one of the 2-dimensional spaces $\mathbf{R}(2,0)$, $\mathbf{R}(1,1)$, or $\mathbf{R}(0,2)$, and that W is the vector space defined in the proof of Lemma 11.17. Recall from Lemma 11.17 that

$$\text{Cl}(W) \cong \text{Cl}(r,s) \otimes \text{Cl}(V) \quad \text{(as algebras)}.$$

Prove that the extra structure in $\text{Cl}(W)$ of the three involutions and the inner product is given by

$$(1) \quad (a \otimes b)\tilde{} = \tilde{a} \otimes \tilde{b},$$
$$(2) \quad (a \otimes b)^{\wedge} = \breve{a} \otimes \hat{b},$$
$$(3) \quad (a \otimes b)^{\vee} = \hat{a} \otimes \breve{b},$$
$$(4) \quad \langle a \otimes b, c \otimes d \rangle = \langle \tilde{a}, c \rangle \, \langle b, d \rangle.$$

(b) Show that, as associative algebras with unit,

$$\text{Cl}(p-1, q) \cong \text{Cl}(q-1, p) \quad \text{(symmetry about the line } q = p+1\text{)}.$$

Moreover, show that this isomorphism can be chosen to preserve the hat involutions and the inner products.

12. The Split Case Cl(p,p)

The case Cl(r,s) with $r = s$ is particularly easy to analyze, independent of the inductive method used in the last chapter. Furthermore, the split case can be used to determine information about the complex Clifford algebra $\text{Cl}_{\mathbf{C}}(2p) = \text{Cl}(p,p) \otimes_{\mathbf{R}} \mathbf{C}$ and the other real Clifford algebras Cl(r,s) ($r + s = 2p$) as subalgebras of Cl(p,p) $\otimes_{\mathbf{R}} \mathbf{C}$. This chapter is independent of Chapter 11 and may be read first.

A MODEL FOR Cl(p,p)

Let $\mathbf{R}(p,p) \equiv \mathbf{R}(p,0) \times \mathbf{R}(p,0)$ with $z \equiv (u,v) \in \mathbf{R}(p,p)$ and $\|z\| = \|u\| - \|v\|$. Let $E_u x \equiv u \wedge x$ denote left exterior multiplication by $u \in \mathbf{R}(p,0)$ for $x \in \Lambda\mathbf{R}(p,0)$. Let $I_u x \equiv u \llcorner x$ denote left interior multiplication by $u \in \mathbf{R}(p,0)$ for $x \in \Lambda\mathbf{R}(p,0)$. Then $L_u^+ \equiv E_u - I_u$ denotes left Clifford multiplication by u in $\text{Cl}(p,0) \cong \Lambda\mathbf{R}^p$, and $L_u^- \equiv E_u + I_u$ denotes left Clifford multiplication by u in $\text{Cl}(0,p) \cong \Lambda\mathbf{R}^p$.

Lemma 12.1. L_u^+ and L_v^- anticommute for all $u,v \in \mathbf{R}^p = \Lambda^1\mathbf{R}^p$.

Proof 1:

$$
\begin{aligned}
L_u^+ L_v^- + L_v^- L_u^+ &= (E_u - I_u)(E_v + I_v) + (E_v + I_v)(E_u - I_u)\\
&= E_u I_v + I_v E_u - (I_v E_u + E_v I_u)\\
&= \langle u,v \rangle - \langle v,u \rangle = 0.
\end{aligned}
$$

Here we use the fact that

$$E_u I_v + I_v E_u = \langle u, v \rangle,$$

obtained from (9.3) by polarization (i.e., replace w by $u + v$ in (9.3)). ∎

Proof 2: By Problem 1, $L_v^- x = R_v^+ \tilde{x}$. Therefore

$$\left(L_v^+ L_v^- + L_v^- L_u^+ \right) x = u\tilde{x}v + \tilde{u}\,\tilde{x}v = 0$$

for all x. ∎

Lemma 12.1. *Let $e_1, \ldots, e_p$ denote the standard orthonormal basis for* $\mathbf{R}(p, 0)$. *The algebra generated by $E_{e_1}, \ldots, E_{e_p}, I_{e_1}, \ldots, I_{e_p}$ is all of* $\mathrm{End}_{\mathbf{R}}(\Lambda \mathbf{R}^p)$.

Proof: The proof is by induction on p. If $p = 1$, then computing the four 2×2 matrices for the operators $I_e E_e, I_e, E_e, E_e I_e$ with respect to the basis $1, e$ for $\Lambda \mathbf{R}$ yields the result. Now assume the lemma is true for p and let $A_{IJ} \in \mathrm{End}_{\mathbf{R}}(\Lambda \mathbf{R}^p)$ denote the linear map that sends e_J to e_I and e_K to zero for $K \neq J$ (i.e., the matrix with a single 1 in the I, J^{th} position). Let $e \equiv e_{p+1}$ and observe that

(12.3)
$$
\begin{aligned}
I_e A_{IJ} E_e \text{ maps } e_J &\text{ to } e_I, \\
A_{IJ} I_e \text{ maps } e \wedge e_J &\text{ to } e_I, \\
E_e A_{IJ} \text{ maps } e_J &\text{ to } e \wedge e_I, \\
E_e A_{IJ} I_e \text{ maps } e \wedge e_J &\text{ to } e \wedge e_I,
\end{aligned}
$$

while mapping all other e_K or $e \wedge e_K$ to zero.

Theorem 12.4. *There exists an algebra isomorphism*

(12.5)
$$\mathrm{Cl}(p, p) \cong \mathrm{End}_{\mathbf{R}}(\Lambda \mathbf{R}^p),$$

extending the map $\phi : \mathbf{R}(p, p) \to \mathrm{End}_{\mathbf{R}}(\Lambda \mathbf{R}^p)$ defined by

(12.6)
$$\phi(u, v) \equiv E_{u+v} - I_{u-v} = L_u^+ + L_v^-.$$

Proof: Because of Lemma 12.1,

$$\left(L_u^+ + L_v^- \right)\left(L_u^+ + L_v^- \right) = \left(L_u^+ \right)^2 + \left(L_v^- \right)^2 + L_u^+ L_v^- + L_v^- L_u^+ = -(\|u\| - \|v\|)$$

for all $u, v \in \mathbf{R}^p$. The Fundamental Lemma of Clifford Algebras implies that ϕ extends to an algebra homomorphism in the following diagram:

$$
\begin{array}{ccc}
\mathbf{R}(p, p) & \xrightarrow{\phi} & \mathrm{End}_{\mathbf{R}}\left(\Lambda \mathbf{R}^p\right) \\
\cap & \nearrow & \\
\mathrm{Cl}(p, p) & \phi &
\end{array}
$$

(12.7)

Because of Lemma 12.1, the image of the map $\phi : \mathrm{Cl}(p, p) \to \mathrm{End}_{\mathbf{R}}(\Lambda \mathbf{R}^p)$ must be all of $\mathrm{End}_{\mathbf{R}}(\Lambda \mathbf{R}^p)$. Therefore, by a dimension count, ϕ must be an isomorphism. ∎

     Consider the map $\phi : \mathrm{Cl}(p, p) \to \mathrm{End}_{\mathbf{R}}(\Lambda \mathbf{R}^p)$ as an identification.

Corollary 12.8. *Let $e_1, \ldots, e_p$ denote the standard orthonormal basis for $\mathbf{R}(p, 0)$.*

(12.9) $$L_{e_1}^+, \ldots, L_{e_p}^+, L_{e_1}^-, \ldots, L_{e_p}^-$$

provides the standard basis (of γ matrices) for $\mathbf{R}(p, p) \subset \mathrm{End}_{\mathbf{R}}(\Lambda \mathbf{R}^p) \cong \mathrm{Cl}(p, p)$.

Definition 12.10 (The Split Case). *The vector space $\mathbf{P}(p, p)$ of pinors is defined to be $\Lambda \mathbf{R}^p$, and the isomorphism*

(12.11) $$\mathrm{Cl}(p, p) \cong \mathrm{End}_{\mathbf{R}}(\mathbf{P})$$

is called the pinor representation of $\mathrm{Cl}(p, p)$.

     Since the dimension $n \equiv 2p$ is even, the canonical automorphism of $\mathrm{Cl}(p, p)$ is given by

(12.12) $$\tilde{a} = \lambda a \lambda^{-1} \quad \text{for all } a \in \mathrm{Cl}(p, p),$$

where λ is a unit volume element for $\mathbf{R}(p, p)$. Recall that there are two choices for λ and that making a choice of λ is equivalent to choosing an orientation for $\mathbf{R}(p, p)$.

Definition 12.13 (The Split Case). *Assume that an orientation for $\mathbf{R}(p, p)$ is given, and let λ denote the positive unit volume for $\mathbf{R}(p, p)$.*

(12.14) $$\mathbf{S}_+ \equiv \{x \in \mathbf{P} : \lambda x = x\}$$

is the space of positive spinors, and

(12.15) $\mathbf{S}_- \equiv \{x \in \mathbf{P} : \lambda x = -x\}$

is the space of *negative spinors.*

The map $\tilde{x} \equiv \lambda x$ for $x \in \mathbf{P}$ is sometimes called the canonical pinor involution. This notation $x \to \tilde{x} \equiv \lambda x$ is consistent with the notation $a \to \tilde{a}$ for $a \in \mathrm{Cl}(p,p)$ because

(12.16) $\widetilde{ax} = \lambda ax = \lambda a \lambda^{-1} \lambda x = \tilde{a}\,\tilde{x}.$

The decomposition

(12.17) $\mathbf{P} = \mathbf{S}_+ \oplus \mathbf{S}_-$

determines a 2 × 2 blocking

(12.18) $A = \begin{pmatrix} a & b \\ c & d \end{pmatrix}$ for each $A \in \mathrm{End}_{\mathbf{R}}(\mathbf{P}).$

Note that

(12.19) $\lambda = \begin{pmatrix} 1 & 0 \\ 0 & -1 \end{pmatrix}.$

Lemma 12.20.

(12.21) $\mathrm{Cl}^{\mathrm{even}}(p,p) \cong \left\{ \begin{pmatrix} a & 0 \\ 0 & b \end{pmatrix} : a \in \mathrm{End}_{\mathbf{R}}(\mathbf{S}_+),\ b \in \mathrm{End}_{\mathbf{R}}(\mathbf{S}_-) \right\}$

(12.22)

$\mathrm{Cl}^{\mathrm{odd}}(p,p) \cong \left\{ \begin{pmatrix} 0 & a \\ b & 0 \end{pmatrix} : a \in \mathrm{End}_{\mathbf{R}}(\mathbf{S}_-, \mathbf{S}_+),\ b \in \mathrm{End}_{\mathbf{R}}(\mathbf{S}_+, \mathbf{S}_-) \right\}.$

Proof: Consult (12.12) and (12.19). ∎

Corollary 12.23. $\dim \mathbf{S}_+ = \dim \mathbf{S}_-.$

Proof: If $s_\pm \equiv \dim \mathbf{S}_\pm$, then

$s_+^2 + s_-^2 = \dim \mathrm{Cl}^{\mathrm{even}}(p,p) = \dim \mathrm{Cl}^{\mathrm{odd}}(p,p) = 2s_+ s_-.$ ∎

PINOR INNER PRODUCTS FOR $\mathrm{Cl}(p,p) \cong \mathrm{End}_{\mathbf{R}}(\mathbf{P}(p,p))$

The pinor space has inner products $\hat{\varepsilon}$ and $\check{\varepsilon}$ that determine the hat and check involutions as adjoints with respect to these inner products.

Let $(\ ,\ )$ denote the inner product on $\mathbf{P} \equiv \Lambda\mathbf{R}(p,0)$ induced by the standard inner product on $\mathbf{R}(p,0)$. Let $\sigma \equiv e_1 \wedge \cdots \wedge e_p$ denote the unit volume element for $\mathbf{R}(p,0)$. Utilizing the Clifford multiplication on $\mathbf{P} \equiv \Lambda\mathbf{R}(p,0) \cong \mathrm{Cl}(p,0)$, given $x \in \mathbf{P}$, let $\sigma x \in \mathbf{P}$ denote the Clifford product of σ and x in $\mathrm{Cl}(p,0)$.

Definition 12.24. *The hat bilinear form $\hat\varepsilon$ on* $\mathbf{P} \equiv \Lambda\mathbf{R}(p,0)$ *is defined by*

$$(12.25) \qquad (p \text{ odd}) \quad \hat\varepsilon(x,y) \equiv (x,\sigma y) \quad \text{for all } x,y \in \mathbf{P},$$

$$(12.26) \qquad (p \text{ even}) \quad \hat\varepsilon(x,y) \equiv (\tilde x,\sigma y) \quad \text{for all } x,y \in \mathbf{P}.$$

The check bilinear form $\check\varepsilon$ on $\mathbf{P} \equiv \Lambda\mathbf{R}(p,0)$ *is defined by*

$$(12.27) \qquad \check\varepsilon(x,y) \equiv \hat\varepsilon(\tilde x, y).$$

Hat/Check Theorem 12.28. *Both $\hat\varepsilon$ and $\check\varepsilon$ are nondegenerate bilinear forms on the space $\mathbf{P}$ of pinors. The Clifford hat anti-automorphism, $A \mapsto \hat A$, is equal to the adjoint on $\mathrm{End}_{\mathbf{R}}(\mathbf{P})$ with respect to the bilinear form $\hat\varepsilon$. That is, given $A \in \mathrm{Cl}(p,p) \cong \mathrm{End}(\mathbf{P})$, $\hat A$ is determined by*

$$(12.29) \qquad \hat\varepsilon(Ax,y) = \hat\varepsilon(x,\hat Ay) \quad \text{for all } x,y \in \mathbf{P}.$$

Similarly,

$$(12.29') \qquad \check\varepsilon(Ax,y) = \check\varepsilon(x,\check Ay) \quad \text{for all } x,y \in \mathbf{P}.$$

The bilinear forms $\hat\varepsilon$ and $\check\varepsilon$ on $\mathbf{P}$ are either symmetric (with split signature) or skew, as indicated in the following table.

Table 12.30

$p \bmod 4$	$\hat\varepsilon$	$\check\varepsilon$
0	symmetric	symmetric
1	skew	symmetric
2	skew	skew
3	symmetric	skew

Moreover,

$$(12.31) \quad \mathbf{S}_+ \text{ and } \mathbf{S}_- \text{ are nondegenerate and orthogonal if } p \text{ is even, and}$$

$$(12.32) \qquad \mathbf{S}_+ \text{ and } \mathbf{S}_- \text{ are totally null if } p \text{ is odd.}$$

Proof: Since $(\,,\,)$ is nondegenerate on $\mathbf{P} \equiv \Lambda\mathbf{R}(p,0)$, and both $x \mapsto \tilde x$ and $x \mapsto \sigma x$ are invertible, the bilinear forms $\hat\varepsilon$ and $\check\varepsilon$ are nondegenerate. Suppose (12.29) is true. Then (12.29') follows directly:

$$\check\varepsilon(Ax,y) = \hat\varepsilon(\widetilde{Ax},y) = \hat\varepsilon(\tilde A \tilde x, y) = \hat\varepsilon(\tilde x, \check Ay) = \check\varepsilon(x, \check Ay).$$

Let $A \to A^\dagger$ denote the anti-automorphism of $\text{End}(\mathbf{P})$ determined by $\hat{\varepsilon}$. Suppose we can show that

(12.33)   if $A \in \mathbf{R}(p,p) \subset \text{Cl}(p,p) \cong \text{End}(\mathbf{P})$, then $A^\dagger = -A$.

Then $(A^\dagger)^\wedge$ is an automorphism of $\text{Cl}(p,p)$ that extends the identity on $\mathbf{R}(p,p)$, so that by the uniqueness part of the fundamental lemma of Clifford algebras $(A^\dagger)^\wedge = A$ for all $A \in \text{Cl}(p,p)$. Therefore, $A^\dagger = A^\wedge$ for all $A \in \text{Cl}(p,p)$.

The next lemma implies, as a corollary that $A^\dagger = -A$ for $A \in \mathbf{R}(p,p)$, completing the proof of (12.33).

Lemma 12.34. *Suppose* $u \in \mathbf{R}(p,0)$ *and* $x \in \Lambda \mathbf{R}(p,0)$.

(12.35)   *(p odd)* $u \wedge (\sigma x) = -\sigma(u \llcorner x)$ *and* $u \llcorner (\sigma x) = -\sigma(u \wedge x)$.

(12.36)   *(p even)* $u \wedge (\sigma x) = \sigma(u \llcorner x)$ *and* $u \llcorner (\sigma x) = \sigma(u \wedge x)$.

Note that Lemma 12.34 implies

(12.37)   *(p odd)* $\begin{aligned} (E_u - I_u)\sigma &= \sigma(E_u - I_u), \\ (E_u + I_u)\sigma &= -\sigma(E_u + I_u), \end{aligned}$

(12.38)   *(p even)* $\begin{aligned} (E_u - I_u)\sigma &= -\sigma(E_u - I_u), \\ (E_u + I_u)\sigma &= \sigma(E_u + I_u). \end{aligned}$

Corollary 12.39.
$$(E_u - I_u)^\dagger = -(E_u - I_u)$$

and

$$(E_u + I_u)^\dagger = -(E_u + I_u)$$

for all $u \in \mathbf{R}(p,0)$.

Proof: Suppose p is odd. Then

$$\begin{aligned}
\hat{\varepsilon}((E_u - I_u)x, y) &= ((E_u - I_u)x, \sigma y) = -(x, (E_u - I_u)\sigma y) \\
&= -(x, \sigma(E_u - I_u)y) = -\hat{\varepsilon}(x, (E_u - I_u)y),
\end{aligned}$$

and

$$\begin{aligned}
\hat{\varepsilon}((E_u + I_u)x, y) &= ((E_u + I_u)x, \sigma y) = (x, (E_u + I_u)\sigma y) \\
&= -(x, \sigma(E_u + I_u)y) = -\hat{\varepsilon}(x, (E_u + I_u)y).
\end{aligned}$$

The proof for p even is Problem 2(a). ∎

Proof of Lemma 12.34: Suppose $u \wedge x = 0$. Then x is of the form $x = u \wedge y$ and hence $u \llcorner \sigma x = 0$. Therefore, $u \wedge (\sigma x) = u\sigma x$ and $\sigma(u \llcorner x) = -\sigma u x$. Suppose $u \llcorner x = 0$. Then $u \wedge \sigma x = 0$. Therefore, $u \llcorner (\sigma x) = -u\sigma x$ and $\sigma(u \wedge x) = \sigma u x$. If p is odd, then $u\sigma = \sigma u$ and (12.35) follows. If p is even, then $u\sigma = -\sigma u$ and (12.36) follows.

The proof of the Hat/Check Theorem 12.28 is completed as follows:

If either $\hat{\varepsilon}$ or $\check{\varepsilon}$ has a signature (i.e., is symmetric), then by Problem 9.10 the signature must be split.

Table 12.30 is deduced as follows. First, suppose p is even so that $\hat{\varepsilon}(x,y) \equiv (\tilde{x}, \sigma y)$. Then

$$(12.40) \qquad \hat{\varepsilon}(y,x) \equiv (\tilde{y}, \sigma x) = (\widetilde{\sigma x}, y) = (\sigma \tilde{x}, y) = (\tilde{x}, \hat{\sigma} y).$$

Here $\hat{\sigma}$ is the Cl$(p,0)$ hat of the unit volume element σ for $\mathbf{R}(p,0)$. Since $\hat{\sigma} = \sigma$ if $p = 0 \bmod 4$, and $\hat{\sigma} = -\sigma$ if $p = 2 \bmod 4$, this verifies that $\hat{\varepsilon}$ is symmetric if $p = 0 \bmod 4$ and that $\hat{\varepsilon}$ is skew if $p = 2 \bmod 4$. The analysis of $\hat{\varepsilon}$ for p odd is Problem 2 (b).

Using the facts that

$$(12.41) \qquad \lambda^2 = 1,$$

$$(12.42) \qquad \hat{\lambda} = \check{\lambda} = \lambda, \text{ for } p \text{ even},$$

$$(12.43) \qquad \hat{\lambda} = \check{\lambda} = -\lambda, \text{ for } p \text{ odd},$$

both (12.31) and (12.32) follow. For example, if p is even and $x \in \mathbf{S}_+, y \in \mathbf{S}_-$, then

$$\hat{\varepsilon}(x,y) = -\hat{\varepsilon}(\lambda x, \lambda y) = -\hat{\varepsilon}(x, \hat{\lambda}\lambda y) = -\hat{\varepsilon}(x,y),$$

proving that $\mathbf{S}_+$ and $\mathbf{S}_-$ are orthogonal. ∎

The bilinear form $\check{\varepsilon}$ on $\mathbf{P}$ can also be understood in terms of $(\, , \,)$ and σ. Alternatively, $\check{\varepsilon}$ can be analyzed in terms of $\hat{\varepsilon}$ using the relationship

$$(12.44) \qquad \check{\varepsilon}(x,y) = \hat{\varepsilon}(\lambda x, y), \text{ with } \lambda = \begin{pmatrix} 1 & 0 \\ 0 & -1 \end{pmatrix}.$$

The details are left as Problem 3. ∎

THE COMPLEX CLIFFORD ALGEBRAS (CONTINUED FROM CHAPTER 9)

The complex Clifford algebra $\mathrm{Cl}_{\mathbf{C}}(2p)$ is obtained from the complex vector space $\mathbf{C}^{2p}$ equipped with a square norm (see (9.74)).

Theorem 12.45. *As complex associative algebras with unit,*

(12.46) $\text{Cl}_{\mathbf{C}}(2p) \cong M_N(\mathbf{C})$ *with* $N = 2^p$.

Also, given an isomorphism

$$\text{Cl}_{\mathbf{C}}(2p) \cong \text{End}_{\mathbf{C}}(\mathbf{P}_{\mathbf{C}}),$$

there exist complex inner products $\hat{\varepsilon}_{\mathbf{C}}$ *and* $\check{\varepsilon}_{\mathbf{C}}$ *on* $\mathbf{P}_{\mathbf{C}}$, *so that*

(12.47) $\hat{\varepsilon}_{\mathbf{C}}(ax, y) = \hat{\varepsilon}_{\mathbf{C}}(x, \hat{a}y)$

and

(12.48) $\check{\varepsilon}_{\mathbf{C}}(ax, y) = \check{\varepsilon}_{\mathbf{C}}(x, \check{a}y)$

for all $a \in \text{Cl}_{\mathbf{C}}(2p)$ *and* $x, y \in \mathbf{P}_{\mathbf{C}}$. *Both* $\hat{\varepsilon}_{\mathbf{C}}$ *and* $\check{\varepsilon}_{\mathbf{C}}$ *are* **C**-*symmetric or* **C**-*skew exactly when their counterparts,* $\hat{\varepsilon}$ *and* $\check{\varepsilon}$ *in the split case, are* **R**-*symmetric or* **R**-*skew as indicated in Table 12.30.*

Proof: It is convenient to use the square norm on $\mathbf{C}^{2p} \cong \mathbf{C}(p, p)$ paralleling the real split case. Let

$$z \equiv (u, v) \in \mathbf{C}(p, 0) \times \mathbf{C}(p, 0) \cong \mathbf{C}^{2p}$$

have square norm

$$\|z\| = \|u\| - \|v\| = u_1^2 + \cdots + u_p^2 - v_1^2 - \cdots - v_p^2.$$

Let

(12.49) $L_u^+ \equiv E_u - I_u \in \text{End}_{\mathbf{C}}(\Lambda \mathbf{C}^p)$

(12.50) $L_u^- \equiv E_u + I_u \in \text{End}_{\mathbf{C}}(\Lambda \mathbf{C}^p)$

with interior multiplication I_u determined by the square norm $\|u\| = u_1^2 + \cdots + u_p^2$ on $\mathbf{C}^p$.
    Exactly as in Lemma 12.1,

$$L_u^+ L_v^- + L_v^- L_u^+ = 0$$

since L_u^+ and L_v^- acting on $\Lambda \mathbf{C}^p$ are just the complexifications of the real operators L_u^+ and L_v^- acting on $\Lambda \mathbf{R}^p$. The map $\phi : \mathbf{C}(p, p) \to \text{End}_{\mathbf{C}}(\Lambda \mathbf{C}^p)$ defined by $\phi(z) \equiv \phi(u, v) \equiv L_u^+ + L_v^-$ has a unique extension

$$\phi : \text{Cl}_{\mathbf{C}}(2p) \longrightarrow \text{End}_{\mathbf{C}}(\Lambda \mathbf{C}^p)$$

that is a complex algebra isomorphism. This proves that $\mathrm{Cl}_{\mathbf{C}}(2p) \cong M_N(\mathbf{C})$.

Because of the results of Chapter 8, any other isomorphism $\mathrm{Cl}_{\mathbf{C}}(2p) \cong \mathrm{End}_{\mathbf{C}}(\mathbf{P}_{\mathbf{C}})$ of complex algebras is equivalent, via a complex linear intertwining operator, to the isomorphism

$$(12.51) \qquad \mathrm{Cl}_{\mathbf{C}}(2p) \cong \mathrm{End}_{\mathbf{C}}(\Lambda\mathbf{C}^p).$$

Thus, we need only construct $\hat{\varepsilon}_{\mathbf{C}}$ and $\check{\varepsilon}_{\mathbf{C}}$ on $\Lambda\mathbf{C}^p$. Now simply take $\hat{\varepsilon}_{\mathbf{C}}$ (and $\check{\varepsilon}_{\mathbf{C}}$) to be the complexifications of the real inner products $\hat{\varepsilon}$ and $\check{\varepsilon}$ on $\Lambda\mathbf{R}^p$. ■

Corollary 12.52. *The complex Clifford algebras* $\mathrm{Cl}_{\mathbf{C}}(2p)$, *in even dimensions, contain no proper nontrivial two-sided ideals.*

Proof: Because of (12.46), it suffices to show that $M_N(\mathbf{C})$ has no proper nontrivial two-sided ideals. This was done in Chapter 8. ■

Corollary 12.53. *The real Clifford algebras* $\mathrm{Cl}(r,s)$, *in even dimensions* $r + s = 2p$, *contain no proper nontrivial two-sided ideals.*

Proof: Since $\mathrm{Cl}_{\mathbf{C}}(2p) \cong \mathrm{Cl}(r,s) \otimes_{\mathbf{R}} \mathbf{C}$ is the complexification of the real algebra $\mathrm{Cl}(r,s)$, Corollary 12.53 is an immediate consequence of Corollary 12.52. ■

$\mathrm{Cl}(r,s)(r+s \equiv 2p)$ AS A SUBALGEBRA OF $\mathrm{Cl}_{\mathbf{C}}(2p) \equiv \mathrm{Cl}(p,p) \otimes_{\mathbf{R}} \mathbf{C}$

Each Clifford algebra $\mathrm{Cl}(r,s)$ of even dimension $n \equiv r + s \equiv 2p$ is of the form $\mathrm{Cl}(p+k, p-k)$ with $r - s \equiv 2k$, $k \in \mathbf{Z}$. These Clifford algebras are real subalgebras of $\mathrm{Cl}_{\mathbf{C}}(2p)$. By Proposition 9.76,

$$(12.54) \qquad \mathrm{Cl}(p+k, p-k) \otimes_{\mathbf{R}} \mathbf{C} = \mathrm{Cl}_{\mathbf{C}}(2p).$$

The conjugation or reality operator fixing $\mathrm{Cl}(p+k, p-k)$ in $\mathrm{Cl}_{\mathbf{C}}(2p)$ will be denoted $\mathcal{R}$, when the signature $r, s = p+k, p-k$ has been prescribed.

No matter the signature, the model taken for $\mathrm{Cl}_{\mathbf{C}}(2p) \cong \mathrm{Cl}(p+k, p-k) \otimes_{\mathbf{R}} \mathbf{C}$ will be the complexification of the split case $\mathrm{Cl}(p,p)$. Thus,

$$(12.55) \qquad \mathrm{Cl}_{\mathbf{C}}(2p) \cong \mathrm{Cl}(p,p) \otimes_{\mathbf{R}} \mathbf{C}$$

is realized as the complexification of the matrix algebra

$$(12.56) \qquad \mathrm{Cl}(p,p) \cong \mathrm{End}_{\mathbf{R}}(\Lambda\mathbf{R}^p).$$

Namely,

$$(12.57) \qquad \text{Cl}_{\mathbf{C}}(2p) \cong \text{End}_{\mathbf{C}}(\Lambda \mathbf{C}^p),$$

where the pinor space for $\text{Cl}_{\mathbf{C}}(2p)$,

$$(12.58) \qquad \Lambda \mathbf{C}^p \cong \Lambda \mathbf{R}^p \otimes_{\mathbf{R}} \mathbf{C},$$

is the complexification of the pinor space $\Lambda \mathbf{R}^p$ for $\text{Cl}(p, p)$.

The conjugation, or reality operator for $\text{Cl}(p, p)$, contained in $\text{Cl}_{\mathbf{C}}(2p)$ $\cong \text{Cl}(p, p) \otimes_{\mathbf{R}} \mathbf{C}$ will be denoted $\mathcal{C}$. The complexification $\Lambda \mathbf{C}^p \cong \Lambda \mathbf{R}^p \otimes_{\mathbf{R}}$ $\mathbf{C}$ also comes equipped with a conjugation, the natural extension of the conjugation on

$$(12.59) \qquad \mathbf{C}^p \cong \mathbf{R}^p \otimes \mathbf{C}.$$

This conjugation on $\Lambda \mathbf{C}^p$ that fixes $\Lambda \mathbf{R}^p$ will be denoted C (or $\overline{x} \equiv Cx$ for $x \in \Lambda \mathbf{C}^p$). In turn, the conjugation C on $\Lambda \mathbf{C}^p$ induces a conjugation on $\text{End}_{\mathbf{C}}(\Lambda \mathbf{C}^p)$:

$$(12.60) \qquad \overline{a} \equiv CaC \quad \text{for all } a \in \text{End}_{\mathbf{C}}(\Lambda \mathbf{C}^p),$$

which fixes $\text{End}_{\mathbf{R}}(\Lambda \mathbf{R}^p)$ in $\text{End}_{\mathbf{C}}(\Lambda \mathbf{C}^p)$. Here $\text{End}_{\mathbf{R}}(\Lambda \mathbf{R}^p)$ is naturally considered a subspace of $\text{End}_{\mathbf{C}}(\Lambda \mathbf{C}^p)$ by extending each real linear map of $\Lambda \mathbf{R}^p$ to a complex linear map of $\Lambda \mathbf{C}^p \cong \Lambda \mathbf{R}^p \otimes \mathbf{C}$. Thus,

$$(12.61) \qquad \mathcal{C}a = \overline{a} \quad \text{for all } a \in \text{Cl}_{\mathbf{C}}(2p) \cong \text{End}_{\mathbf{C}}(\Lambda \mathbf{C}^p).$$

Remark 12.62. Note that the map $a \to \mathcal{C}a = \overline{a}$ is a real algebra automorphism of $\text{Cl}_{\mathbf{C}}(2p) = \text{End}_{\mathbf{C}}(\Lambda \mathbf{C}^p)$, that is complex antilinear.

Also note that each element $A \in \text{End}_{\mathbf{R}}(\Lambda \mathbf{C}^p)$ has a unique decomposition

$$(12.63) \qquad A = a + bC$$

into a complex linear map $a \in \text{End}_{\mathbf{C}}(\Lambda \mathbf{C}^p)$ and an anticomplex linear map bC, where $b \in \text{End}_{\mathbf{C}}(\Lambda \mathbf{C}^p)$ is complex linear.

Let $e^+, \ldots, e_p^+, e_1^-, \ldots, e_p^-$ denote the standard basis for $\mathbf{R}(p, p)$ considered as a subspace of $\text{End}_{\mathbf{R}}(\Lambda \mathbf{R}^p) \cong \text{Cl}(p, p)$. That is,

$$(12.64) \qquad \begin{aligned} e_j^+ &\equiv L_{e_j}^+, \quad \text{left Clifford multiplication by } e_j \\ &\text{on } \text{Cl}(p, 0) \cong \Lambda \mathbf{R}^p, \end{aligned}$$

and

$$(12.65) \qquad \begin{aligned} e_j^- &\equiv L_{e_j}^-, \quad \text{left Clifford multiplication by } e_j \\ &\text{on } \text{Cl}(0, p) \cong \Lambda \mathbf{R}^p. \end{aligned}$$

The signature of $\mathbf{R}(p, p)$, considered as a subspace of

$$\mathbf{C}(p, p) \subset \text{End}_{\mathbf{C}}(\Lambda \mathbf{C}^p) \cong \text{Cl}_{\mathbf{C}}(2p)$$

can easily be modified.

Definition 12.66. *Let*

(12.69)   (k positive) $e_1^+, \ldots, e_p^+, ie_1^-, \ldots, ie_k^- \; ; \; e_{k+1}^-, \ldots, e_p^-,$

(12.70)   (k negative) $e_1^+, \ldots, e_{p+k}^+ \; ; \; ie_{p+k+1}^+, \ldots, ie_p^+, e_1^-, \ldots, e_p^-$

define a standard basis for $\mathbf{R}(p+k, p-k)$ *as a subspace of* $\mathbf{C}(p,p) \subset$ $\mathrm{End}_{\mathbf{C}}(\Lambda \mathbf{C}^p) \cong \mathrm{Cl}_{\mathbf{C}}(2p)$.

Proposition 12.69. *The real subalgebra of* $\mathrm{End}_{\mathbf{C}}(\Lambda \mathbf{C}^p)$ *generated by* $\mathbf{R}(p+k, p-k)$ *is isomorphic to* $\mathrm{Cl}(p+k, p-k)$.

Proof: The inner product $\langle \, , \, \rangle$ on $\mathbf{C}(p,p)$, restricted to $\mathbf{R}(p+k, p-k)$ is real-valued with signature $p+k, p-k$. Thus the Fundamental Lemma of Clifford Algebras applies, yielding an algebra homomorphism

(12.70) $\phi : \mathrm{Cl}(p+k, p-k) \longrightarrow \mathrm{End}_{\mathbf{C}}(\Lambda \mathbf{C}^p)$

that extends the inclusion

$$\mathbf{R}(p+k, p-k) \subset \mathrm{End}_{\mathbf{C}}(\Lambda \mathbf{C}^p)$$

given by Definition 12.66. Since the dimension $n \equiv 2p$ is even, $\mathrm{Cl}(p+k, p-k)$ contains no proper nontrivial two-sided ideals, by Corollary 12.53. Therefore, the kernel of ϕ is trivial. ∎

THE PINOR REALITY MAP

In order to identify $\mathrm{Cl}(p+k, p-k)$ as a matrix algebra, it is useful to compute the centralizer of $\mathrm{Cl}(p+k, p-k)$ in $\mathrm{End}_{\mathbf{R}}(\Lambda \mathbf{C}^p)$, i.e., all real linear maps of $\Lambda \mathbf{C}^p$ that commute with $\mathrm{Cl}(p+k, p-k)$. Of course, the centralizer always contains $\mathbf{C}$, all complex multiples of the identity. Note that $\mathrm{End}_{\mathbf{R}}(\Lambda \mathbf{C}^p)$ is larger yet than $\mathrm{Cl}_{\mathbf{C}}(2p) \cong \mathrm{End}_{\mathbf{C}}(\Lambda \mathbf{C}^p)$.

Theorem 12.71. *The centralizer of* $\mathrm{Cl}(p+k, p-k)$ *in* $\mathrm{End}_{\mathbf{R}}(\Lambda \mathbf{C}^p)$, *which canonically contains* $\mathbf{C}$, *is the algebra*

(12.72) $M_2(\mathbf{R})$ *if* $k = 0, 3 \bmod 4$ $(r - s = 0, 6 \bmod 8)$,

(12.73) $\mathbf{H}$ *if* $k = 1, 2 \bmod 4$ $(r - s = 2, 4 \bmod 8)$.

Remark 12.74. Recall from Chapter 6 that both $M_2(\mathbf{R})$ and $\mathbf{H}$ are normed algebras and that the square norm $\| \; \|$ is uniquely determined by the algebraic structure.

Theorem 12.71 will be proved after some lemmas are established.

Lemma 12.75. $A \in \mathrm{End}_{\mathbf{R}}(\Lambda \mathbf{C}^p)$ *commutes with* $\mathrm{Cl}(p+k, p-k)$ *if and only if* A *is of the form* $A = a + b\mu C$, *with* $a, b \in \mathbf{C}$ *and* μ *defined as follows:*

(12.76)   (k even positive)$\mu \equiv e_1^- \cdots e_k^-$,

(12.77)   (k odd positive) $\mu \equiv e_1^+ \cdots e_p^+ e_{k+1}^- \cdots e_p^-$,

(12.78)   (k even negative) $\mu \equiv e_{p+k+1}^+ \cdots e_p^+$,

(12.79)   (k odd negative) $\mu \equiv e_1^+ \cdots e_{p+k}^+ e_1^- \cdots e_p^-$.

Proof: Given $A \in \mathrm{End}_{\mathbf{R}}(\Lambda \mathbf{C}^p)$, it has a unique decomposition (see (12.63))

$$A = a + bC, \quad \text{with } a, b \in \mathrm{End}_{\mathbf{C}}(\Lambda \mathbf{C}^p).$$

Suppose A commutes with all $u \in \mathrm{Cl}(p+k, p-k)$, or equivalently, with all $u \in \mathbf{R}(p+k, p-k)$. Since $Cu = \overline{u}C$, this is equivalent to

(12.80) $au = ua$ and $b\overline{u} = ub$ for all $u \in \mathbf{R}(p+k, p-k)$.

Since a commutes with i and

$$\mathrm{End}_{\mathbf{C}}(\Lambda \mathbf{C}^p) \cong \mathrm{Cl}_{\mathbf{C}}(2p) \cong \mathrm{Cl}(p+k, p-k) \oplus i\,\mathrm{Cl}(p+k, p-k),$$

(12.81) a belongs to the center of $\mathrm{Cl}_{\mathbf{C}}(2p)$.

Thus, $a \in \mathbf{C}$ is a complex scalar (because of either the complex version of Lemma 9.49 or the fact that $\mathrm{Cl}_{\mathbf{C}}(2p) \cong M_N(\mathbf{C})$).

Referring back to the choice (Definition 12.66) of basis for

$$\mathbf{R}(p+k, p-k) \subset \mathrm{Cl}(p+k, p-k) \subset \mathrm{End}_{\mathbf{C}}(\Lambda \mathbf{C}^p) \cong \mathrm{Cl}_{\mathbf{C}}(2p),$$

the condition $b\overline{u} = ub$ becomes

(12.82)   (k positive) b commutes with $e_1^+, \ldots, e_p^+, e_{k+1}^-, \ldots, e_p^-$
    and anticommutes with $e_1^-, \ldots, e_k^-$.

(12.83)   (k negative) b commutes with $e_1^+, \ldots, e_{p+k}^+, e_1^-, \ldots, e_p^-$
    and anticommutes with $e_{p+k+1}^+, \ldots, e_p^+$.

Since these conditions are valid for b if and only if they are valid for both the even part of b and the odd part of b, it suffices to consider the cases of b even and b odd separately.

Suppose k is positive and b is even. Then, by (9.53), b does not involve $e_1^+, \ldots, e_p^+, e_{k+1}^-, \ldots, e_p^-$, while by (9.54), b does involve $e_1^-, \ldots, e_k^-$. Thus, b is a scalar multiple of $e_1^- \ldots e_k^-$, and hence k must be even.

Suppose k is positive and b is odd. Then by (9.56), b does involve $e_1^+, \ldots, e_p^+, e_{k+1}^-, \ldots, e_p^-$, while by (9.55), b does not involve $e_1^-, \ldots, e_k^-$. Thus, b is a scalar multiple of $e_1^+ \cdots e_p^+ e_{k+1}^- \cdots e_p^-$, and hence $2p - k$ and k must be odd.

The proofs for k negative are similar and so are omitted. ∎

Definition 12.84. *Suppose μ is defined as in Lemma 12.75. Then*

$$(12.85) \qquad R \equiv e^{i\theta}\mu C \in \text{End}_{\mathbf{R}}(\Lambda \mathbf{C}^p), \quad \text{for } \theta \in \mathbf{R}$$

is called a choice of pinor reality map for Cl($p+k, p-k$).

Lemma 12.86. *Suppose R is a pinor reality map for* Cl($p+k, p-k$). *Then*

$$(12.87) \qquad Ri = -iR, \text{ i.e., } R \text{ is complex antilinear.}$$

$$(12.88) \qquad \mathcal{R}a = RaR^{-1} \text{ for all } a \in \text{Cl}_{\mathbf{C}}(2p) \cong \text{Cl}(p+k, p-k) \otimes_{\mathbf{R}} \mathbf{C},$$

$$(12.89) \qquad \begin{aligned} R^2 &= 1 \text{ for } k = 0, 3 \bmod 4 \ (r - s = 0, 6 \bmod 8), \\ R^2 &= -1 \text{ for } k = 1, 2 \bmod 4 \ (r - s = 2, 4 \bmod 8). \end{aligned}$$

Proof: Since $e^{i\theta}\mu \in \text{End}_{\mathbf{C}}(\Lambda \mathbf{C}^p) \cong \text{Cl}_{\mathbf{C}}(2p)$, $R \equiv e^{i\theta}\mu C$ is complex antilinear. By Lemma 12.75, $RaR^{-1} = a$ if $a \in \text{Cl}(p+k, p-k)$, and since $Ri = -iR$, $RbR^{-1} = -b$ if $b \in i\,\text{Cl}(p+k, p-k)$. Therefore (12.88) is valid. Finally, to prove (12.89), note that $R^2 = e^{i\theta}\mu C e^{i\theta}\mu C = \mu^2$, and that μ is a unit volume element for

$$\begin{aligned} (k \text{ even positive}) \quad &\mathbf{R}(0,k) \subset \text{Cl}(0,k) \cong \Lambda\mathbf{R}(0,k), \\ (k \text{ odd positive}) \quad &\mathbf{R}(p,p-k) \subset \text{Cl}(p,p-k) \cong \Lambda\mathbf{R}(p,p-k), \\ (k \text{ even negative}) \quad &\mathbf{R}(k,0) \subset \text{Cl}(-k,0) \cong \Lambda\mathbf{R}(-k,0), \\ (k \text{ odd negative}) \quad &\mathbf{R}(p+k,p) \subset \text{Cl}(p+k,p) \cong \Lambda\mathbf{R}(p+k,p). \end{aligned}$$

Now (12.89) follows immediately from Proposition 9.57. ∎

Proof of Theorem 12.71. The centralizer of Cl($p + k, p - k$) in $\text{End}_{\mathbf{R}}(\Lambda\mathbf{C}^p)$ is $\mathbf{C} \oplus R\mathbf{C}$, where R anticommutes with i and $R^2 = \pm 1$ depending on $k \bmod 4$, so the theorem follows. ∎

It is convenient to choose the reality map to have an additional property. Recall the hat bilinear form $\hat{\varepsilon}$ on the pinor space $\Lambda\mathbf{R}^p$ for Cl(p,p) (the split case), as well as the complex extension, denoted $\hat{\varepsilon}_{\mathbf{C}}$, which is a complex inner product on the pinor space $\Lambda\mathbf{C}^p \cong \Lambda\mathbf{R}^p \otimes \mathbf{C}$ for Cl$_{\mathbf{C}}(2p)$.

Lemma 12.90. *The pinor reality map R for $\mathrm{Cl}(p+k, p-k)$ can be chosen so that*

$$\hat{\varepsilon}_{\mathbf{C}}(Rx, Ry) = \overline{\hat{\varepsilon}_{\mathbf{C}}(x, y)} \quad \text{for all } x, y \in \Lambda \mathbf{C}^p.$$

Proof: Let $R \equiv e^{i\theta}\mu C$ denote a reality map. Then

$$\hat{\varepsilon}(Rx, Ry) = \hat{\varepsilon}\left(Ce^{-i\theta}\mu x, Ce^{-i\theta}\mu y\right)$$
$$= \overline{\hat{\varepsilon}\left(e^{-i\theta}\mu x, e^{-i\theta}\mu y\right)} = e^{2i\theta}\hat{\varepsilon}(\mu x, \mu y)$$
$$= e^{2i\theta}\|\mu\|\hat{\varepsilon}(x, y).$$

Since μ is the product of unit simple vectors, $\|\mu\| = \pm 1$. If $\|\mu\| = 1$, take $R = \pm\mu C$, while if $\|\mu\| = -1$ take $R = \pm i\mu C$. ∎

A SECOND PROOF OF THE CLASSIFICATION THEOREM

A second proof of the important Theorem 11.3 can be given using the reality map R on the pinor space $\Lambda \mathbf{C}^p$ for $\mathrm{Cl}_{\mathbf{C}}(2p) \cong \mathrm{Cl}(r, s) \otimes_{\mathbf{R}} \mathbf{C}$.

Theorem 12.91.

If $r - s = 0, 6 \bmod 8$, then $\mathrm{Cl}(r, s) \cong \mathrm{End}_{\mathbf{R}}\left(\Lambda_R \mathbf{C}^p\right)$.

If $r - s = 2, 4 \bmod 8$, then $\mathrm{Cl}(r, s) \cong \mathrm{End}_{\mathbf{H}}\left(\Lambda \mathbf{C}^p\right)$.

Remark 12.92. Combined with Corollary 11.65 this theorem provides a second proof of Theorem 11.3.

Proof:

    Case 1 $(r - s = 0, 6 \bmod 8)$: Choose a reality map R for $\mathrm{Cl}(r, s)$. Then $R^2 = 1$. Therefore, R is a conjugation. In general, conjugations on a complex vector space V are characterized by two properties:

$$(12.93) \qquad\qquad Ri = -iR \quad \text{and} \quad R^2 = 1.$$

This is because the negative eigenspace for R is just i times the positive eigenspace for R, which implies that these two eigenspaces are of the same dimension. Let

$$(12.94) \qquad\qquad \Lambda_R \mathbf{C}^p = \{x \in \Lambda \mathbf{C}^p : Rx = x\}$$

denote the set fixed by the reality map R. Since R is a conjugation and $\mathcal{R}a \equiv RaR^{-1}$ (for all $a \in \mathrm{End}_{\mathbf{C}}(\Lambda \mathbf{C}^p)$) fixes $\mathrm{Cl}(r, s)$,

$$(12.95) \qquad\qquad \mathrm{Cl}(r, s) \cong \mathrm{End}_{\mathbf{R}}(\Lambda_R \mathbf{C}^p).$$

Case 2 ($r - s = 2, 4 \bmod 8$): The centralizer of Cl(r, s) in End$_\mathbf{R}(\Lambda \mathbf{C}^p)$ $\subset$ End$_\mathbf{C}(\Lambda \mathbf{C}^p) \cong$ Cl$_\mathbf{C}(2p)$ is $\mathbf{H} \cong \mathbf{C} \oplus R\mathbf{C}$. Therefore, $\mathbf{H}$ (the centralizer) defines a right quaternion structure on $\Lambda \mathbf{C}^p$, and since Cl(r, s) is exactly the subset of End$_\mathbf{C}(\Lambda \mathbf{C}^p)$ that commutes with R,

$$(12.96) \qquad \text{Cl}(r, s) \cong \text{End}_\mathbf{H}(\Lambda \mathbf{C}^p). \quad \blacksquare$$

PURE SPINORS

In general, the orbit structure of Pin acting on the pinor space $\mathbf{P}$ is extremely complicated. However, there is one particular orbit that warrants special attention, the so-called *pure spinors*. We shall examine the split signature case and the complex case. See Lawson–Micheleson [12] for a slightly different treatment of the complex case and some nice applications to geometry. Also consult Proposition 13.78 and Problem 13.8 in the next chapter for a further elaboration of the material in this section in terms of squares of spinors.

Let $\mathbf{N}$ denote the space of all p-dimensional totally null vector subspaces of $\mathbf{R}(p, p)$. Recall, by Problem 2.4, that the maximal dimension possible for a totally null subspace of $\mathbf{R}(p, p)$ is p. Consider a pinor representation:

$$(12.97) \qquad \text{Cl}(p, p) \cong \text{End}_\mathbf{R}(\mathbf{P}).$$

Given a pinor $x \in \mathbf{P}$, let

$$(12.98) \qquad N_x \equiv \{z \in \mathbf{R}(p, p) \subset \text{Cl}(p, p) : z(x) = 0\}.$$

Note that for $x \neq 0$, N_x is always totally null, since $0 = z^2(x) = -\|z\|x$ for all $z \in N_x$.

Definition 12.99. *A pinor $x \in \mathbf{P}(p, p)$ is said to be pure if* $\dim N_x = p$, *i.e., $N_x \in \mathbf{N}$. Let PURE(p) or PURE denote the set of all pure pinors.*

Consider Pin(p) $\equiv$ Pin($0, p$) as a subgroup of Pin(p, p), where $O(p)$ is the subgroup of $O(p, p)$, which acts as the identity on the first factor $\mathbf{R}(p, 0) \subset \mathbf{R}(p, p)$. That is, $A \in O(p)$ sends (u, v) to (u, Av), and Pin(p) is generated by the unit sphere in $\mathbf{R}(0, p) \subset \mathbf{R}(p, p)$.

See Proposition 13.78 for a further elaboration of the next theorem in terms of the square of pure spinors.

Theorem 12.100. $\mathrm{Pin}(p)$ *acts transitively on* $\mathrm{PURE}/\mathbf{R}^*$ *with trivial isotropic subgroup.* $O(p)$ *acts transitively on* $\mathbf{N}$ *with trivial isotropic subgroup. The map sending* $x \in \mathrm{PURE}$ *to* $N_x \in \mathbf{N}$ *induces an equivariant isomorphism* $\chi : \mathrm{PURE}/\mathbf{R}^* \to \mathbf{N}$:
(12.101)
$$\chi_a(N_x) \equiv N_{ax} = a N_x a^{-1} \quad \text{for all } a \in \mathrm{Pin}(p) \text{ and all } x \in \mathrm{PURE}.$$

The orbit PURE *consists of two connected components of dimension* $\frac{1}{2}p(p-1)$; *one, denoted* PURE^+, *is a subset of* $\mathbf{S}_+$, *the space of positive spinors, while the other, denoted* PURE^-, *is a subset of* $\mathbf{S}_-$, *the space of negative spinors.*

Proof: First we show that $O(p)$ acts transitively on $\mathbf{N}$ with trivial isotropy subgroup. Suppose $N \in \mathbf{N}$ is totally null of dimension p. Then $N \cap (\{0\} \times \mathbf{R}^p) = \{0\}$ implies that $N = \{(u, Au) : u \in \mathbf{R}^p\}$ is the graph over $\mathbf{R}^p \times \{0\}$ of a linear map $A \in \mathrm{End}(\mathbf{R}^p)$. A graph $N = \mathrm{graph}\, A$ is totally null if and only if
$$\langle (u, Au), (v, Av) \rangle = \langle u, v \rangle - \langle Au, Av \rangle = 0$$
for all $u, v \in \mathbf{R}(p, 0)$, i.e., if and only if $A \in O(p)$. Now it is obvious that $O(p)$ acts transitively on $\mathbf{N}$ with trivial isotropy subgroup. In particular, since $O(p)$ has exactly two components, this proves that $\mathbf{N}$ consists of two connected components, each of dimension $\frac{1}{2}p(p-1)$. The p-planes in one component are called α -*planes*, while the p-planes in the other component are called β -*planes*.

Suppose $a \in \mathrm{Pin}(p, p)$ and $x \in \mathrm{PURE}$ are given. Let $z, w \in \mathbf{R}(p, p)$ be related by $z = a^{-1}wa$ or $w = aza^{-1}$. Then the following are equivalent: $w \in N_{ax}$, $wax = 0$, $azx = 0$, $zx = 0$, $z \in N_x$, $w \in aN_xa^{-1}$. This proves that $\mathrm{Pin}(p) \subset \mathrm{Pin}(p, p)$ maps the set of pure pinors PURE into itself and that the map from $\mathrm{PURE}/\mathbf{R}^*$ to $\mathbf{N}$ is equivariant, i.e., $N_{ax} = aN_xa^{-1}$.

Next we shall show that there exists a positive spinor s_+, which is pure, and that if $N_x = N_{s_+}$ for any other pure pinor, then $x = \lambda\, x_+$ with $\lambda \in \mathbf{R}^*$, is a scalar multiple of s_+. Assuming this fact, the proof of the theorem is completed as follows. Each totally null plane $N \in \mathbf{N}$ is of the form $N = aN_{s_+}a^{-1} = N_{as_+}$ for some $a \in \mathrm{Pin}(p)$, since $O(p)$ acts transitively on $\mathbf{N}$. Therefore, the map from PURE to $\mathbf{N}$ is surjective. The equivariance, plus the fact that $N_x = N_{s_+}$ if and only if $x = \lambda s_+$ with $\lambda \in \mathbf{R}^*$, implies that $\mathrm{PURE}/\mathbf{R}^* \cong \mathbf{N}$.

Since $\mathrm{Spin}(p) \subset \mathrm{Spin}(p, p)$ has two connected components, one $\mathrm{Spin}^0(p) = \mathrm{Spin}(p) \cap \mathrm{Cl}^{\mathrm{even}}(p, p)$ consisting of even Clifford elements and the other $\mathrm{Spin}(p) \cap \mathrm{Cl}^{\mathrm{odd}}(p, p)$ consisting of odd Clifford elements, the orbit of $\mathrm{Spin}(p)$ through $s_+ \in \mathbf{S}_+$ must consist of a component PURE^+ in $\mathbf{S}^+$ and a component PURE^- in $\mathbf{S}_-$.

To prove the existence of a pure positive spinor s_+, we use the model $\mathrm{Cl}(p,p) \cong \mathrm{End}_{\mathbf{R}}(\mathbf{P})$ of this chapter with $\mathbf{P} \equiv \Lambda\mathbf{R}(p,0)$. Consider the positive spinor $s_+ \equiv 1 \in \Lambda^{\mathrm{even}}\mathbf{R}(p,0)$. A vector $z \equiv (u,v) \in \mathbf{R}(p,p) \subset \mathrm{Cl}(p,p) \cong \mathrm{End}(\mathbf{P})$ acts on s_+ by

$$z(s_+) = (E_{u+v} - I_{u-v})(s_+) = u + v.$$

Therefore, $N_{s_+} = \{(u,-u) : u \in \mathbf{R}^p\}$, which is p-dimensional (i.e., $s_+ \in$ PURE). If $x \in$ PURE and $N_x = N_{s_+}$, then for all $z = (u,-u) \in N_{s_+}$, $z(x) = -I_{2u}(x) = -2u \lrcorner x$ vanishes, i.e., $u \lrcorner x = 0$ for all $u \in \mathbf{R}(p,0)$. This proves that the element $x \in \Lambda\mathbf{R}(p,0) \equiv \mathbf{P}$ must belong to $\Lambda^0\mathbf{R}(p,0)$, i.e., $x = \lambda s_+$ for some $\lambda \in \mathbf{R}^*$, completing the proof of the theorem. ∎

Now we consider the complex case. Suppose $V_{\mathbf{C}}$ is a complex inner product space and assume that the dimension $n \equiv \dim_{\mathbf{C}} V_{\mathbf{C}} = 2p$ is even. Let $\mathbf{N}_{\mathbf{C}}$ denote the space of all totally null p-planes through the origin in $V_{\mathbf{C}}$. (Note that $\mathbf{N}_{\mathbf{C}}$ is a complex submanifold of $G_{\mathbf{C}}(p, V_{\mathbf{C}})$ and hence is projective algebraic.) Assume that $\mathrm{Cl}_{\mathbf{C}}(V_{\mathbf{C}}) \cong \mathrm{End}_{\mathbf{C}}(\mathbf{P}_{\mathbf{C}})$ is a pinor representation. A pinor $x \in \mathbf{P}_{\mathbf{C}}$ is said to be *pure* if

(12.102) $\qquad N_x \equiv \{z \in V_{\mathbf{C}} \subset \mathrm{Cl}(V_{\mathbf{C}}) : z(x) = 0\}$

is of complex dimension p, i.e., $N_x \in \mathbf{N}_{\mathbf{C}}$. Let $\mathrm{PURE}_{\mathbf{C}}$ denote the set of pure spinors.

The complex inner product space $V_{\mathbf{C}}$ may be expressed as $V_{\mathbf{C}} \cong V(p,p) \otimes_{\mathbf{R}} \mathbf{C}$ with $V(p,p) \cong \mathbf{R}(p,p)$ split. The model

$$\mathrm{Cl}(p,p) \cong \mathrm{End}_{\mathbf{R}}(\Lambda\mathbf{R}(p,0))$$

can be complexified, and some parts of the proof of Theorem 12.100 carry over to the complex case. For example,

(12.103)   (a)   there exists a pure positive spinor $s_+ \in \mathbf{S}^+$, and

               (b) $N_x = N_{s_+}$ if and only if $x = \lambda s_+$ for some $\lambda \in \mathbf{C}^*$

are proved exactly as in the split case. However, just because $N \in \mathbf{N}_{\mathbf{C}}$ is totally null $N \cap (\{0\} \times \mathbf{C}^p)$ need not be equal to $\{0\}$. That is, it is no longer true that each $N \in \mathbf{N}_{\mathbf{C}}$ can be graphed over $\mathbf{C}^p \times \{0\} \subset \mathbf{C}^p \times \mathbf{C}^p$, as in the split case.

Consequently, the model

(12.104) $V_{\mathbf{C}} \cong V(2p) \otimes_{\mathbf{R}} \mathbf{C}$,   with $V(2p) \cong \mathbf{R}(2p)$ positive definite, is more convenient. Now each $N \in \mathbf{N_C}$ can be graphed over $V \equiv V(2p) \subset V \oplus iV = V_{\mathbf{C}}$, since iV cannot contain any null vectors other than 0. Suppose $N \equiv \operatorname{graph} A \equiv \{u + iAu : u \in V\}$ is the graph of $A \in \operatorname{End}_{\mathbf{R}}(V)$. Then N is totally null if and only if $\langle u + iAu, v + iAv \rangle = \langle u, v \rangle - \langle Au, Av \rangle + i(\langle Au, v \rangle + \langle u, Av \rangle) = 0$. That is, $A \in O(2p)$ and $A^t = -A$, or equivalently, $A \in O(2p)$ and $A^2 = -\operatorname{Id}$. Thus, $A \in \operatorname{Cpx}(2p)$ is an orthogonal complex structure. This proves that

$$\mathbf{N_C} \cong \operatorname{Cpx}(2p).$$

Recall that $\operatorname{Cpx}(2p) \cong O(2p)/U(p)$ with $g \in O(2p)$ sending $J \in \operatorname{Cpx}(2p)$ to gJg^{-1}. Therefore, $\mathbf{N_C} \cong O(2p)/U(p)$ with $g \in O(2p)$ sending $N \in \mathbf{N_C}$ to gNg^{-1}. The space

$$(12.105) \qquad \mathbf{N_C} \cong \operatorname{Cpx}(2p) \cong O(2p)/U(p)$$

is called the *twistor space* (at a point on a manifold), i.e., the *twistor fiber*.

Remark 12.106. Suppose $J \in \operatorname{Cpx}(V)$. Note that $N \equiv \operatorname{graph} J \equiv \{u + iJu : u \in V\}$ is just $V^{1,0}(J)$, the $+i$ eigenspace of J (extended to $V_{\mathbf{C}}$). As in the split case, it follows easily that $N_{ax} = aN_x a^{-1}$ for $x \in \operatorname{PURE}_{\mathbf{C}}$ and $a \in \operatorname{Pin}(2p, \mathbf{C})$. The group $\operatorname{Pin}(2p)$ is naturally a subgroup of $\operatorname{Pin}(2p, \mathbf{C})$, since $V(2p) \subset V_{\mathbf{C}}$. This completes the proof of the following result.

Theorem 12.107. $\operatorname{Pin}(2p) \subset \operatorname{Pin}(2p, \mathbf{C})$ *acts transitively on* $\operatorname{PURE}_{\mathbf{C}}/\mathbf{C}^*$. *The map* χ *sending* $x \in \operatorname{PURE}_{\mathbf{C}}$ *to* $N_x \in \mathbf{N_C} \cong \operatorname{Cpx}(2p)$ *is an equivariant isomorphism* χ:

$$(12.108) \qquad \operatorname{PURE}_{\mathbf{C}}/\mathbf{C}^* \cong \mathbf{N_C} \cong \operatorname{Cpx}(2p) \cong O(2p)/U(p).$$

Also, $\operatorname{PURE}_{\mathbf{C}} = \operatorname{PURE}_{\mathbf{C}}^+ \cup \operatorname{PURE}_{\mathbf{C}}^-$ *with* $\operatorname{PURE}^{\pm}$ *a connected subset of* $\mathbf{S}_{\mathbf{C}}^{\pm}$.

Remark 12.109. In particular, given $s_+ \in \operatorname{PURE}_{\mathbf{C}}^+$ (with associated complex structure J) and an element $a \in \operatorname{Pin}(2p)$:

$$(12.110) \qquad \chi_a(s_+) \in U(p) \text{ if and only if } as_+ = cs_+ \text{ with } c \in \mathbf{C}^*.$$

Proposition 13.79 has a complex analogue which says that

$$(12.111) \qquad \text{if } s \in \operatorname{PURE}_{\mathbf{C}} \text{ then } s \,\hat{\circ}\, s = c\, dz_1 \wedge \cdots \wedge dz_p,$$

with $c \in \mathbf{C}^*$ and $z_1, \ldots, z_p$ a choice of complex linear coordinates for $V_{\mathbf{C}}(2p)$. The *square of s* is defined by $(s \,\hat{\circ}\, s)(x) = s\hat{\varepsilon}(s, x)$. It then follows easily that

$$(12.112) \qquad \{a \in \operatorname{Pin}(2p) : as = s\} \cong \operatorname{SU}(p),$$

with the isomorphism given by the vector representation χ. That is, the isotropy subgroup of Pin$(2p)$ at a pure spinor is SU(p) (cf. Problem 8).

PROBLEMS

1. Given $u \in V$ and $x \in \text{Cl}(V)$, show that

$$x \cdot u = u \wedge \tilde{x} + u \lrcorner \tilde{x}.$$

This formula has the following interesting interpretation: Right Clifford multiplication on x by a vector u is the same as first replacing x by $\tilde{x}$ and then left multiplying by u using Clifford multiplcation based on the inner product $-\langle , \rangle$ of signature s,r rather than r,s.

2. (a) Give the proof of Corollary 12.39 if p is even.

   (b) Complete the proof that $\hat{\varepsilon}$ is skew if $p = 1$ mod 4, $\hat{\varepsilon}$ is symmetric if $p = 3$ mod 4.

3. Complete the proof of the $\check{\varepsilon}$ portion of Table 12.30.

4. (a) Prove that, for $r - s = 1$ or 5 mod 8, Cl(r,s) contains no proper nontrivial two-sided ideals.

   (b) Prove that, for $r - s = 3$ or 7 mod 8, the only proper nontrivial two-sided ideals of Cl(r,s) are Cl$^{\pm}(r,s)$.

   (c) Suppose

$$\begin{array}{ccc} \mathbf{R}(r,s) & \xrightarrow{\phi} & A \text{ (nontrivial)} \\ \cap & \nearrow \phi & \\ \text{Cl}(r,s) & & \end{array}$$

is given exactly as in the Fundamental Lemma of Clifford Algebras. Show that ϕ is injective if $r - s \neq 3, 7$ mod 8, while the kernel of ϕ is either $\{0\}$, Cl$(r,s)^{+}$, or Cl$(r,s)^{-}$ if $r - s = 3, 7$ mod 8.

5. In the model $\mathbf{P}(p,p) = \Lambda \mathbf{R}(p,0)$, Cl$(p,p) \cong \text{End}_{\mathbf{R}}(\mathbf{P})$ of this chapter, show that the set of pure positive spinors PURE^{+} is equal to the subset $\mathbf{R}^{*} \cdot \text{Spin}(0,p)$ of $\Lambda^{\text{even}}\mathbf{R}^{p} = \mathbf{S}_{+}$.

6. Prove that the span of PURE^{+} is $\mathbf{S}^{+}(p,p)$.

7. Show that a positive spinor $x \in \mathbf{S}_{+}(4,4)$ is pure if and only if x is null, i.e., $\varepsilon(x,x) = 0$.

8. Suppose $J \in \text{Cpx}(2p)$ is an orthogonal complex structure on the positive euclidean space $\mathbf{R}(2p)$. Suppose $e_1, Je_1, \ldots, e_p, Je_p$, is an orthonormal basis for $\mathbf{R}(2p)$ and let $\alpha^1, \beta^1, \ldots, \alpha^p, \beta^p$ denote the dual basis. Then $\omega = \alpha^1 \wedge \beta^1 + \ldots + \alpha^p \wedge \beta^p$ is the standard Kähler

form. Let χ : $\text{Spin}(2p) \to \text{SO}(2p)$ denote the vector representation. Identify $\wedge\mathbf{R}(2p)$ and $\wedge\mathbf{R}(2p)^*$ using the inner product. Let $\lambda \equiv \alpha^1 \wedge \beta^1 \wedge \ldots \wedge \alpha^p \wedge \beta^p$ denote the unit volume.

(a) Show that

$$\phi \equiv 2^{-p/2}(1 + \alpha^1 \wedge \beta^1) \cdots (1 + \alpha^p \wedge \beta^p) \in \text{Spin}(2p) \subset \text{Cl}(2p)$$

is independent of the orthonormal basis.

(b) Show that $\chi_\phi = J$, $\phi^2 = \lambda$, and that $\phi^{-1} = \phi\lambda = \hat{\phi}$.

(c) Define $\tilde{U}(p) \equiv \{a \in \text{Pin}(2p) : \chi_a \in U(p)\}$. Show that

$$\tilde{U}(p) = \{a \in \text{Spin}(2p) : a\phi = \phi a\}$$

and

$$\tilde{U}(p) = \{a \in \text{Spin}(2p) : a\omega = \omega a\}.$$

(d) Using (12.111), show that the isotropy subgroup of $\text{Pin}(2p)$ at a pure spinor is $SU(p)$.

(e) Show that $\chi^{-1}(SU(p))$ has two connected components.

9. (a) Suppose $N \in \mathbf{N}$ is a p-dimensional totally null plane in $\mathbf{R}(p,p)$ and $a \in \text{GL}_\mathbf{R}(N)$. Show that there exists $A \in O(p,p)$ with $A|_N = a$; i.e., show that a extends to an isometry of $\mathbf{R}(p,p)$.

Hint: Use the model $N = \mathbf{R}^p \times \{0\} \subset \mathbf{R}^p \times \mathbf{R}^p$ with $\|(u,v)\| = uv$.

(b) Show that any linear isomorphism $a : N_1 \to N_2$ of totally null p-planes $N_1, N_2 \in \mathbf{N}$ can be extended to an isometry A of $\mathbf{R}(p,p)$, i.e., $A \in O(p,p)$.

(c) Prove Witt's Theorem: Suppose $\overline{V}$ and $\overline{W}$ are isometric euclidean vector spaces (i.e., same dimension and signature). Then each isometry $a : V \to W$, from a subspace V of $\overline{V}$ onto a subspace W of $\overline{W}$ extends to an isometry A from $\overline{V}$ to $\overline{W}$.

13. Inner Products on the Spaces of Spinors and Pinors

The hat and the check anti-automorphism of $\mathrm{Cl}(r,s)$ can be described in terms of extra structure on the space $\mathbf{P}(r,s)$ of pinors. These two anti-automorphisms agree on $\mathrm{Cl}(r,s)^{\mathrm{even}}$ and will be referred to as the *canonical involution* of $\mathrm{Cl}(r,s)^{\mathrm{even}}$. This canonical involution on $\mathrm{Cl}(r,s)^{\mathrm{even}}$ can be described in terms of extra structure on the space $\mathbf{S}(r,s)$ of spinors.

THE SPINOR INNER PRODUCT

The *spinor inner product*, denoted ε, is an inner product on the space $\mathbf{S}(r,s)$ of spinors with the property that the adjoint with respect to ε is the canonical involution of $\mathrm{Cl}(r,s)^{\mathrm{even}}$. The next result establishes the existence of spinor inner products and classifies the types. It is interesting to note that all of the different types of inner products introduced in Chapter 2 can occur as spinor inner products.

Theorem 13.1. *There exists an inner product ε on the space of spinors $\mathbf{S}(r,s)$ called the* spinor inner product *with the property that, given $a \in \mathrm{Cl}(r,s)^{\mathrm{even}}$,*

$$(13.2) \qquad \varepsilon(ax, y) = \varepsilon(x, \hat{a}y) \quad \text{for all } x, y \in \mathbf{S}.$$

247

Table 13.3. Spinor Inner Products

r − s = 0 mod 8: p mod 4 ε

$S \equiv S_+ \oplus S_-$ real

$\mathrm{Cl}(r,s)^{\mathrm{even}} \equiv$
$\mathrm{End}_{\mathbf{R}}(S_+) \oplus \mathrm{End}_{\mathbf{R}}(S_-)$

p mod 4	ε
0	$\varepsilon \equiv \varepsilon_+ \oplus \varepsilon_-$, both $\varepsilon_\pm$ **R**-symmetric
1	ε reflective, $S_- \cong S_+^*$
2	$\varepsilon \equiv \varepsilon_+ \oplus \varepsilon_-$, both $\varepsilon_\pm$ **R**-skew
3	ε reflective, $S_- \cong S_+^*$

r − s = 1, 7 mod 8: p mod 4 ε

$\mathrm{Cl}(r,s)^{\mathrm{even}} \equiv \mathrm{End}_{\mathbf{R}}(S)$

p mod 4	ε
0	**R**-symmetric
1	**R**-skew
2	**R**-skew
3	**R**-symmetric

r − s = 2, 6 mod 8: p mod 4 ε

$\mathrm{Cl}(r,s)^{\mathrm{even}} \equiv \mathrm{End}_{\mathbf{C}}(S)$

p mod 4	ε
0	**C**-symmetric
1	**C**-hermitian (symmetric)
2	**C**-skew
3	**C**-hermitian (skew)

r − s = 3, 5 mod 8: p mod 4 ε

$\mathrm{Cl}(r,s)^{\mathrm{even}} \equiv \mathrm{End}_{\mathbf{H}}(S)$

p mod 4	ε
0	**H**-hermitian skew
1	**H**-hermitian symmetric
2	**H**-hermitian symmetric
3	**H**-hermitian skew

r − s = 4 mod 8: p mod 4 ε

$S \equiv S_+ \oplus S_-$

$\mathrm{Cl}(r,s) \cong$
$\mathrm{End}_{\mathbf{H}}(S_+) \oplus \mathrm{End}_{\mathbf{H}}(S_-)$

p mod 4	ε
0	$\varepsilon \equiv \varepsilon_+ \oplus \varepsilon_-$, both $\varepsilon_\pm$ **H**-hermitian skew
1	ε reflective, $S_- \cong S_+^*$
2	$\varepsilon \equiv \varepsilon_+ \oplus \varepsilon_-$, both $\varepsilon_\pm$ **H**-hermitian symmetric
3	ε reflective, $S_- \cong S_+^*$

The type of the spinor inner product ε is described in the Table 13.3. Here $n \equiv r + s \equiv 2p$ defines p if the dimension n is even, and $n \equiv r + s \equiv 2p + 1$ defines p if the dimension n is odd. If the signature of $\mathbf{R}(r, s)$ is definite (i.e., positive, in case $s = 0$, or negative in case $r = 0$), then ε is definite. (In particular, ε must be of the type that has a signature.) In all other cases, where ε has a signature, the signature of ε is split.

Remark. Since the spinor inner product ε is only determined up to a change of scale (see Theorem 8.33), it is *not* the signature r', s' of ε that is an invariant but the absolute value $|r' - s'|$. However, we continue to use the terminology "ε has a signature." In the cases $\varepsilon \equiv \varepsilon_+ \oplus \varepsilon_-$, another abuse of terminology occurs in Theorem 13.1. Here either ε_+ or ε_- may be changed by a nonzero real constant, so that, strictly speaking, $\varepsilon \equiv \varepsilon_+ \oplus \varepsilon_-$ does not have a signature even in the absolute value sense, $|r' - s'|$, discussed above. However, if $\varepsilon_\pm$ has a signature $r_\pm, s_\pm$, then $|r_\pm - s_\pm|$ is a true invariant.

The proof of Theorem 13.1 (constructing the spinor inner products ε) will be given as a corollary of Theorem 13.17 (constructing the pinor inner products $\hat{\varepsilon}$ and $\breve{\varepsilon}$).

The next remark applies to the algebra $A = \mathrm{Cl}(r, s)^{\mathrm{even}}$ for either $r - s = 0 \bmod 8$ (with $F \equiv \mathbf{R}$), or $r - s = 4 \bmod 8$ (with $F \equiv \mathbf{H}$). It gives the definition of the term ε-reflective used in Table 13.3.

Remark 13.4. Suppose that the algebra A is of the form

$$A = M_N(F) \times M_N(F) \cong \mathrm{End}_F(V_+) \oplus \mathrm{End}_F(V_-).$$

Suppose $\varepsilon : V_+ \times V_- \to F$ is a nondegenerate hermitian F-bilinear form (or equivalently, $\flat(u)(v) \equiv \varepsilon(u, v)$ for all $u \in V_+, v \in V_-$ defines an F-linear isomorphism $\flat : V_+ \to V_-^*$). Then ε is said to be *reflective*, and the anti-automorphism of the algebra A sending $(a, b) \in A$ to $(b^*, a^*) \in A$, defined by

$$(13.5) \qquad \begin{aligned} \varepsilon(au, v) &= \varepsilon(u, a^*v) \quad \text{and} \\ \varepsilon(u, bv) &= \varepsilon(b^*u, v) \quad \text{for all } u \in V_+, v \in V_-, \end{aligned}$$

is called the (*reflective*) ε *involution* of A.

See Problem 1.7 for the definition of W^*, when W is a right $\mathbf{H}$-space.

Suppose a reflective $\varepsilon : V_+ \times V_- \to F$ is given. Then there exists an (unique up to a change of scale) F-hermitian symmetric inner product on $V \equiv V_+ \oplus V_-$ (also denoted ε) with the property that the reflective ε involution is the adjoint with respect to this inner product on V. Namely, (13.6)

$$\varepsilon(z_1, z_2) \equiv \varepsilon(x_1, y_2) + \overline{\varepsilon(x_2, y_1)}$$

$$\text{for all } z_1 = (x_1, y_1),\ z_2 = (x_2, y_2) \in V \equiv V_+ \oplus V_-,$$

defines the inner product on V. Note that we could have defined this inner product on V by

$$\varepsilon(z_1, z_2) \equiv \varepsilon(x_1, y_2) - \overline{\varepsilon(x_2, y_1)}$$

so that ε would be F-hermitian skew. Thus, the property of the inner product on V being symmetric or skew is of no significance, and we simply say ε is *reflective* on $V \equiv V_+ \oplus V_-$.

THE SPIN REPRESENTATION AND THE (REDUCED) CLASSICAL COMPANION GROUP $\mathrm{Cp}^0(r, s)$

As noted after (10.38), the reduced spin group $\mathrm{Spin}^0(r, s)$ is a subgroup of the *classical companion* group

(13.7) $\mathrm{Cp}(r, s) \equiv \{a \in \mathrm{Cl}^{\mathrm{even}}(r, s) : a\hat{a} = 1\}.$

Since $\hat{a}$ is the adjoint of a with respect to the spinor inner product ε, this group $\mathrm{Cp}(r, s)$ is indeed a classical group, namely, the group that fixes the inner product ε. Thus Theorem 13.1 provides a classical description of this companion group $\mathrm{Cp}(r, s)$ (see Problem 1).

The subgroup of the classical companion group, equal to the connected component of the identity element in $\mathrm{Cp}(r, s)$, is called the *reduced classical companion* to $\mathrm{Spin}^0(r, s)$ and denoted by $\mathrm{Cp}^0(r, s)$.

Recall from (10.53) and Problem 10.7 that for the spin representations ρ in dimension $n \geq 3$, the determinant (real or complex) of $\rho(a)$, with $a \in \mathrm{Spin}$, is in the finite set $\{\pm 1, \pm i\}$. Therefore, for $a \in \mathrm{Spin}^0(r, s)$, each such determinant must be one, because $\mathrm{Spin}^0(r, s)$ is connected.

Each of these reduced classical companion groups (is a classical group that) can be read off from the next theorem. For example, if $r - s = 0 \bmod 8$ and $n = r + s = 2p$ with $p = 0 \bmod 4$ then $\mathrm{Cp}^0(r, s) = \mathrm{SO}^\dagger(\mathbf{S}_+) \times \mathrm{SO}^\dagger(\mathbf{S}_-)$ is the reduced classical companion of $\mathrm{Spin}^0(r, s)$ (cf. Problem 5). The main result of the section *Cartan's Isomorphisms* in Chapter 14 is obtained by reading off the (reduced) classical companion group $Cp^0(r, s)$ in low dimensions.

Theorem 13.8 (The Spin Representations).

$r - s = 0 \bmod 8$:

$p = 0 \bmod 4$ $1 \to \mathbf{Z}_2 \equiv \{1, \lambda\} \longrightarrow \mathrm{Spin}^0(r, s) \xrightarrow{\rho^+} \mathrm{SO}^\uparrow(\mathbf{S}_+)$,

                   $1 \to \mathbf{Z}_2 \equiv \{1, -\lambda\} \to \mathrm{Spin}^0(r, s) \xrightarrow{\rho^-} \mathrm{SO}^\uparrow(\mathbf{S}_-)$,

$p = 2 \bmod 4$ $1 \to \mathbf{Z}_2 \equiv \{1, \lambda\} \longrightarrow \mathrm{Spin}^0(r, s) \xrightarrow{\rho^+} \mathrm{Sp}(\mathbf{S}_+, \mathbf{R})$,

                   $1 \to \mathbf{Z}_2 \equiv \{1, -\lambda\} \to \mathrm{Spin}^0(r, s) \xrightarrow{\rho^-} \mathrm{Sp}(\mathbf{S}_-, \mathbf{R})$,

                   except for $\mathrm{Spin}^0(2,2)$

$p = 1, 3 \bmod 4$ $1 \to \mathbf{Z}_2 \equiv \{1, \lambda\} \longrightarrow \mathrm{Spin}^0(r, s) \xrightarrow{\rho^+} \mathrm{SL}(\mathbf{S}_+, \mathbf{R})$,

                   $1 \to \mathbf{Z}_2 \equiv \{1, -\lambda\} \to \mathrm{Spin}^0(r, s) \xrightarrow{\rho^-} \mathrm{SL}(\mathbf{S}_+^*, \mathbf{R})$,

                   except for $\mathrm{Spin}(1,1) \cong \mathrm{GL}(1,\mathbf{R})$.

$r - s = 4 \bmod 8$:

$p = 0 \bmod 4$ $1 \to \mathbf{Z}_2 \equiv \{1, \lambda\} \longrightarrow \mathrm{Spin}^0(r, s) \xrightarrow{\rho^+} \mathrm{SK}(\mathbf{S}_+)$,

                   $1 \to \mathbf{Z}_2 \equiv \{1, -\lambda\} \to \mathrm{Spin}^0(r, s) \xrightarrow{\rho^-} \mathrm{SK}(\mathbf{S}_-)$,

$p = 2 \bmod 4$ $1 \to \mathbf{Z}_2 \equiv \{1, \lambda\} \longrightarrow \mathrm{Spin}^0(r, s) \xrightarrow{\rho^+} \mathrm{HU}(\mathbf{S}_+)$,

                   $1 \to \mathbf{Z}_2 \equiv \{1, -\lambda\} \to \mathrm{Spin}^0(r, s) \xrightarrow{\rho^-} \mathrm{HU}(\mathbf{S}_-)$,

                   except for $\mathrm{Spin}(4,0) = \mathrm{Spin}(0,4)$,

$p = 1, 3 \bmod 4$ $1 \to \mathbf{Z}_2 \equiv \{1, \lambda\} \longrightarrow \mathrm{Spin}^0(r, s) \xrightarrow{\rho^+} \mathrm{SL}(\mathbf{S}_+, \mathbf{H})$,

                   $1 \to \mathbf{Z}_2 \equiv \{1, -\lambda\} \to \mathrm{Spin}^0(r, s) \xrightarrow{\rho^-} \mathrm{SL}(\mathbf{S}_+^*, \mathbf{H})$.

$r - s = 1, 7 \bmod 8$:

$p = 0, 3 \bmod 4$ $\mathrm{Spin}^0(r, s) \subset \mathrm{SO}^\uparrow(\mathbf{S})$,

$p = 1, 2 \bmod 4$ $\mathrm{Spin}^0(r, s) \subset \mathrm{Sp}(\mathbf{S}, \mathbf{R})$.

$r - s = 2, 6 \bmod 8$:

$p = 0 \bmod 4$ $1 \to \mathrm{Spin}^0(r, s) \xrightarrow{\rho} \mathrm{SO}(\mathbf{S}, \mathbf{C})$,

                   $1 \to \mathrm{Spin}^0(r, s) \xrightarrow{\overline{\rho}} \mathrm{SO}(\overline{\mathbf{S}}, \mathbf{C})$,

$p = 1, 3 \bmod 4$ $1 \to \mathrm{Spin}^0(r, s) \xrightarrow{\rho} \mathrm{SU}(\mathbf{S})$,

                   $1 \to \mathrm{Spin}^0(r, s) \xrightarrow{\overline{\rho}} \mathrm{SU}(\overline{\mathbf{S}})$,

                   except for $\mathrm{Spin}(2,0) = \mathrm{Spin}(0,2) = \mathrm{U}(1)$,

$p = 2 \bmod 4$ $1 \to \mathrm{Spin}^0(r, s) \xrightarrow{\rho} \mathrm{Sp}(\mathbf{S}, \mathbf{C})$,

                   $1 \to \mathrm{Spin}^0(r, s) \xrightarrow{\overline{\rho}} \mathrm{Sp}(\overline{\mathbf{S}}, \mathbf{C})$.

$r - s = 3, 5 \bmod 8$:

$p = 0, 3 \bmod 4$ $\mathrm{Spin}^0(r, s) \subset \mathrm{SK}(\mathbf{S})$,

$p = 1, 2 \bmod 4$ $\mathrm{Spin}^0(r, s) \subset \mathrm{HU}(\mathbf{S})$.

Proof: The only part of Theorem 13.8 that does not follow from the spinor inner product theorem and the extra determinant restrictions discussed above is the fact that

$$(13.9) \qquad \ker \rho_+ = \{1, \lambda\}, \quad \text{kernel } \rho_- = \{1, -\lambda\}$$

for $r - s = 0 \mod 4$ and the dimension $n \equiv r + s \geq 6$. This follows from the next theorem (cf. Problem 10.11). ∎

Theorem 13.10. *If the dimension $n \equiv r + s \geq 5$ and*

$$\rho : \mathrm{Spin}^0(r, s) \longrightarrow \mathrm{End}_{\mathbf{R}}(W)$$

is a nontrivial representation of the (reduced) spin group $\mathrm{Spin}^0(r, s)$, then

$$(13.11) \qquad \ker \rho \subset \mathrm{cen}\ \mathrm{Spin}^0(r, s).$$

This is a standard result proved in most textbooks on Lie algebras (the complex Lie algebra $\mathfrak{so}(n, \mathbf{C})$ has no nontrivial two-sided ideals for $n \geq 5$, i.e., $\mathfrak{so}(n, \mathbf{C})$ is *simple* for $n \geq 5$).

THE PINOR INNER PRODUCTS $\check{\varepsilon}$ AND $\hat{\varepsilon}$

The extra structure on $\mathbf{P}(r, s)$ required to describe the hat anti-automorphism, mapping a to $\hat{a}$, is an inner product on $\mathbf{P}(r, s)$, denoted $\hat{\varepsilon}$, with the property that given $a \in \mathrm{Cl}(r, s)$,

$$(13.12) \qquad \hat{\varepsilon}(ax, y) = \hat{\varepsilon}(x, \hat{a}y) \quad \text{for all } x, y \in \mathbf{P}.$$

The extra structure on $\mathbf{P}(r, s)$ required to describe the check anti-automorphism, mapping a to $\check{a}$, is an inner product on $\mathbf{P}(r, s)$, denoted $\check{\varepsilon}$, with the property that given $a \in \mathrm{Cl}(r, s)$,

$$(13.13) \qquad \check{\varepsilon}(ax, y) = \check{\varepsilon}(x, \check{a}y) \quad \text{for all } x, y \in \mathbf{P}.$$

Recall from Definition 11.30 and Theorems 11.34 and 11.66 that there exists a spinor structure map $s \in \mathrm{End}_{\mathbf{R}}(\mathbf{P})$ that determines the canonical automorphism of $\mathrm{Cl}(r, s)$ by

$$(13.14) \qquad \tilde{a} = sas^{-1} \quad \text{for all } a \in \mathrm{Cl}(r, s).$$

Consequently, either one of $\hat{\varepsilon}$ or $\check{\varepsilon}$ can be used to determine (define) the other by the formula:

$$(13.15) \qquad \check{\varepsilon}(x, y) \equiv \hat{\varepsilon}(sx, y) \quad \text{for all } x, y \in \mathbf{P}.$$

For example, if $\hat{\varepsilon}$ satisfying (13.12) is given and $\check{\varepsilon}$ is defined by (13.15), then

$$(13.16) \qquad \check{\varepsilon}(ax, y) \equiv \hat{\varepsilon}(sax, y) = \hat{\varepsilon}(\tilde{a}sx, y) = \hat{\varepsilon}(sx, \check{a}y) \equiv \check{\varepsilon}(x, \check{a}y).$$

Because of Theorem 8.33, $\hat{\varepsilon}$ and $\check{\varepsilon}$ are unique up to a change of scale (as usual, the cases $r - s = 3, 7$ mod 8 will be somewhat different). In particular, if both $\hat{\varepsilon}$ and $\check{\varepsilon}$ are given, then a change of scale may be required in order for (13.15) to be true.

The next theorem describes the *pinor inner products* $\hat{\varepsilon}$ and $\check{\varepsilon}$.

Theorem 13.17. *There exist inner products $\hat{\varepsilon}$ and $\check{\varepsilon}$ on the space of pinors* $\mathbf{P}(r, s)$ *with the property that, for each $a \in \mathrm{Cl}(r, s)$*

$$(13.18) \quad \hat{\varepsilon}(ax, y) = \hat{\varepsilon}(x, \hat{a}y) \quad \text{and} \quad \check{\varepsilon}(ax, y) = \check{\varepsilon}(x, \check{a}y) \quad \text{for all } x, y \in \mathbf{P}.$$

The type of the pinor inner products $\hat{\varepsilon}$ and $\check{\varepsilon}$ is described in Table 13.19. Here $n \equiv r + s \equiv 2p$ defines p if the dimension n is even, while $n \equiv r + s \equiv 2p + 1$ defines p if the dimension n is odd. If $\mathbf{R}(r, s)$ is positive definite (i.e., $s = 0$), then $\hat{\varepsilon}$ is (positive) definite. If $\mathbf{R}(r, s)$ is negative definite (i.e., $r = 0$), then $\check{\varepsilon}$ is (positive) definite. In all other cases, if $\hat{\varepsilon}$ or $\check{\varepsilon}$ has a signature it must be split.

The proofs of the spinor Theorem 13.1 and the pinor Theorem 13.17 are intertwined in the following five steps.

Step 1. Suppose n is even and that the $\hat{\varepsilon}$ portion of the Pinor Theorem is valid. Then the $\check{\varepsilon}$ portion of the Pinor Theorem is valid.

Step 2. Suppose n is even. The inner product $\hat{\varepsilon}$ is constructed from the inner product $\hat{\varepsilon}_{\mathbf{C}}$ of Theorem 12.45 using the pinor reality map R.

Step 3. Suppose n is odd. The Spinor Theorem is deduced from the Pinor Theorem for even dimensions.

Step 4. Suppose n is odd. The Pinor Theorem is deduced from the Spinor Theorem.

Step 5. Suppose n is even. The Spinor Theorem is deduced from the Pinor Theorem for odd dimensions.

Table 13.19. Pinor Inner Products

$r - s \equiv 0 \bmod 8$:

$\mathrm{Cl}(r, s) \cong \mathrm{End}_{\mathbf{R}}(\mathbf{P})$

$\mathbf{P} \equiv \mathbf{S}_+ \oplus \mathbf{S}_-$

$\lambda = s = \begin{pmatrix} 1 & 0 \\ 0 & -1 \end{pmatrix}$

$p \bmod 4$	$\hat{\varepsilon}$	$\check{\varepsilon}$
0	**R**-symmetric	**R**-symmetric
1	**R**-skew	**R**-symmetric
2	**R**-skew	**R**-skew
3	**R**-symmetric	**R**-skew

$r - s \equiv 1 \bmod 8$:

$\mathrm{Cl}(r, s) \cong \mathrm{End}_{\mathbf{C}}(\mathbf{P})$

$\mathbf{P} \equiv \mathbf{S} \otimes_{\mathbf{R}} \mathbf{C}$

$\lambda = i, \; s = \text{conjugation}$

$p \bmod 4$	$\hat{\varepsilon}$	$\check{\varepsilon}$
0	**C**-hermitian (symmetric)	**C**-symmetric
1	**C**-skew	**C**-hermitian (skew)
2	**C**-hermitian (skew)	**C**-skew
3	**C**-symmetric	**C**-hermitian (sym

Table 13.19. Pinor Inner Products (continued)

r − s = 2 mod 8:

$\mathrm{Cl}(r,s) \cong \mathrm{End}_{\mathbf{H}}(\mathbf{P})$
$\mathbf{P} \equiv \mathbf{S} \oplus \overline{\mathbf{S}}$

$\lambda = s = \mathbf{L}_i$
(left multiplication by i)

R_j is intertwining operator

$p \bmod 4$	$\hat{\varepsilon}$	$\check{\varepsilon}$
0	**H**-hermitian skew	**H**-hermitian skew
1	**H**-hermitian symmetric	**H**-hermitian skew
2	**H**-hermitian symmetric	**H**-hermitian symmetric
3	**H**-hermitian skew	**H**-hermitian symmetric

r − s = 3 mod 8:

$\mathrm{Cl}(r,s) \cong \mathrm{End}_{\mathbf{H}}(\mathbf{P}_+) \oplus \mathrm{End}_{\mathbf{H}}(\mathbf{P}_-)$
$\mathbf{P}_\pm = \mathbf{S}$

$\lambda = \begin{pmatrix} 1 & 0 \\ 0 & -1 \end{pmatrix} \quad s = \begin{pmatrix} 0 & 1 \\ 1 & 0 \end{pmatrix}$

$p \bmod 4$	$\hat{\varepsilon}$	$\check{\varepsilon}$
0	$\hat{\varepsilon}(z_1, z_2) = \check{\varepsilon}(sz_1, z_2)$	$\check{\varepsilon} \equiv \varepsilon \oplus \varepsilon$, ske
1	$\hat{\varepsilon} = \varepsilon \oplus \varepsilon$, ε **H**-hermitian symmetric	$\check{\varepsilon}(z_1, z_2)$
2	$\hat{\varepsilon}(z_1, z_2) = \check{\varepsilon}(sz_1, z_2)$	$\check{\varepsilon} = \varepsilon \oplus \varepsilon$, sy
3	$\hat{\varepsilon} = \varepsilon \oplus \varepsilon$, ε **H**-hermitian skew	$\check{\varepsilon}(z_1, z_2)$

Table 13.3. Pinor Inner Products (continued)

$r - s = 4 \bmod 8$: $\mathrm{Cl}(r,s) \cong \mathrm{End}_{\mathbf{H}}(\mathbf{P})$ $\mathbf{P} = \mathbf{S}_+ \oplus \mathbf{S}_-$ $\lambda = s = \begin{pmatrix} 1 & 0 \\ 0 & -1 \end{pmatrix}$	$p \bmod 4$	$\hat{\varepsilon}$	$\check{\varepsilon}$	$\mathbf{S}_+, \mathbf{S}_-$
	0	**H**-hermitian skew	**H**-hermitian skew	orthogo
	1	**H**-hermitian symmetric	**H**-hermitian skew	totally
	2	**H**-hermitian symmetric	**H**-hermitian symmetric	orthogo
	3	**H**-hermitian skew	**H**-hermitian symmetric	totally

$r - s = 5 \bmod 8$: $\mathrm{Cl}(r,s) \cong \mathrm{End}_{\mathbf{C}}(\mathbf{P})$ $\mathbf{S} \equiv \mathbf{P}$ with right **H**-structure $\lambda = \mathbf{R}_i,\ s = \mathbf{R}_j$	$p \bmod 4$	$\hat{\varepsilon}$	$\check{\varepsilon}$
	0	**C**-hermitian (skew)	**C**-symmetri
	1	**C**-skew	**C**-hermitia
	2	**C**-hermitian (symmetric)	**C**-skew
	3	**C**-symmetric	**C**-hermitia

Table 13.3. Pinor Inner Products (continued)

$r - s = 6$ mod 8:

$\mathrm{Cl}(r,s) \cong \mathrm{End}_{\mathbf{R}}(\mathbf{P})$

$\lambda = s = i$

$\mathbf{P} \otimes_{\mathbf{R}} \mathbf{C} = \mathbf{S} \oplus \bar{\mathbf{S}}$

$\mathbf{C}$ (conjugation) is an intertwining operator

p mod 4	$\hat{\varepsilon}$	$\check{\varepsilon}$	$\lambda = i$
0	**R**-symmetric	**R**-symmetric	anti-isometry
1	**R**-skew	**R**-symmetric	isometry
2	**R**-skew	**R**-skew	anti-isometry
3	**R**-symmetric	**R**-skew	isometry

$r - s = 7$ mod 8:

$\mathrm{Cl}(r,s) \cong \mathrm{End}_{\mathbf{R}}(\mathbf{P}_+) \oplus \mathrm{End}_{\mathbf{R}}(\mathbf{P}_-)$

$\mathbf{P}_\pm = \mathbf{S}$

$\lambda = \begin{pmatrix} 1 & 0 \\ 0 & -1 \end{pmatrix} \quad s = \begin{pmatrix} 0 & 1 \\ 1 & 0 \end{pmatrix}$

p mod 4	$\hat{\varepsilon}$	
0	$\hat{\varepsilon}(z_1, z_2) \equiv \check{\varepsilon}(sz_1, z_2)$	$\check{\varepsilon} = \varepsilon$
1	$\hat{\varepsilon} = \varepsilon \oplus \varepsilon,\ \varepsilon\ \mathbf{R}$-skew	$\check{\varepsilon}(z_1, z_2)$
2	$\hat{\varepsilon}(z_1, z_2) = \check{\varepsilon}(sz_1, z_2)$	$\check{\varepsilon} = \varepsilon$
3	$\hat{\varepsilon} = \varepsilon \oplus \varepsilon,\ \varepsilon\ \mathbf{R}$-symmetric	$\check{\varepsilon}(z_1, z_2)$

The Pinor Inner Products $\check{\varepsilon}$ and $\hat{\varepsilon}$

Proof of Theorems 13.1 and 13.17.

Step 1.

Assume $n \equiv r + s \equiv 2p$ is even. Also, for the moment assume that $\hat{\varepsilon}$ has been constructed and satisfies Theorem 13.17. Let λ denote the choice of unit volume element for $\mathbf{R}(r, s)$.

Definition 13.20. (n even). *The (check) pinor inner product $\check{\varepsilon}$ on $\mathbf{P}(r, s)$ is defined by*

$$(13.21) \qquad \check{\varepsilon}(x, y) \equiv \hat{\varepsilon}(\lambda x, y) \quad \text{for all } x, y \in \mathbf{P}.$$

As noted above in 13.16, if $a \in \mathrm{Cl}(r, s)$, then

$$\check{\varepsilon}(ax, y) = \check{\varepsilon}(x, \breve{a}y) \quad \text{for all } x, y \in \mathbf{P},$$

because of the corresponding fact for $\hat{\varepsilon}$. Also note that

$$(13.22) \qquad \hat{\varepsilon} \text{ is } F\text{-bilinear if and only if } \check{\varepsilon} \text{ is } F\text{-blinear}$$

and

$$\hat{\varepsilon} \text{ is } F\text{-hermitian bilinear if and only if}$$

$$(13.23) \qquad \check{\varepsilon} \text{ is } F\text{-hermitian bilinear.}$$

Lemma 13.24. *If $n \equiv r + s \equiv 2p$ is even, then*

(a) (p even)

> $\hat{\varepsilon}$ *is symmetric if and only if $\check{\varepsilon}$ is symmetric, and*
> $\hat{\varepsilon}$ *is skew if and only if $\check{\varepsilon}$ is skew,*

(b) (p odd)

> $\hat{\varepsilon}$ *is symmetric if and only if $\check{\varepsilon}$ is skew, and*
> $\check{\varepsilon}$ *is skew if and only if $\check{\varepsilon}$ is symmetric.*

We give the proof for the nonhermitian cases. The proof for the hermitian cases is essentially the same.

Proof: Because of Proposition 9.27,

$$(13.25) \qquad \hat{\lambda} = \lambda \text{ if } p \text{ is even and } \hat{\lambda} = -\lambda \text{ if } p \text{ is odd.}$$

Assume that $\hat{\varepsilon}(y, x) = \pm \hat{\varepsilon}(x, y)$ with the $+$ if $\hat{\varepsilon}$ is symmetric, and the $-$ if $\hat{\varepsilon}$ is skew. Then

$$\check{\varepsilon}(y, x) \equiv \hat{\varepsilon}(\lambda y, x) = \hat{\varepsilon}(y, \hat{\lambda} x) = \pm \hat{\varepsilon}(\hat{\lambda} x, y),$$

which, because of (13.25), equals $\pm \hat{\varepsilon}(\lambda x, y) = \pm \check{\varepsilon}(x, y)$ if p is even, and equals $\mp \hat{\varepsilon}(\lambda x, y) = \mp \check{\varepsilon}(x, y)$ if p is odd. ∎

If $r - s = 0, 4 \mod 8$, then $\lambda^2 = 1$ and the eigenspaces

$$\mathbf{S}_\pm \equiv \{ x \in \mathbf{P} : \lambda x = \pm x \}$$

are (by definition) the spaces of positive and negative spinors.

Lemma 13.26. *If* $r - s = 0, 4$ mod 8 *(and* $n \equiv r + s \equiv 2p$), *then*

(a) *(p even)* $\mathbf{S}_+$ *and* $\mathbf{S}_-$ *are orthogonal for both* $\hat{\varepsilon}$ *and* $\check{\varepsilon}$,

(b) *(p odd)* $\mathbf{S}_+$ *and* $\mathbf{S}_-$ *are totally null for both* $\hat{\varepsilon}$ *and* $\check{\varepsilon}$.

Proof: Suppose p is even. Then, for $x \in \mathbf{S}_+, y \in \mathbf{S}_-$,

$$\hat{\varepsilon}(x, y) = -\hat{\varepsilon}(\lambda x, \lambda y) = -\hat{\varepsilon}(x, \hat{\lambda} \lambda y) = -\hat{\varepsilon}(x, \lambda^2 y) = -\hat{\varepsilon}(x, y),$$

since $\hat{\lambda} = \lambda$. The proof of (a) for $\check{\varepsilon}$ is identical since $\check{\lambda} = \lambda$.

Suppose p is odd. Then, for either $x, y \in \mathbf{S}_+$ both positive, or for $x, y \in \mathbf{S}_-$ both negative,

$$\hat{\varepsilon}(x, y) = \hat{\varepsilon}(\lambda x, \lambda y) = \hat{\varepsilon}(x, \hat{\lambda} \lambda y) = -\hat{\varepsilon}(x, \lambda^2 y) = -\hat{\varepsilon}(x, y),$$

since $\hat{\lambda} = -\lambda$. The proof of (b) for $\check{\varepsilon}$ is identical since $\check{\lambda} = -\lambda$. ∎

If $r - s = 2, 6$ mod 8 then $\lambda^2 = -1$. Note that

(13.27) λ is an $\hat{\varepsilon}$ isometry if and only if λ is an $\check{\varepsilon}$ isometry.

(13.28) λ is an $\hat{\varepsilon}$ anti-isometry if and only if λ is an $\check{\varepsilon}$ anti-isometry.

Both of these statements are true because; if $\hat{\varepsilon}(\lambda x, \lambda y) = \pm \hat{\varepsilon}(x, y)$, then

$$\check{\varepsilon}(\lambda x, \lambda y) \equiv \hat{\varepsilon}(\lambda^2 x, \lambda y) = \pm \hat{\varepsilon}(\lambda x, y) = \pm \check{\varepsilon}(x, y).$$

Lemma 13.29. *If* $r - s = 2, 6$ mod 8 *(and* $n \equiv r + s \equiv 2p$) *then*

(a) *(p even)* λ *is an anti-isometry for* $\hat{\varepsilon}$ *and* $\check{\varepsilon}$,

(b) *(p odd)* λ *is an isometry for* $\hat{\varepsilon}$ *and* $\check{\varepsilon}$.

Proof: $\hat{\varepsilon}(\lambda x, \lambda y) = \hat{\varepsilon}(x, \hat{\lambda} \lambda y)$, which equals $\hat{\varepsilon}(x, \lambda^2 y) = -\hat{\varepsilon}(x, y)$ if p is even, and which equals $\hat{\varepsilon}(x, \lambda^2 y) = \hat{\varepsilon}(x, y)$ if p is odd, because of (13.25). ∎

Step 2.

In the even dimensional cases, it remains to construct $\hat{\varepsilon}$ satisfying the Pinor Theorem 13.17.

Recall from Chapter 12 that $\mathrm{Cl}(r, s)$ can be realized as a subalgebra of $\mathrm{End}_{\mathbf{C}}(\Lambda \mathbf{C}^p) \cong \mathrm{Cl}_{\mathbf{C}}(n)$ when $n \equiv r + s \equiv 2p$. Also recall that there exists a reality operator R that has the property that

(13.30) $\mathcal{R}a \cong RaR^{-1}$, for all $a \in \mathrm{End}_{\mathbf{C}}(\Lambda \mathbf{C}^p) \cong \mathrm{Cl}(n) = \mathrm{Cl}(r, s) \otimes_{\mathbf{R}} \mathbf{C}$, where $\mathcal{R}$ is the natural conjugation on $\mathrm{Cl}(r,s) \otimes_{\mathbf{R}} \mathbf{C}$. Both the fact that each of the three involutions sending a to $\tilde{a}, \hat{a}$, and $\breve{a}$ on $\mathrm{Cl}_{\mathbf{C}}(n)$ leave $\mathrm{Cl}_{\mathbf{C}}(r,s) \subset \mathrm{Cl}(n)$ fixed, and the fact that each of these involutions when restricted to $\mathrm{Cl}(r,s)$, yield the corresponding involutions on $\mathrm{Cl}(r,s)$, will (implicitly) be assumed in the following.

The complex inner product $\hat{\varepsilon}_{\mathbf{C}}$ on $\Lambda\mathbf{C}^p$ constructed in Chapter 12 will provide the basis for the construction of $\hat{\varepsilon}$ on $\mathbf{R}(r,s)$. The facts that we shall need about $\hat{\varepsilon}_{\mathbf{C}}$ are listed here for convenience.

Given $a \in \mathrm{Cl}_{\mathbf{C}}(n) \cong \mathrm{End}_{\mathbf{C}}(\Lambda\mathbf{C}^p)$,

$$(13.31) \qquad \hat{\varepsilon}_{\mathbf{C}}(ax,y) = \hat{\varepsilon}_{\mathbf{C}}(x,\hat{a}y) \quad \text{for all } x,y \in \Lambda\mathbf{C}^p.$$

The reality operator $R \in \mathrm{End}_{\mathbf{R}}(\Lambda\mathbf{C}^p)$ for $\mathrm{Cl}(r,s)$ may be chosen so that

$$(13.32) \qquad \hat{\varepsilon}_{\mathbf{C}}(Rx,Ry) = \overline{\hat{\varepsilon}_{\mathbf{C}}(x,y)} \quad \text{for all } x,y \in \Lambda\mathbf{C}^p.$$

Also,

$$(13.33) \qquad \begin{aligned} &\text{if } p = 0,3 \bmod 4, \text{ then } \hat{\varepsilon}_{\mathbf{C}} \text{ is } \mathbf{C}\text{-symmetric;} \\ &\text{if } p = 1,2 \bmod 4, \text{ then } \hat{\varepsilon}_{\mathbf{C}} \text{ is } \mathbf{C}\text{-skew.} \end{aligned}$$

Case $\mathbf{r-s = 0,6 \bmod 8}$. Here $R^2 = 1$ and $\mathrm{Cl}(r,s) \cong \mathrm{End}_{\mathbf{R}}(\mathbf{P})$, where the space of pinors is defined to be $\mathbf{P} \equiv \{x \in \Lambda\mathbf{C}^p : Rx = x\}$ (see (12.94) and (12.95)). Note that because of condition (13.32), $\hat{\varepsilon}_{\mathbf{C}}$ restricted to $\mathbf{P}$ is real-valued and nondegenerate.

Definition 13.34. *($r-s = 0,6 \bmod 8$) The hat inner product on* $\mathbf{P}(r,s)$ *is defined to be*

$$\hat{\varepsilon} \equiv \hat{\varepsilon}_{\mathbf{C}}|_{\mathbf{P}}.$$

The condition (13.18) for $\hat{\varepsilon}$ follows immediately from the analogous condition (13.31) on $\hat{\varepsilon}_{\mathbf{C}}$. Finally, because of (13.33), $\hat{\varepsilon}$ is $\mathbf{R}$-symmetric when $p = 0,3 \bmod 4$ and $\hat{\varepsilon}$ is $\mathbf{R}$-skew when $p = 1,2 \bmod 4$.

Case $\mathbf{r-s = 2,4 \bmod 8}$. Here $R^2 = -1$ and $\mathrm{Cl}(r,s) \cong \mathrm{End}_{\mathbf{H}}(\mathbf{P})$, where the space of pinors $\mathbf{P}(r,s)$ is defined to be $\Lambda\mathbf{C}^p$ equipped with the right $\mathbf{H}$-structure given by $I \equiv i$ and $J \equiv R$ (see (12.96)).

Definition 13.35. *($r-s = 2,4 \bmod 8$). The hat inner product $\hat{\varepsilon}$ on* $\mathbf{P}(r,s)$ *is defined to be*

$$(13.36) \qquad \hat{\varepsilon}(x,y) \equiv -\hat{\varepsilon}_{\mathbf{C}}(xJ,y) + j\hat{\varepsilon}_{\mathbf{C}}(x,y) \quad \text{for all } x,y \in \mathbf{P}.$$

The condition (13.18) for $\hat{\varepsilon}$ follows from the analogous condition (13.31) on $\hat{\varepsilon}_{\mathbf{C}}$ because J commutes with all $a \in \mathrm{Cl}(r,s)$.

Lemma 13.37. *Suppose W is a right **H**-space and $\varepsilon_{\mathbf{C}}$ is a complex bilinear form on W with respect to the complex structure I (right multiplication) on W, that satisfies*

$$(13.38) \qquad \varepsilon_{\mathbf{C}}(xJ, yJ) = \overline{\varepsilon_{\mathbf{C}}(x, y)}.$$

Let

$$(13.39) \qquad h(x, y) \equiv -\varepsilon_{\mathbf{C}}(xJ, y) + j\varepsilon(x, y) \quad \text{for all } x, y \in W.$$

Then

$(13.40) \qquad h$ *is **H**-hermitian symmetric if and only if $\varepsilon_{\mathbf{C}}$ is **C**-skew;*

$(13.41) \qquad h$ *is **H**-hermitian skew if and only if $\varepsilon_{\mathbf{C}}$ is **C**-symmetric.*

Proof: Since $\varepsilon_{\mathbf{C}}$ is complex bilinear and I, J as well as i, j anticommute, it is easy to show that

$$(13.42) \qquad h(xI, y) = -ih(x, y) \quad \text{and} \quad h(x, yI) = h(x, y)i.$$

Also,

$$(13.43) \qquad h(xJ, y) = \varepsilon_{\mathbf{C}}(x, y) + j\varepsilon_{\mathbf{C}}(xJ, y) = -jh(x, y)$$

is automatic.

Finally, to prove that

$$(13.44) \qquad h(x, yJ) = h(x, y)j,$$

the hypothesis (13.38) must be used:

$$h(x, yJ) = -\varepsilon_{\mathbf{C}}(xJ, yJ) + j\varepsilon_{\mathbf{C}}(x, yJ) = -\overline{\varepsilon_{\mathbf{C}}(x, y)} - j\overline{\varepsilon_{\mathbf{C}}(xJ, y)}$$
$$= j\varepsilon_{\mathbf{C}}(x, y)j - \varepsilon_{\mathbf{C}}(xJ, y)j = h(x, y)j.$$

Thus, h is **H**-hermitian.

The proofs of the remainder of (13.40) and (13.41) are similar and so are omitted (also cf. Lemmas 2.72 and 2.78). ■

Because of Lemma 13.37 and (13.33), $\hat{\varepsilon}$ is **H**-hermitian skew when $p = 0, 3 \bmod 4$ and $\hat{\varepsilon}$ is **H**-hermitian symmetric when $p = 1, 2 \bmod 4$.

Combining Step 1 and Step 2, this completes the proof of the Pinor Theorem 13.17 when the dimension $n \equiv r + s$ is even.

Step 3.

Next we give the proof of the Spinor Theorem 13.1 when $n \equiv r + s \equiv 2p + 1$ is odd. The proof is based on two ingredients. The first ingredient is the pair of isomorphism:

$$(13.45) \qquad \mathrm{Cl}(r,s)^{\mathrm{even}} \cong \mathrm{Cl}(r-1,s) \quad \text{if } r \geq 1,$$

and

$$(13.45') \qquad \mathrm{Cl}(r,s)^{\mathrm{even}} \cong \mathrm{Cl}(s-1,r) \quad \text{if } s \geq 1,$$

from Theorem 9.38. (The important fact is that these isomorphisms preserve the hat anti-automorphisms.) The second ingredient is the $\hat{\varepsilon}$ portion of the Pinor Theorem 13.17 for even dimensions.

Case $r - s = 1, 7 \bmod 8$. Then $r - 1 - s = 0, 6 \bmod 8$, so that $\mathrm{Cl}(r,s)^{\mathrm{even}} \cong \mathrm{Cl}(r-1,s) \cong \mathrm{End}_{\mathbf{R}}(\mathbf{P}(r-1,s))$, and we may take $\mathbf{S}(r,s) \equiv \mathbf{P}(r-1,s)$. The *spinor inner product* ε on $\mathbf{S}(r,s)$ is, by definition, the hat inner product $\hat{\varepsilon}$ on $\mathbf{P}(r-1,s)$. Finally, ε is **R**-symmetric if $p = 0, 3 \bmod 4$, and ε is **R**-skew if $p = 1, 2 \bmod 4$ by Theorem 13.17 applied to $\hat{\varepsilon}$ on $\mathbf{P}(r-1,s)$. If $r = 0$, consider the isomorphism $\mathrm{Cl}(r,s)^{\mathrm{even}} \cong \mathrm{Cl}(s-1,r) \cong \mathrm{End}_{\mathbf{R}}(\mathbf{P}(s-1,r))$ instead.

Case $r - s = 3, 5 \bmod 8$. Then $r - 1 - s = 2, 4 \bmod 8$, so that $\mathrm{Cl}(r,s)^{\mathrm{even}} \cong \mathrm{Cl}(r-1,s) \cong \mathrm{End}_{\mathbf{H}}(\mathbf{P}(r-1,s))$, and we may take $\mathbf{S}(r,s)$ to be the right **H**-space $\mathbf{P}(r-1,s)$. The *spinor inner product* ε on $\mathbf{S}(r,s)$ is, by definition, the hat inner product $\hat{\varepsilon}$ on $\mathbf{P}(r-1,s)$. Finally, ε is **H**-hermitian symmetric if $p = 1, 2 \bmod 4$, and ε is **H**-hermitian skew if $p = 0, 3 \bmod 4$ by Theorem 13.17 applied to $\mathbf{P}(r-1,s)$. If $r = 0$, consider the isomorphism $\mathrm{Cl}(r,s)^{\mathrm{even}} \cong \mathrm{Cl}(s-1,r) \cong \mathrm{End}_{\mathbf{H}}(\mathbf{P}(s-1,r))$ instead.

Step 4.

Next the Pinor Theorem 13.17 for odd dimensions will be deduced from the Spinor Theorem 13.1 for odd dimensions.

Case $r - s = 1 \bmod 8$. Recall from Theorem 11.66 that $\mathbf{P}(r,s) \equiv \mathbf{S}(r,s) \otimes_{\mathbf{R}} \mathbf{C}$ and that $\mathrm{Cl}(r,s) = \mathrm{End}_{\mathbf{C}}(\mathbf{P})$; and $\mathrm{Cl}(r,s)^{\mathrm{even}} \cong \mathrm{End}_{\mathbf{R}}(\mathbf{S})$, where $\lambda \equiv i$ is a unit volume element and s, the spinor structure map, is the natural conjugation sending x to $\bar{x}$ on $\mathbf{P} \equiv \mathbf{S} \otimes_{\mathbf{R}} \mathbf{C}$. Let $\varepsilon_{\mathbf{C}}$ denote the complexification of the spinor inner product ε on $\mathbf{S}$. Consulting the $r - s = 1 \bmod 8$ entry in Table 13.3 associated with Theorem 13.1, proves that

$$(13.46) \qquad \begin{aligned} &\varepsilon_{\mathbf{C}} \text{ is } \mathbf{C}\text{-symmetric for } p = 0, 3 \bmod 4, \\ &\varepsilon_{\mathbf{C}} \text{ is } \mathbf{C}\text{-skew for } p = 1, 2 \bmod 4. \end{aligned}$$

It is natural to require that $\hat{\varepsilon}$ and $\breve{\varepsilon}$ be related by $\breve{\varepsilon}(x, y) = \hat{\varepsilon}(\overline{x}, y)$, for all $x, y \in \mathbf{P}$, because of (13.15).

Definition 13.47. $(r - s = 1 \bmod 8)$ Let $n \equiv r + s \equiv 2p + 1$ define p. The pinor inner products $\hat{\varepsilon}$ and $\breve{\varepsilon}$ on $\mathbf{P}(r, s)$ are defined by: First, if p is even

$$(13.48) \qquad \hat{\varepsilon}(x, y) \equiv \hat{\varepsilon}_{\mathbf{C}}(\overline{x}, y), \quad \breve{\varepsilon}(x, y) \equiv \hat{\varepsilon}_{\mathbf{C}}(x, y) \quad \text{for all } x, y \in \mathbf{P},$$

and, second if p is odd

$$(13.49) \qquad \hat{\varepsilon}(x, y) \equiv \hat{\varepsilon}_{\mathbf{C}}(x, y), \quad \breve{\varepsilon}(x, y) \equiv \hat{\varepsilon}_{\mathbf{C}}(\overline{x}, y), \quad \text{for all } x, y \in \mathbf{P}.$$

Now we prove that

$$(13.50) \qquad \hat{\varepsilon}(ax, y) = \hat{\varepsilon}(x, \hat{a}y) \quad \text{for all } x, y \in \mathbf{P}.$$

If $a \in \mathrm{Cl}(r, s)^{\text{even}}$, then (13.49) is automatic. Suppose $ia \in \mathrm{Cl}(r, s)^{\text{odd}}$, then $a \in \mathrm{Cl}(r, s)^{\text{even}}$. Then

$$(13.51) \qquad \begin{aligned} (p \text{ even}) \quad \hat{\varepsilon}(iax, y) &\equiv \hat{\varepsilon}_{\mathbf{C}}(\overline{iax}, y) = -i\hat{\varepsilon}_{\mathbf{C}}(a\overline{x}, y) \\ &= -i\hat{\varepsilon}_{\mathbf{C}}(\overline{x}, \hat{a}y) = -\hat{\varepsilon}_{\mathbf{C}}(\overline{x}, i\hat{a}y) = -\hat{\varepsilon}(\overline{x}, i\hat{a}y), \end{aligned}$$

where $-i\hat{a} = ia$ since $\hat{i} = -i$ (the dimension $n \equiv 2p + 1 = 1 \bmod 4$).

$$(13.52) \qquad (p \text{ odd}) \quad \hat{\varepsilon}(iax, y) = \hat{\varepsilon}_{\mathbf{C}}(iax, y) = \hat{\varepsilon}_{\mathbf{C}}(x, i\hat{a}y) = \hat{\varepsilon}(x, i\hat{a}y),$$

where $i\hat{a} = ia$ since $\hat{i} = i$ (the dimension $n \equiv 2p + 1 = 3 \bmod 4$).

The type of $\hat{\varepsilon}$ and $\breve{\varepsilon}$ listed in the $r - s = 1 \bmod 8$ portion of Theorem 13.17 follows from (13.46) and the definitions of $\hat{\varepsilon}$ and $\breve{\varepsilon}$.

Case $r - s = 3 \bmod 8$. Recall from Theorem 11.66 that $\mathbf{S}(r, s)$ is a right $\mathbf{H}$-space and that $\mathbf{P}(r, s) \equiv \mathbf{P}_+(r, s) \oplus \mathbf{P}_-(r, s)$ with $\mathbf{P}_\pm(r, s) \equiv \mathbf{S}(r, s)$. The unit volume element is $\lambda = \begin{pmatrix} 1 & 0 \\ 0 & -1 \end{pmatrix}$ and the spinor structure map is $s = \begin{pmatrix} 0 & 1 \\ 1 & 0 \end{pmatrix}$. Also recall that $\mathrm{Cl}(r, s) \cong \mathrm{End}_{\mathbf{H}}(\mathbf{P}_+) \oplus \mathrm{End}_{\mathbf{H}}(\mathbf{P}_-)$ and $\mathrm{Cl}(r, s)^{\text{even}} \cong \mathrm{End}_{\mathbf{H}}(\mathbf{S})$ embedded diagonally in $\mathrm{End}_{\mathbf{H}}(\mathbf{P}_+) \oplus \mathrm{End}_{\mathbf{H}}(\mathbf{P}_-)$. Let ε denote the spinor inner product on $\mathbf{S}$. Because of Theorem 13.1,

$$(13.53) \qquad \begin{aligned} &\varepsilon \text{ is } \mathbf{H}\text{-hermitian skew if } p = 0, 3 \bmod 4, \text{ and} \\ &\varepsilon \text{ is } \mathbf{H}\text{-hermitian symmetric if } p = 1, 2 \bmod 4. \end{aligned}$$

Defintion 13.54. *The pinor inner products $\hat{\varepsilon}$ and $\check{\varepsilon}$ on $\mathbf{P} \equiv \mathbf{S} \oplus \mathbf{S}$ are defined by*

$$(13.55) \qquad (p \text{ even}) \quad \hat{\varepsilon}(z_1, z_2) = \check{\varepsilon}(sz_1, z_2) \quad \text{and} \quad \check{\varepsilon} = \varepsilon \oplus \varepsilon.$$

$$(13.56) \qquad (p \text{ odd}) \quad \hat{\varepsilon} = \varepsilon \oplus \varepsilon \quad \text{and} \quad \check{\varepsilon}(z_1, z_2) = \hat{\varepsilon}(sz_1, z_2).$$

It remains to prove that

$$(13.57) \qquad (p \text{ even}) \quad \check{\varepsilon}(Az_1, z_2) = \check{\varepsilon}(z_1, \breve{A}z_2)$$

and

$$(13.58) \qquad (p \text{ odd}) \quad \hat{\varepsilon}(Az_1, z_2) = \hat{\varepsilon}(z_1, \hat{A}z_2)$$

for $A = (a, b)$ with $a, b \in \text{End}_{\mathbf{H}}(\mathbf{S})$ and $z_1 = (x_1, y_1), z_2 = (x_2, y_2) \in \mathbf{P} \equiv \mathbf{S} \oplus \mathbf{S}$.

Both cases are automatic if $A \equiv (a, a) \in \text{Cl}(r, s)^{\text{even}}$. Each $B \equiv (a, -a) \in \text{Cl}(r, s)^{\text{odd}}$ is of the form $B = \lambda A$ with $A = (a, a) \in \text{Cl}(r, s)^{\text{even}}$. If p is even, the dimension $2p + 1 = 1 \mod 4$ and $\hat{\lambda} = -\lambda$, while if p is odd then $2p + 1 = 3 \mod 4$ and $\hat{\lambda} = \lambda$. Now (13.57) and (13.58) follow easily.

Case $r - s = 5 \mod 8$. Recall from Theorem 11.66 that $\mathbf{S}(r, s)$ is a right $\mathbf{H}$-space and that $\mathbf{P}(r, s)$ equals $\mathbf{S}(r, s)$ equipped with the complex structure I. The spinor structure map on $\mathbf{P}$ is $s \equiv J$ and the unit volume element is $\lambda \equiv I$. Let ε denote the spinor inner product on $\mathbf{S}$. Because of Theorem 13.1,

$$(13.59) \qquad \begin{array}{l} \varepsilon \text{ is } \mathbf{H}\text{-hermitian skew if } p = 0, 3 \mod 4, \text{ and} \\ \varepsilon \text{ is } \mathbf{H}\text{-hermitian symmetric if } p = 1, 2 \mod 4. \end{array}$$

Definition 13.60. *The pinor inner products $\hat{\varepsilon}$ and $\check{\varepsilon}$ are defined to be complex-valued inner products by*

$$(13.61) \qquad (p \text{ even}) \quad \varepsilon(x, y) \equiv \hat{\varepsilon}(x, y) + j\check{\varepsilon}(x, y),$$

and

$$(13.62) \qquad \begin{array}{l} (p \text{ odd}) \quad \varepsilon(x, y) \equiv \check{\varepsilon}(x, y) - j\hat{\varepsilon}(x, y) \\ \qquad (\text{or } j\varepsilon(x, y) \equiv \hat{\varepsilon}(x, y) + j\check{\varepsilon}(x, y)) \end{array}$$

for all $x, y \in \mathbf{P} \equiv \mathbf{S}$.

Note that in both cases,

$$(13.63) \qquad \check{\varepsilon}(x, y) = \hat{\varepsilon}(xJ, y) \quad \text{for all } x, y \in \mathbf{P}.$$

Now we prove that

(13.64) $\hat{\varepsilon}(ax, y) = \hat{\varepsilon}(x, \hat{a}y)$ for all $x, y \in \mathbf{P}$.

If $a \in \mathrm{Cl}(x, s)^{\text{even}}$, then (13.64) is automatic. Each $b \in \mathrm{Cl}(r, s)^{\text{odd}}$ is of the form $b = a\lambda$ with $a \in \mathrm{Cl}(r, s)^{\text{even}}$. Therefore,

$$\varepsilon(bx, y) = \varepsilon(a\lambda x, y) = \varepsilon(\lambda x, \hat{a}y) = \varepsilon(xI, \hat{a}y) = -i\varepsilon(x, \hat{a}y).$$

If p is even, this implies that $\hat{\varepsilon}(bx, y) = -i\hat{\varepsilon}(x, \hat{a}y) = -\hat{\varepsilon}(x, \lambda\hat{a}y)$, which (since the dimension $n = 2p + 1 = 1$ mod 4 implies $\hat{\lambda} = -\lambda$) equals $\hat{\varepsilon}(x, \hat{\lambda}\hat{a}y) = \hat{\varepsilon}(x, \hat{b}y)$. If p is odd, this implies that $\hat{\varepsilon}(a\lambda x, y) = i\hat{\varepsilon}(x, \hat{a}y)$, which, since the dimension $2p + 1 = 3$ mod 4, equals $\hat{\varepsilon}(x, \lambda\hat{a}y) = \hat{\varepsilon}(x, a\lambda y)$.

The type of $\hat{\varepsilon}$ and $\check{\varepsilon}$ follows from the type of ε listed in (13.59) because of the next lemma.

Lemma 13.65. *Suppose $h \equiv \alpha + j\beta$ is an $\mathbf{H}$-hermitian form and α and β are complex valued. If h is $\mathbf{H}$-hermitian symmetric, then α is $\mathbf{C}$-hermitian (symmetric) and β is $\mathbf{C}$-skew. If h is $\mathbf{H}$-hermitian skew, then α is $\mathbf{C}$-hermitian (skew) and β is $\mathbf{C}$-symmetric.*

Remark 13.66. The inner product α is said to be the *first complex part* of h while β is said to be the *second complex part* of h (cf. Chapter 2).

Case $r - s = 7$ mod 8. This case is so similar to the case $r - s = 3$ mod 8 that the proof is omitted.

This completes the proof of the Pinor Theorem 13.17. It remains to give the proof of the Spinor Theorem 13.1 in the even dimensional cases.

Step 5.

Using the hat preserving isomorphisms

$$\mathrm{Cl}(r, s)^{\text{even}} \cong \mathrm{Cl}(r - 1, s)$$

$$\mathrm{Cl}(r, s)^{\text{even}} \cong \mathrm{Cl}(s - 1, r),$$

the $r - s$ odd mod 8 portion of the Spinor Theorem 13.1 can be read off from the even portion of the Pinor Theorem 13.17 (Problem 3). ∎

This completes the description of the types for the spinor inner products ε and the pinor inner products $\hat{\varepsilon}$ and $\check{\varepsilon}$. The discussion of the signatures of these inner products will be given later in the section entitled *Signature*, as an application of pinor multiplication.

PINOR MULTIPLICATION

The inner products $\hat{\varepsilon}$ and $\check{\varepsilon}$ on $\mathbf{P}(r,s)$ enable one to multiply pinors, as described in Definition 8.42.

An application to calibrations is given in Chapter 14, and an application to pure spinors in the split case is given at the end of this section. The product $x \odot y$ induced by $\hat{\varepsilon}$ will be denoted $x \,\hat{\circ}\, y$, and the product $x \odot y$ induced by $\check{\varepsilon}$ will be denoted $x \,\check{\circ}\, y$. Therefore, if $x, y \in \mathbf{P}$ are given, then $x \,\hat{\circ}\, y$ and $x \,\check{\circ}\, y \in \mathrm{End}_F(\mathbf{P})$ are defined by

$$(13.67) \qquad\qquad (x \,\hat{\circ}\, y)(z) \equiv x\hat{\varepsilon}(y,z)$$

and

$$(13.68) \qquad\qquad (x \,\check{\circ}\, y)(z) \equiv x\check{\varepsilon}(y,z)$$

for all $z \in \mathbf{P}$.

For all signatures r, s, the pinor representation enables us to consider $\mathrm{Cl}(r,s)$ as a subalgebra of $\mathrm{End}_F(\mathbf{P}(r,s))$, with equality

$$\mathrm{Cl}(r,s) \cong \mathrm{End}_F(\mathbf{P})$$

unless $r - s = 3$ mod 4. If $r - s = 3$ mod 4, then

$$\mathrm{Cl}(r,s) \cong \mathrm{End}_F(\mathbf{P}_+) \oplus \mathrm{End}_F(\mathbf{P}_-) \subset \mathrm{End}_F(\mathbf{P}).$$

However, in all cases, $\rho(\lambda)$ (which we also denote by λ in the present context) has real trace zero. Therefore Theorem 9.65 implies that

Lemma 13.69. *Consider* $\mathrm{Cl}(r,s) \subset \mathrm{End}_F(\mathbf{P}(r,s))$. *The Clifford inner product* $\langle\,,\,\rangle$ *can be expressed as a real trace:*

$$(13.70) \qquad \langle a, b \rangle = (\dim_{\mathbf{R}} \mathbf{P})^{-1} \, \mathrm{trace}_{\mathbf{R}} \, \hat{a}b \quad \text{for all } a, b \in \mathrm{Cl}(r,s).$$

Remark 13.71. Note that this implies that the twisted Clifford inner product is also given by a trace:

$$(13.72) \qquad \langle \tilde{a}, b \rangle = (\dim_{\mathbf{R}} \mathbf{P})^{-1} \, \mathrm{trace}_{\mathbf{R}} \, \check{a}b.$$

This lemma states that the natural $\mathbf{R}$-symmetric inner product on $\mathrm{End}_F(\mathbf{P})$ induced by $\hat{\varepsilon}$ (see Defintion 8.38) is exactly the same as the Clifford inner product on $\mathrm{Cl}(r,s)$. Therefore, all the results of the section *Inner Products* in Chapter 8 are applicable to $\mathbf{P}, \hat{\varepsilon}$ and $\mathrm{End}_F(\mathbf{P}), \langle\,,\,\rangle$. These results are summarized in the next two theorems. Let $N \equiv \dim_F \mathbf{P}$.

Theorem 13.73. $\langle x \; \hat{\circ} \; y, a \rangle = \frac{1}{N} \mathrm{Re}\, \hat{\varepsilon}(ay, x)$, *for all* $a \in \mathrm{Cl}(r, s)$ *and* $x, y \in \mathbf{P}$.

Note that $\hat{\varepsilon}(ay, x)$ has the order of x and y reversed.

Theorem 13.74. *For signatures where* $\hat{\varepsilon}$ *is* F-*hermitian on* $\mathbf{P}(r, s)$,

$$(13.75) \qquad \langle x \; \hat{\circ} \; y, z \; \hat{\circ} \; w \rangle = \frac{1}{N} \, \mathrm{Re} \left(\overline{\hat{\varepsilon}(x, z)} \hat{\varepsilon}(y, w) \right).$$

For signatrues where $\hat{\varepsilon}$ *is* F-*symmetric or* F-*skew on* $\mathbf{P}(r, s)$,

$$(13.76) \qquad \langle x \; \hat{\circ} \; y, z \; \hat{\circ} \; w \rangle = \frac{1}{N} \, \mathrm{Re} \left(\hat{\varepsilon}(x, z) \hat{\varepsilon}(y, w) \right).$$

Here $x, y, z, w \in \mathbf{P}(r, s)$.

Remark 13.77. Because of (13.72), if the Clifford inner product $\langle a, b \rangle$ is replaced by the twisted Clifford inner product $\langle \tilde{a}, b \rangle$, the hat product $x \; \hat{\circ} \; y$ by the check product $x \; \check{\circ} \; y$, and $\hat{\varepsilon}$ by $\check{\varepsilon}$, then both Theorem 13.73 and Theorem 13.74 remain valid.

Other properties of pinor multiplication are easily read off Lemma 8.44.

Proposition 13.78. *For all* $a \in \mathrm{Cl}(r, s)$

(i) $(ax) \; \hat{\circ} \; y = a(x \; \hat{\circ} \; y)$, *and* $x \; \hat{\circ} \; (ay) = (x \; \hat{\circ} \; y)\hat{a}$,

(ii) $(x \; \hat{\circ} \; y)(z \; \hat{\circ} \; w) = (x \hat{\varepsilon}(y, z)) \; \hat{\circ} \; w$,

(iii) $(x \; \hat{\circ} \; y)\hat{} = y \; \hat{\circ} \; x$ *if* ε *is symmetric (hermitian or pure)*,

(iv) $(x \; \hat{\circ} \; y)\hat{} = -y \; \hat{\circ} \; x$ *if* ε *is skew (hermitian or pure)*.

Recall the notion of a pure spinor in the split case $\mathbf{P}(p, p)$. The square of a pure spinor represents the associated null plane in $\Lambda \mathbf{R}(p, p)$.

Proposition 13.79. *Suppose* $s \in \mathrm{PURE} \subset \mathbf{P}(p, p)$ *is a pure spinor. Then the square* $s \; \hat{\circ} \; s \in \mathrm{Cl}(p, p) \cong \Lambda \mathbf{R}(p, p)$ *is of the form* $s \; \hat{\circ} \; s = z_1 \wedge \cdots \wedge z_p$ *with* $N_s = \mathrm{span}\{z_1, \ldots, z_p\} = \{z \in \mathbf{R}(p, p) : z(s) = 0\}$ *totally null. Conversely, each* $\xi \in \mathrm{Cl}(p, p) = \Lambda \mathbf{R}(p, p)$ *of the form* $\xi = z_1 \wedge \cdots \wedge z_p$ *with* $N = \mathrm{span}\{z_1, \cdots, z_p\}$ *totally null is the square of a pure spinor.*

Proof: Since, for all $a \in \mathrm{Pin}(p)$, $a(s \; \hat{\circ} \; s)a^{-1} = (as) \; \hat{\circ} \; (as)$, we need only show that for some $s_+ \in \mathrm{PURE}^+, s_+ \; \hat{\circ} \; s_+ = z_1 \wedge \ldots \wedge z_p$ with $z_1, \cdots, z_p \in N_{s_+}$ (see Theorem 12.100). As in the proof of Theorem 12.100, note that $s_+ \equiv 1 \in \Lambda \mathbf{R}(p, 0) \equiv \mathbf{P}(p, p)$ is a pure positive spinor. By (13.67) and Definition 12.24,

$$(13.80) \qquad (s_+ \; \hat{\circ} \; s_+)(x) = (1 \; \hat{\circ} \; 1)(x) = \hat{\varepsilon}(1, x) \cdot 1 = (1, \sigma x) \cdot 1.$$

Therefore, $s_+ \hat{\circ} s_+ = P\sigma$, where P is orthogonal projection onto the line through $1 \in \Lambda\mathbf{R}(p,0) = \mathbf{P}$. Let $z_j \equiv -\frac{1}{2}(e_j, -e_j)$ so that $z_1, \ldots, z_p$ is a basis for N_{s_+}. Recall that $z_j = I_{e_j}$. Therefore, $\xi \equiv z_1 \cdots z_p = I_{e_1} \circ \cdots \circ I_{e_p}$ maps the volume element $\sigma \equiv e_1 \wedge \cdots \wedge e_p \in \Lambda\mathbf{R}(p,0) = \mathbf{P}$ to ± 1 and the orthogonal complement of σ to zero. This proves that $s_+ \hat{\circ} s_+ = \pm\xi$. ∎

SIGNATURE

First note that, by Problem 9.10,

(13.81)   If $s \geq 1$ (i.e., $\mathbf{R}(r,s)$ not positive definite), then $\hat{\varepsilon}$ must have split signature (if it has a signature).

(13.82)   If $r \geq 1$ (i.e., $\mathbf{R}(r,s)$ not negative definite), then $\breve{\varepsilon}$ must have split signature (if it has a signature).

Consequently, unless $r = 0$ or $s = 0$, the spinor inner product ε must have split signature (if it has a signature)

Case s = 0 (i.e., R(r,0) positive definite). Then the Clifford inner product $\langle\, ,\, \rangle$ on $\mathrm{Cl}(r,0) \cong \Lambda\mathbf{R}^r$ is positive definite. Suppose x is a null vector in $\mathbf{P}$, i.e., $\hat{\varepsilon}(x,x) = 0$. By Theorem 13.74,

$$\langle x \hat{\circ} y, x \hat{\circ} y \rangle = 0 \quad \text{for all } y \in \mathbf{P}.$$

Since $\langle\, ,\, \rangle$ is positive definite, $x \hat{\circ} y = 0$ for all y. Thus $x = 0$. This proves that $\hat{\varepsilon}$ has no nonzero null vectors, so that $\hat{\varepsilon}$ must be definite (and by adjusting the scale we may assume that $\hat{\varepsilon}$ is positive definite).

Case r = 0 (i.e., R(0,s) negative definite). Let $\langle\, ,\, \rangle$ denote the Clifford inner product on $\mathrm{Cl}(0,s) \cong \Lambda\mathbf{R}^s$. The twisted Clifford inner product $(a,b) \equiv \langle \tilde{a}, b \rangle$ is positive definite. Using the product $x \breve{\circ} y$ and Theorem 13.74 it follows that $\breve{\varepsilon}$ cannot have nonzero null vectors.

The signature part of Theorem 13.1 (cf. the Remark that follows this Theorem), for the spinor inner product ε on $\mathbf{S}(r,s)$, follows because of the hat preserving isomorphism

$$\mathrm{Cl}(r,0)^{\mathrm{even}} \cong \mathrm{Cl}(r-1,0).$$

Table 13.19 can be used to determine exactly when $\hat{\varepsilon}$ and $\breve{\varepsilon}$ have a signature.

Proposition 13.83.

(i) *The pinor inner product* $\hat{\varepsilon}$ *has a signature if and only if either* $s = 0 \bmod 4$ *or* $r = 3 \bmod 4$.

(ii) *The pinor inner product* $\check{\varepsilon}$ *has a signature if and only if either* $r = 0 \bmod 4$ *or* $s = 1 \bmod 4$.

PROBLEMS

1. List the classical companions $\mathrm{Cp}(r, s)$, defined by (13.7), as classical groups.

2. Deduce the $r - s = 0, 4 \bmod 8$ cases of the Spinor Theorem 13.1 from the $r - s = 0, 4 \bmod 8$ cases of the Pinor Theorem 13.17.

3. (a) Deduce the $r - s = 0 \bmod 8$ case of the Spinor Theorem 13.1 from the $r - s = 7 \bmod 8$ cases of the Pinor Theorem 13.17.

 (b) Deduce the $r - s = 2, 6 \bmod 8$ case of the Spinor Theorem 13.1 from the $r - s = 1, 5 \bmod 8$ cases of the Pinor Theorem 13.17.

 (c) Deduce the $r - s = 4 \bmod 8$ case of the Spinor Theorem 13.1 from the $r - s = 3 \bmod 8$ case of the Pinor Theorem 13.17.

4. Suppose $r \neq 0 \bmod 4$. Show that the check inner product $\check{\varepsilon}$ on $\mathbf{P}(r, 0)$ never has a signature, using Table 13.19.

5. Suppose $r - s = 0 \bmod 8$ and $r + s = 2p$ with $p = 1$ or $3 \bmod 4$. Show that the (reduced) classical companion ($p > 1$) is

$$\mathrm{SL}(p, \mathbf{R}) \cong \left\{ \begin{pmatrix} a & 0 \\ 0 & (a^t)^{-1} \end{pmatrix} : a \in \mathrm{SL}(p, \mathbf{R}) \right\}.$$

6. Suppose $\mathrm{Cl}(V) \cong \mathrm{End}_{\mathbf{F}}(\mathbf{P})$ is the pinor representation for a positive definite inner product space V.

 (a) Prove that $G \equiv \{a \in \mathrm{Cl}(V) : a\hat{a} = 1\}$ is a subgroup of $O(\mathrm{Cl}(V))$, and hence compact.

 (b) Pick any positive definite inner product on $\mathbf{P}$ and define $\hat{\varepsilon}$ to be the average of this inner product over the compact group G. Show that $\hat{\varepsilon}$ is the pinor inner product (up to a scale).

7. A real vector space V equipped with a linear map J, with $J^2 = 1$ and $\dim V_+ = \dim V_-$, where $V_\pm = \{x \in V : Jx = \pm x\}$ is called a **L**-vector space. Assume $r - s = 0 \bmod 8$.

 (a) Show that $\mathbf{P}(r, s)$ is an **L**-vector space.

(b) Combine $\hat{\varepsilon}$ and $\check{\varepsilon}$ to obtain an "**L**-valued inner product" on $\mathbf{P}(r,s)$. Define **L**-hermitian symmetric, etc., and describe the "type" of this inner product (depending on p mod 4 with $r + s \equiv 2p$ defining p).

8. State and prove a complex analogue of Proposition 13.79 concerning squares of pure spinors.

14. Low Dimensions

The results of the previous chapters are particularly interesting in the low dimensional cases. These special cases are examined in some detail in this chapter.

CARTAN'S ISOMORPHISMS

Recall from Chapter 1 some low dimensional isomorphisms of the groups defined in that chapter (see Proposition 1.40):

$$\mathrm{SO}(2) \cong U(1) \cong \mathrm{SK}(1) \cong S^1 \text{ and } \mathrm{CSO}(2) \cong \mathrm{GL}(1, \mathbf{C}) \cong \mathrm{SO}(2, \mathbf{C}),$$

$$\mathrm{SO}(4) \cong \mathrm{HU}(1) \cdot \mathrm{HU}(1) \text{ and } \mathrm{CSO}(4) = \mathrm{GL}(1, \mathbf{H}) \cdot \mathbf{H}^*,$$

$$\mathrm{Sp}(1, \mathbf{R}) \cong \mathrm{SL}(2, \mathbf{R}) \cong \mathrm{SU}(1, 1) \text{ and } \mathrm{SL}(2, \mathbf{C}) \cong \mathrm{Sp}(1, \mathbf{C}),$$

$$\mathrm{SU}(2) \cong \mathrm{HU}(1) \cong \mathrm{SL}(1, \mathbf{H}) \cong S^3 \text{ and } \mathrm{SO}^\uparrow(3, 1) \cong \mathrm{SO}(3, \mathbf{C}).$$

The final isomorphism $\mathrm{SO}^\uparrow(3, 1) \cong \mathrm{SO}(3, \mathbf{C})$, which was proved in Chapter 3, is also a consequence of two of the spin isomorphisms presented below—see $\mathrm{Spin}^0(3, 1)$ and $\mathrm{Spin}(3, \mathbf{C})$, and note that both are $\mathrm{SL}(2, \mathbf{C})$, which has center $\mathbf{Z}_2 \equiv \{\pm 1\}$. Also $\mathrm{SU}(2) \cong \mathrm{HU}(2)$ follows from $\mathrm{Spin}(3, 0) = \mathrm{HU}(1)$ and $\mathrm{Spin}(0, 3) = \mathrm{SU}(2)$ (see Remark 14.113).

In low dimensions, the spin groups Spin^0 are not new but are isomorphic to classical groups. Recall from Problem 10.2(d) that $\text{Spin}^0(r,s) \equiv \{a \in \text{Spin}(r,s) : a\hat{a} = 1\}$ is the connected component of the identity in $\text{Spin}(r,s)$ ($n \geq 2$), except when both $r = 1$ and $s = 1$.

Theorem 14.1 (The Spin Isomorphisms).

$n \equiv r + s \equiv 2$:

$$\text{Spin } (2) \quad \cong U(1)$$

$$\text{Spin}^0(1,1) = \left\{ \pm \begin{pmatrix} e^t & 0 \\ 0 & e^{-t} \end{pmatrix} : t \in \mathbf{R} \right\} \cong \mathbf{R} \oplus \mathbf{R}$$

$n \equiv r + s \equiv 3$:

$$\text{Spin } (3) \quad = \text{HU}(1) \ (\cong \text{SU}(2))$$
$$\text{Spin}^0(2,1) \cong \text{SL}(2,\mathbf{R})$$

$n \equiv r + s \equiv 4$:

$$\text{Spin } (4) \quad \cong \text{HU}(1) \times \text{HU}(1)$$
$$\text{Spin}^0(3,1) \cong \text{SL}(2,\mathbf{C})$$
$$\text{Spin}^0(2,2) \cong \text{SL}(2,\mathbf{R}) \times \text{SL}(2,\mathbf{R})$$

$n = r + s \equiv 5$:

$$\text{Spin } (5) \quad \cong \text{HU}(2)$$
$$\text{Spin}^0(4,1) \cong \text{HU}(1,1)$$
$$\text{Spin}^0(3,2) \cong \text{Sp}(2,\mathbf{R})$$

$n \equiv r + s \equiv 6$:

$$\text{Spin } (6) \qquad \cong \text{SU}(4)$$
$$\text{Spin}^0(5,1) \qquad \cong \text{SL}(2,\mathbf{H})$$
$$\text{Spin}^0(4,2) \qquad \cong \text{SU}(2,2)$$
$$\rho_\pm : \text{Spin}^0(3,3)/\mathbf{Z}_2 \cong \text{SL}(4,\mathbf{R})$$

$n \equiv r + s \equiv 8$:

$$\rho_\pm : \text{Spin}^0(6,2)/\mathbf{Z}_2 \cong \text{SK}(4).$$

Proof: These results can be read off from Theorem 13.8 by comparing dimensions and connectivity. Recall the type of the spinor inner product implies:

$$\text{Spin}^0(6) \subset U(4)$$
$$\text{Spin}^0(5,1) \subset \text{GL}(2,\mathbf{H}),$$
$$\text{Spin}^0(4,2) \subset U(2,2),$$
$$\text{Spin}^0(3,3)/\mathbf{Z}_2 \subset \text{GL}(4,\mathbf{R}).$$

The extra information required in these cases is that each element of Spin^0 has determinant equal to one. This follows easily from the fact that

Spin^0 is connected and the fact that the determinant takes on a finite number of values for any representation of Pin (see the section *Determinants* in Chapter 10).

In summary, using the spinor inner products (and in the four cases listed above, information about the determinants), the group Spin^0 is a subgroup of the particular classical group (the *(reduced) classical companion*) described in Theorem 14.1. To complete the proof, equality is obtained by counting dimensions and using connectivity of this (reduced) classical companion associated with the group Spin^0. The last case, signature $6, 2$ in dimension 8 is a special case of Theorem 14.3 presented below. ∎

Theorem 14.2 (Dimension Seven).

$$\text{Spin}(7) \subset \text{SO}(8) \subset M_8(\mathbf{R})$$
$$\text{Spin}^0(6, 1) \subset \text{SK}(4) \subset M_4(\mathbf{H})$$
$$\text{Spin}^0(5, 2) \subset \text{SK}(4) \subset M_4(\mathbf{H})$$
$$\text{Spin}^0(4, 3) \subset \text{SO}^\dagger(4, 4).$$

Proof: This result follows from Theorem 13.8 exactly parallel to the proof of the previous theorem. ∎

Theorem 14.3 (Dimension Eight). *First,* $\text{Spin}^0(7, 1) \subset \text{SO}(8, \mathbf{C})$ *and* $\text{Spin}^0(5, 3) \subset \text{SO}(8, \mathbf{C})$.

Second, **$\text{Spin}(8) \subset \text{SO}(8) \times \text{SO}(8)$**. *Moreover, the*

the positive spin representation

$$1 \to \mathbf{Z}_2 \equiv \{1, \lambda\} \to \text{Spin}(8) \xrightarrow{\rho+} \text{SO}(8) \to 1,$$

the negative spin representation

$$1 \to \mathbf{Z}_2 \equiv \{1, -\lambda\} \to \text{Spin}(8) \xrightarrow{\rho-} \text{SO}(8) \to 1,$$

and the vector representation

$$1 \to \mathbf{Z}_2 \equiv \{1, -1\} \to \text{Spin}(8) \xrightarrow{\chi} \text{SO}(8) \to 1$$

are exact sequences of groups.

Also, **$\text{Spin}^0(4,4) \subset \text{SO}(8) \times \text{SO}(8)$**. *Moreover,*

the positive spin representation

$$1 \to \mathbf{Z}_2 \equiv \{1, \lambda\} \to \text{Spin}^0(4, 4) \xrightarrow{\rho+} \text{SO}^\dagger(4, 4) \to 1,$$

the negative spin representation

$$1 \to \mathbf{Z}_2 \equiv \{1, -\lambda\} \to \text{Spin}^0(4, 4) \xrightarrow{\rho-} \text{SO}^\dagger(4, 4) \to 1,$$

and the vector representation

$$1 \to \mathbf{Z}_2 \equiv \{1, -1\} \to \text{Spin}^0(4, 4) \xrightarrow{\chi} \text{SO}^\dagger(4, 4) \to 1$$

are exact sequences of groups.

In addition, $\mathbf{Spin}^0(6,2) \subset \mathbf{SK}(4) \times \mathbf{SK}(4)$. *Moreover,*

the positive spin representation

$$1 \to \mathbf{Z}_2 \equiv \{1, \lambda\} \to \mathrm{Spin}^0(6,2) \xrightarrow{\rho_+} \mathrm{SK}(4) \to 1$$

and the negative spin representation

$$1 \to \mathbf{Z}_2 \equiv \{1, -\lambda\} \to \mathrm{Spin}^0(6,2) \xrightarrow{\rho_-} \mathrm{SK}(4) \to 1$$

are exact sequences of groups.

For each one of these Spin groups in dimension eight, all of the group representations listed above are inequivalent.

Proof: Most of the information in this theorem follows from Theorem 13.8 in a parallel manner to the proof of Theorem 14.1. It remains to prove that the representations listed above are not equivalent. For example, suppose

$$\rho_+ : \mathrm{Spin}(8) \to M_8(\mathbf{R}) \cong \mathrm{End}_{\mathbf{R}}(\mathbf{S}_+)$$

and

$$\rho_- : \mathrm{Spin}(8) \to M_8(\mathbf{R}) \cong \mathrm{End}_{\mathbf{R}}(\mathbf{S}_-)$$

were equivalent, with intertwining operator $f : \mathbf{S}_- \to \mathbf{S}_+$. Then $\rho_+(g) = f \circ \rho_-(g) \circ f^{-1}$ for all $g \in \mathrm{Spin}(8)$. In particular, this implies that $\ker \rho_+ = \ker \rho_-$ which is false. ∎

Theorem 14.1 has a complex analogue.

Theorem 14.4.

$$\mathrm{Spin}(2, \mathbf{C}) \cong \mathbf{C}^*$$
$$\mathrm{Spin}(3, \mathbf{C}) \cong \mathrm{SL}(2, \mathbf{C})$$
$$\mathrm{Spin}(4, \mathbf{C}) \cong \mathrm{SL}(2, \mathbf{C}) \times \mathrm{SL}(2, \mathbf{C})$$
$$\mathrm{Spin}(5, \mathbf{C}) \cong \mathrm{Sp}(2, \mathbf{C})$$
$$\mathrm{Spin}(6, \mathbf{C}) \cong \mathrm{SL}(4, \mathbf{C}).$$

Hint of Proof: For $n = 2p$ even, complexify the split case, $\mathrm{Cl}(2p) \cong \mathrm{Cl}(p,p) \otimes_{\mathbf{R}} \mathbf{C}$, and refer to the split cases $\mathrm{Spin}^0(p,p)$, $p = 2$ and 3, in Theorem 14.1. For $n = 2p + 1$ odd, refer to the $\mathrm{Spin}^0(p+1,p)$ case in Theorem 14.1. The details are omitted.

There is an alternate elementary proof of this theorem. This proof for $\mathrm{Spin}(6, \mathbf{C})$ is outlined in Problem 1. ∎

In addition to the isomorphisms presented above:

Theorem 14.5. *The following groups are isomorphic.*

$$\mathrm{SK}(2) \cong \mathrm{SU}(2) \times \mathrm{SU}(1,1)/\mathbf{Z}_2$$
$$\mathrm{SK}(3) \cong \mathrm{SU}(3,1)/\mathbf{Z}_2.$$

Proof: See problem 2. ∎

TRIALITY

In dimension eight, for both the positive definite case and the split case, the spin group Spin^0 has three inequivalent 8-dimensional representations: the positive spin representation ρ_+, the negative spin representation ρ_-, and the vector representation χ. There are automorphisms of the group Spin^0, called the *triality* automorphisms, which interchange these three representations. Of course, they cannot be inner automorphisms since the three representations are distinct (Theorem 14.3). Thus, by definition, the automorphisms are *outer*.

There is an elegant description of these triality automorphisms in terms of the octonians $\mathbf{O}$ for $\mathrm{Spin}(8)$ and the split octonians $\tilde{\mathbf{O}}$ for $\mathrm{Spin}^0(4,4)$. We shall pursue the positive definite case $\mathrm{Spin}(8)$ using $\mathbf{O}$. However, except for notational changes, the development will apply to $\mathrm{Spin}^0(4,4)$ and $\tilde{\mathbf{O}}$ as well.

As an added bonus, we discuss a very useful concrete model for $\mathrm{Cl}(8,0)$ (or alternatively, $\mathrm{Cl}(4,4)$). As will be showm, all three of the 8-dimensional euclidean spaces $V \equiv \mathbf{R}(8), \mathbf{S}_+$, and $\mathbf{S}_-$ can be identified with $\mathbf{O}$.

First, the space of vectors, $V(8) \subset \mathrm{Cl}(8,0)$, is identified with $\mathbf{O}$ as follows.

Definition 14.6. *Let* $V(8) \subset \mathrm{End}_{\mathbf{R}}(\mathbf{O} \oplus \mathbf{O})$ *denote the positive definite 8-dimensional euclidean space defined by*

$$V(8) \equiv \left\{ \begin{pmatrix} 0 & -R_u \\ -R_{\overline{u}} & 0 \end{pmatrix} : u \in \mathbf{O} \right\}.$$

The square norm on $V(8)$ *is defined by* $\|A(u)\| \equiv \|u\|$, *where*

$$A(u) = \begin{pmatrix} 0 & R_u \\ -R_{\overline{u}} & 0 \end{pmatrix}.$$

Note that

$$(14.7) \qquad A(u)A(v) = \begin{pmatrix} -R_u R_{\overline{v}} & 0 \\ 0 & -R_{\overline{u}} R_v \end{pmatrix}$$

so that

$$A(u)A(u) = -\|u\| \cdot 1 \text{ for all } u \in \mathbf{O}.$$

Thus, by the Fundamental Lemma of Clifford Algebras, the map A : $\mathbf{O} \to \text{End}_{\mathbf{R}}(\mathbf{O} \oplus \mathbf{O})$ extends to an algebra homomorphism $A : \text{Cl}(\mathbf{O}) \to \text{End}_{\mathbf{R}}(\mathbf{O} \oplus \mathbf{O})$. By Lemma 8.6 and Theorem 11.3, $\text{Cl}(8) \cong \text{Cl}(\mathbf{O})$ has no two-sided nontrivial ideals. Thus the algebra homomorphism is injective. A dimension count shows that this map A is onto, so that A is a isomorphism:

$$(14.8) \qquad\qquad \text{Cl}(\mathbf{O}) \cong \text{End}_{\mathbf{R}}(\mathbf{O} \oplus \mathbf{O}).$$

Therefore, the space of pinors may be taken as

$$\mathbf{P} \equiv \mathbf{O} \oplus \mathbf{O}.$$

The isomorphism (14.8) will frequently be composed with the natural isomorphism $\text{Cl}(\mathbf{O}) \cong \Lambda(\mathbf{O})$ (as vector spaces).

Lemma 14.9. *A unit volume element λ for $V(8) \equiv \mathbf{O}$ is given by:*

$$\lambda \equiv \begin{pmatrix} 1 & 0 \\ 0 & -1 \end{pmatrix} \in \text{End}_{\mathbf{R}}(\mathbf{O} \oplus \mathbf{O}) \cong \text{Cl}(\mathbf{O}) \cong \Lambda(\mathbf{O}).$$

Proof 1: Consider the standard orthonormal basis

$$e_0 \equiv 1, \ e_1 \equiv i, \ e_2 \equiv j, \ e_3 \equiv k, \ e_4 \equiv e, \ e_5 \equiv ie, \ e_6 \equiv je, \ e_7 \equiv ke$$

for $\mathbf{O}$. Using octonion multiplication compute

$$A(e_0) \cdots A(e_7) = \begin{pmatrix} 1 & 0 \\ 0 & -1 \end{pmatrix}. \quad \blacksquare$$

Proof 2: Note λ belongs to the twisted center of $\text{Cl}(\mathbf{O})$ by $(9.48')$. Lemma 9.49 implies that $\lambda \in \Lambda^8 \mathbf{O}$ and Theorem 9.65 implies that λ is of unit length.

Remark. Right multiplictaion by i, denoted R_i, induces a complex structure on $\mathbf{O}$. The orientation induced by this complex structure is the same as the orientation induced by this standard basis.

Corollary 14.10. *If $\mathbf{O}$ is identified with $V(8) \subset \text{End}_{\mathbf{R}}(\mathbf{O} \oplus \mathbf{O})$ by*

$$(14.11) \qquad\qquad V(8) \equiv \left\{ \begin{pmatrix} 0 & R_u \\ -R_{\overline{u}} & 0 \end{pmatrix} : u \in \mathbf{O} \right\},$$

then

$$\text{Cl}(\mathbf{O}) \cong \text{End}_{\mathbf{R}}(\mathbf{S}_+ \oplus \mathbf{S}_-)$$

and

$$Cl^{even}(\mathbf{O}) \cong End_{\mathbf{R}}(\mathbf{S}_+) \oplus End_{\mathbf{R}}(\mathbf{S}_-) \cong \left\{ \begin{pmatrix} a & 0 \\ 0 & b \end{pmatrix} : a, b \in End_{\mathbf{R}}(\mathbf{O}) \right\},$$

where $\mathbf{S}_+ \cong \mathbf{O}$ is the first copy of $\mathbf{O}$ and $\mathbf{S}_- \cong \mathbf{O}$ is the second copy of $\mathbf{O}$ in $\mathbf{P} \equiv \mathbf{O} \oplus \mathbf{O}$.

Proof: Note that $\lambda \equiv \begin{pmatrix} 1 & 0 \\ 0 & -1 \end{pmatrix}$ commutes with $\begin{pmatrix} a & c \\ d & b \end{pmatrix}$ if and only if $c = d = 0$. ∎

Lemma 14.12. *The standard inner product on* $\mathbf{O}$,

$$(14.13) \qquad \langle x, y \rangle \equiv Re \, \bar{x} y \quad \text{for all } x, y \in \mathbf{O},$$

when applied to $\mathbf{P} \equiv \mathbf{O} \oplus \mathbf{O}$ *so that the two factors* $\mathbf{S}_+ \cong \mathbf{O}$ *and* $\mathbf{S}_- \cong \mathbf{O}$ *are orthogonal, can be adopted as the pinor inner product* $\hat{\varepsilon}$ *on* $\mathbf{P} \cong \mathbf{S}_+ \oplus \mathbf{S}_-$.

Proof: It suffices to show that, for $z \equiv (x, y)$ and $z' \equiv (x', y') \in \mathbf{O} \oplus \mathbf{O}$,

$$(14.14) \qquad \hat{\varepsilon}(A(u)z, z') = -\hat{\varepsilon}(z, A(u)z') \quad \text{for all } u \in \mathbf{O},$$

where $\hat{\varepsilon}$ is defined by

$$(14.15) \qquad \hat{\varepsilon}(z, z') \equiv \langle x, x' \rangle + \langle y, y' \rangle.$$

Since $A(u)z = (yu, -x\bar{u})$ and $-A(u)z' = (-y'u, x'\bar{u})$, the result (14.14) follows from $R_u^* = R_{\bar{u}}$. ∎

Corollary 14.16. *Under the isomorphism* $Cl(\mathbf{O}) \cong End_{\mathbf{R}}(\mathbf{O} \oplus \mathbf{O})$ *determined by Definition 14.6, the hat anti-automorphism is equal to the adjoint (transpose) with respect to the standard norm on* $\mathbf{O} \oplus \mathbf{O}$. *In particular, Theorem 14.3 for* Spin(8) *yields*

$$(14.17) \qquad Spin(8) \subset SO(\mathbf{O}) \times SO(\mathbf{O}) \cong SO(\mathbf{S}_+) \times SO(\mathbf{S}_-).$$

Note that

$$\{ g \in Cl(\mathbf{O})^{even} : g\hat{g} = 1 \} \cong O(\mathbf{O}) \times O(\mathbf{O}).$$

We shall adopt both the notation

$$(14.18) \qquad \begin{aligned} g &\equiv (g_+, g_-) \in Spin(8), \text{ as well as} \\ g &\equiv (g_0, g_+, g_-) \in Spin(8), \end{aligned}$$

where by definition $g_+ \equiv \rho_+(g)$ and $g_- \equiv \rho_-(g)$, while $g_0 \equiv \chi_g$ denotes the vector representation of g. The vector representation χ will also be denoted by $\rho_0 = \chi$.

Octonian multiplication can be used to give an important characterization of Spin(8).

The Triality Theorem 14.19. *Suppose* (g_0, g_+, g_-) *is a triple of orthogonal linear maps on* $\mathbf{O}$. *Then* $(g_+, g_-) \in \mathrm{Spin}(8)$, *with* g_0 *the vector representation of* (g_+, g_-) *if and only if*

$$(14.20) \qquad g_+(xy) = g_-(x)g_0(y) \quad \text{for all } x, y \in \mathbf{O}.$$

Remark 14.21. If (g_0, g_+, g_-) is a triple of orthogonal transformations on $\mathbf{O}$ satisfying (14.20), then each of the three must have determinant one, because the theorem implies that $g \equiv (g_0, g_+, g_-) \in \mathrm{Spin}(8)$ and $\mathrm{Spin}(8)$ is connected.

Proof: Given

$$A \equiv \begin{pmatrix} a & 0 \\ 0 & b \end{pmatrix} \in \mathrm{End}_{\mathbf{R}}(\mathbf{O}) \oplus \mathrm{End}_{\mathbf{R}}(\mathbf{O}) \cong \mathrm{Cl}(\mathbf{O})^{\mathrm{even}},$$

then, by Corollary 10.50, $A \in \mathrm{Spin}(8)$ if and only if

$$A\hat{A} = 1 \quad (\text{i.e., } a, b \in O(\mathbf{O})), \text{ and } \chi_A(u) \in \mathbf{O} \text{ for all } u \in \mathbf{O},$$

where

$$\chi_A(u) \equiv A \begin{pmatrix} 0 & R_u \\ -R_{\overline{u}} & 0 \end{pmatrix} \hat{A}$$

is the vector representation. Thus, for $A \equiv \begin{pmatrix} a & 0 \\ 0 & b \end{pmatrix}$ to belong to $\mathrm{Spin}(8)$, we must have that $a, b \in O(\mathbf{O})$ and that: for each $u \in \mathbf{O}$, there exists a $v \in \mathbf{O}$ such that

$$(14.22) \qquad \begin{pmatrix} 0 & R_v \\ -R_{\overline{v}} & 0 \end{pmatrix} = A \begin{pmatrix} 0 & R_u \\ -R_{\overline{u}} & 0 \end{pmatrix} \hat{A}.$$

That is,

$$(14.22') \qquad\qquad R_v = a R_u b^t.$$

Considering this as a map sending u to v, it must be the vector representation $g_0 = \rho_0(A)$ of A. Applying (14.22') to $1 \in \mathbf{O}$, we have

$$v = a(b^t(1)u).$$

Therefore, if $g \equiv (g_+, g_-) \in \mathrm{Spin}(8)$, then

$$(14.23) \qquad v = g_0(u) \equiv g_+(g_-^t(1)u) \quad \text{for all } u \in \mathbf{O}$$

defines the vector represenation $g_0 \equiv \chi_g$. Now (14.22′) with $a \equiv g_+$ and $b \equiv g_-$ can be rewritten as

$$(14.24) \qquad wg_0(y) = g_+(g^t_-(w)y) \quad \text{for all } w, y \in \mathbf{O}.$$

Setting $x \equiv g^t_-(w)$ yields the desired equation (14.20).

Conversely, suppose (14.20) is satisfied by (g_0, g_-, g_+). Set

$$A \equiv \begin{pmatrix} g_+ & 0 \\ 0 & g_- \end{pmatrix}.$$

If $u \in \mathbf{O}$ is given, then (with $v \equiv g_0(u)$) the equation (14.22) is satisfied, since the equations (14.22), (14.22′), and (14.24) are all equivalent. ∎

Since conjugation, $c(x) \equiv \overline{x}$, is an orthogonal transformation, the representation $\rho'_0 = c \circ \rho_0 \circ c$ is $O(8)$ equivalent to ρ_0. Similarly, $\rho'_\pm = c \circ \rho_\pm \circ c$ is $O(8)$ equivalent to $\rho_\pm$. Note that if $h \equiv R_u$ and $h' \equiv c \circ h \circ c$ then $h' = L_{\overline{u}}$. The identity $g_+(xy) = g_-(x)g_0(y)$, for $g \in \mathrm{Spin}(8)$, implies that

$$g'_+(xy) = \overline{g_+(\overline{xy})} = \overline{g_+(\overline{y}\,\overline{x})} = \overline{g_-(\overline{y})g_0(\overline{x})} = g'_0(x)g'_-(y).$$

Therefore,

$$(14.25) \qquad \alpha(g_0, g_+, g_-) = (g'_-, g'_+, g'_0)$$

defines an automorphism $\alpha : \mathrm{Spin}(8) \to \mathrm{Spin}(8)$. Similarly,

$$(14.26) \qquad \beta(g_0, g_+, g_-) = (g'_0, g_-, g_+)$$

defines an automorphism $\beta : \mathrm{Spin}(8) \to \mathrm{Spin}(8)$.

Note that the product $\tau = \alpha\beta$ is given by

$$(14.27) \qquad \tau(g_0, g_+, g_-) = (g'_+, g'_-, g_0),$$

and is of order three. τ is called the *triality automorphism* of $\mathrm{Spin}(8)$.

Since $\alpha^2 = \beta^2 = \tau^3$ is the identity homomorphism of $\mathrm{Spin}(8)$ and $\alpha\beta = \tau, \beta\alpha = \tau^2$, it follows that α and β generate a group of automorphisms of $\mathrm{Spin}(8)$ that is isomorphic to S_3, the symmetric group on three letters. These automorphisms are all outer since they act nontrivially on the center, $\{\pm 1, \pm \lambda\} = \{(1, 1, 1), (1, -1, -1), (-1, 1, -1), (-1, -1, 1)\}$, of $\mathrm{Spin}(8)$.

This concrete realization of $\mathrm{Spin}(8)$ using the octonians has many applications. We give one example involving complex structures on $\mathbf{R}^8$ (see Proposition 7.174). Let Ref_ξ denote the orthogonal reflection along the plane ξ (in order to avoid confusion with right octonian multiplication).

Proposition 14.28. Spin(8) *is generated by triples*

$$(\mathrm{Ref}_\xi, J_\xi^+, J_\xi^-) \quad \text{with } \xi \in G_\mathbf{R}(2, \mathbf{O}),$$

where the complex structures J_ξ^+ *and* J_ξ^- *are defined by*

$$J_\xi^+ \equiv \frac{1}{2}(R_v R_{\overline{u}} - R_u R_{\overline{v}})$$

and

$$J_\xi^- \equiv \frac{1}{2}(R_{\overline{v}} R_u - R_{\overline{u}} R_v),$$

if $\xi \equiv u \wedge v$.

Proof: Proposition 10.21 states that $G_\mathbf{R}(2, V) \subset \mathrm{Spin}(8)$ generates Spin(8). Given $g \equiv \xi \in G_\mathbf{R}(2, V) \subset \mathrm{Spin}(8)$, we must show that

$$(14.29) \qquad g_0 = \mathrm{Ref}_\xi, \quad g_+ \equiv \rho_+(\xi) = J_\xi^+, \quad \text{and} \quad g_- \equiv \rho_-(\xi) = J_\xi^-.$$

Given two unit vectors $u, v \in V \equiv \mathbf{O}$, identify u with

$$\begin{pmatrix} 0 & R_u \\ R_{\overline{u}} & 0 \end{pmatrix},$$

and v with

$$\begin{pmatrix} 0 & R_v \\ -R_{\overline{v}} & 0 \end{pmatrix},$$

as in the above discussion. Then the Clifford product is given by

$$(14.30) \qquad u \cdot v = \begin{pmatrix} -R_u R_{\overline{v}} & 0 \\ 0 & -R_{\overline{u}} R_v \end{pmatrix}.$$

Note $g \equiv u \cdot v \in \mathrm{Spin}(8)$, so that (14.30) can be rewritten as

$$(14.31) \qquad (\mathrm{Ref}_u \cdot \mathrm{Ref}_v, -R_u R_{\overline{v}}, -R_{\overline{u}} R_v) \in \mathrm{Spin}(8) \quad \text{for all } u, v \in \mathbf{O}.$$

Finally, given $\xi \in G_\mathbf{R}(2, \mathbf{O})$, choose u, v orthonormal with $\xi = u \wedge v = u \cdot v$ and apply (14.31) and (6.14), to complete the proof of (14.29). ∎

Now we can prove Proposition 7.174 ($n = 4$):

$$(14.32) \qquad \mathrm{Cpx}^+(4) = \left\{ J_\xi^+ : \xi \in G_\mathbf{R}(2, \mathbf{O}) \right\} \cong G_\mathbf{R}(2, \mathbf{O}).$$

Proof of Proposition 7.174 ($n = 4$): First note that the complex structure $R_i = J^+_{1 \wedge i}$ induces the same orientation on $\mathbf{O}$ as the standard basis $e_0, \ldots, e_7$ defined above. Therefore, $R_i \in \mathrm{Cpx}^+(4)$. Since $G_{\mathbf{R}}(2, \mathbf{O})$ is connected, each complex structure J^+_ξ induces the same orientation on $\mathbf{O}$.

It remains to show that the map sending $\xi \in G_{\mathbf{R}}(2, \mathbf{O})$ to $J^+_\xi \in \mathrm{Cpx}^+(4)$ is an isomorphism. This map is just the positive spinor representation ρ_+ restricted to $G_{\mathbf{R}}(2, \mathbf{O})$. Since $\ker \rho_+ = \{1, \lambda\}$ and $\lambda \notin G_{\mathbf{R}}(2, \mathbf{O})$, ρ_+ restricted to $G_{\mathbf{R}}(2, \mathbf{O})$ is injective.

Consider the action of $\mathrm{Spin}(8)$ on $G_{\mathbf{R}}(2, V)$ given by sending ξ to $g\xi g^{-1}$ for each $g = (g_0, g_+, g_-) \in \mathrm{Spin}(8)$. Let K denote the isotropy subgroup at a point ξ_0. Then considering the vector representation and the positive spinor representation yields an equivarient isomorphism between the quotient

$$(14.33) \qquad G_{\mathbf{R}}(2, \mathbf{O}) \cong \mathrm{SO}(8)/(\mathrm{SO}(2) \times \mathrm{SO}(6))$$

and the quotient

$$(14.34) \qquad \mathrm{Cpx}^+(4) \cong \mathrm{SO}(8)/U(4). \quad \blacksquare$$

In the following Remark this isotropy subgroup K of $\mathrm{Spin}(8)$ at ξ_0 is computed explicitly for $\xi_0 = 1 \wedge i$ (see (14.41)).

Remark 14.35. The octonian description of $\mathrm{Spin}(8)$ presented in the Triality Theorem 14.19 naturally leads to octonian descriptions of the two isomorphisms

$$(14.36) \qquad \mathrm{Spin}(6) \cong \mathrm{SU}(4) \quad \text{and} \quad \mathrm{Spin}(5) \cong \mathrm{HU}(2)$$

as follows.

Recall (Problem 10.6) that the subgroup of $\mathrm{Spin}(n)$ that fixes a vector in the vector representation is $\mathrm{Spin}(n - 1)$. Thus

$(14.37) \quad \mathrm{Spin}(7) = \{g \in \mathrm{Spin}(8) : g_0(1) = 1\}$

$(14.38) \quad \mathrm{Spin}(6) = \{g \in \mathrm{Spin}(8) : g_0(1) = 1 \text{ and } g_0(i) = i\}$

$(14.39) \quad \mathrm{Spin}(5) = \{g \in \mathrm{Spin}(8) : g_0(1) = 1, \ g_0(i) = i, \text{ and } g_0(j) = j\}$.

Also, let $\mathrm{Spin}(2)$ denote the following subgroup of $\mathrm{Spin}(8)$:

$$(14.40) \qquad \mathrm{Spin}(2) = \{g \in \mathrm{Spin}(8) : g_0(x) = x \text{ if } x \perp \mathrm{span}\{1, i\}\}.$$

Then it follows that

$$(14.41) \qquad K = \mathrm{Spin}(2) \times \mathrm{Spin}(6).$$

The triality identity (14.20) can be used to compute $\rho_+(G)$ and $\rho_-(G)$ where G is one of the above subgroups of Spin(8). First, we obtain

(14.42) $\mathrm{Spin}(7) = \{g \in \mathrm{Spin}(8) : g_+ = g_-\}$

from (14.20) by setting $y = 1$.

Next, note that each element g of the group Spin(2) defined by (14.40) must have g_0 a rotation Rot_θ through an angle θ in the $1, i$ plane. Thus, g_0 is the product of two reflections along lines in span$\{1, i\}$. For example, $g_0 = \mathrm{Ref}_1 \circ \mathrm{Ref}_{e^{-i\theta/2}}$. Consequently, either g or $-g$ is equal to the Clifford product of $1 \in V(8)$ with $e^{-i\theta/2} \in V(8)$, i.e.,

$$\begin{pmatrix} 0 & 1 \\ -1 & 0 \end{pmatrix} \begin{pmatrix} 0 & R_{e^{-i\theta/2}} \\ -R_{e^{i\theta/2}} & 0 \end{pmatrix} = \begin{pmatrix} -R_{e^{i\theta/2}} & 0 \\ 0 & -R_{e^{-i\theta/2}} \end{pmatrix}.$$

Therefore, with $\psi \equiv \frac{\theta}{2} + \pi$,

(14.43) $\mathrm{Spin}(2) = \{(\mathrm{Rot}_{2\psi}, R_{e^{i\psi}}, R_{e^{-i\psi}}) : \psi \in \mathbf{R}\}.$

Next we show that the positive spinor representation ρ_+ yields an isomorphism

(14.44) $(\mathrm{Spin}(2) \times \mathrm{Spin}(6))/\mathbf{Z}_2 \cong U(4),$

when $\mathbf{S}_+ = \mathbf{O}$ is given the complex structure R_i, and $\mathbf{Z}_2 = \{1, \lambda\}$. If $g \in \mathrm{Spin}(6) \subset \mathrm{Spin}(7)$, then the triality identity (14.20) with $y \equiv i$ implies that $g_+ = g_-$ commutes with the complex structure R_i, so that $g_+ = g_- \in U(4)$. Now a dimension count combined with the connectivity of $U(4)$ completes the proof, showing ρ_+ from Spin(2) $\times$ Spin(6) to $U(4)$ is surjective.

Using Problem 10.7 and the fact that ker $\rho_+ = \{1, \lambda\}$ proves that

(14.45)
$$\mathrm{Spin}(6) = \{(h_0, h, h) : h \in \mathrm{SU}(4), \text{ and } h_0 \text{ is given by}$$
$$h_0(x) \equiv \overline{h(1)}h(x), \text{ for all } x \in V(8) \equiv \mathbf{O}\}.$$

In particular, Spin(6) $\cong$ SU(4).

Finally, the triality identity implies that if $g \in \mathrm{Spin}(5)$, then $g_+ = g_-$ commutes with both R_i and R_j. Thus, $g_+ = g_- \in \mathrm{HU}(2)$, where $\mathbf{O} \cong \mathbf{H}^2$ is provided with the (right) quaternionic structure inducted by $I \equiv R_i, J \equiv R_j$, and $K \equiv R_i R_j$. Again connectivity and a dimension count show that

(14.46) $\rho_+ = \rho_- : \mathrm{Spin}(5) \cong \mathrm{HU}(2).$

TRANSITIVE ACTIONS ON SPHERES

In low dimensions, $n \leq 6$, the spin groups agree with their (larger) classical companions—see Theorem 14.1. For slightly larger dimensions, vestiges of these isomorphisms remain. Consider, for example, $\mathrm{Spin}(7) \subset \mathrm{SO}(8)$. In this case, $\mathrm{SO}(8)$ is the larger classical companion (determined by the spinor inner product ε on $\mathbf{S} = \mathbf{R}^8$). The group $\mathrm{Spin}(7)$ is "almost" as big as its classical companion $\mathrm{SO}(8)$. One way of making this precise is to show that $\mathrm{Spin}(7)$ also acts transitively on spheres in $\mathbf{R}^8$.

This action as well as the other transitive actions to be discussed in this section are listed below for convenient reference.

(14.47) $\mathrm{Spin}(7)/G_2 \cong S^7$.

(14.48) $\mathrm{Spin}(8)/G_2 \cong S^7 \times S^7$.

(14.49) $\mathrm{Spin}(8)/\mathrm{Spin}(7) \cong S^7$.

(14.50) $\mathrm{Spin}(9)/\mathrm{Spin}(7) \cong S^{15}$.

Also, recall

(14.51) $$G_2/\mathrm{SU}(3) \cong S^6$$

from Problem 6.9(a).

First, a model for $\mathrm{Cl}(7)$ is constructed, using the octonians.

A Model for $\mathrm{Cl}(7)$

Identify $V(7) \equiv \mathrm{Im}\,\mathbf{O}$ with the following subspace of $\mathrm{End}_\mathbf{R}(\mathbf{O}) \oplus \mathrm{End}_\mathbf{R}(\mathbf{O})$:

$$V(7) \cong \left\{ \begin{pmatrix} R_u & 0 \\ 0 & -R_u \end{pmatrix} : u \in \mathrm{Im}\,\mathbf{O} \right\}.$$

Let

(14.52) $$A(u) \equiv \begin{pmatrix} R_u & 0 \\ 0 & -R_u \end{pmatrix} \quad \text{for } u \in \mathrm{Im}\,\mathbf{O}.$$

This map $A : V(7) = \mathrm{Im}\,\mathbf{O} \to \mathrm{End}_\mathbf{R}(\mathbf{O}) \oplus \mathrm{End}_\mathbf{R}(\mathbf{O})$ extends to an algebra isomorphism

(14.53) $\mathrm{Cl}(7) \cong \mathrm{End}_\mathbf{R}(\mathbf{O}) \oplus \mathrm{End}_\mathbf{R}(\mathbf{O}) = \mathrm{End}_\mathbf{R}(\mathbf{P}_+) \oplus \mathrm{End}_\mathbf{R}(\mathbf{P}_-)$

by the Fundamental Lemma of Clifford Algebras because $A(u)A(u) = -\|u\| \cdot 1$.

The natural inner product on $\mathbf{P} \equiv \mathbf{O} \oplus \mathbf{O}$ can be adopted as the pinor inner product $\hat{\varepsilon} = \varepsilon \oplus \varepsilon$, since $R_u^* = -R_u$ for $u \in \operatorname{Im}\mathbf{O}$ (one must verify that $\hat{\varepsilon}(A(u)z, z') = -\hat{\varepsilon}(z, A(u)z')$). The unit volume element λ for $V(7,0) \cong \operatorname{Im}\mathbf{O}$ can be chosen to be $\left(\begin{smallmatrix} 1 & 0 \\ 0 & -1 \end{smallmatrix}\right) = A(e_1) \cdots A(e_7)$, consistent with the orientation on $\operatorname{Im}\mathbf{O}$ determined by the basis $e_1, \ldots, e_7$ defined above (see the proof of Lemma 14.9). Obviously, $\operatorname{Cl}(7)^{\text{even}}$ is contained in $\operatorname{End}_{\mathbf{R}}(\mathbf{O})$, embedded diagonally in $\operatorname{End}_{\mathbf{R}}(\mathbf{O}) \oplus \operatorname{End}_{\mathbf{R}}(\mathbf{O}) \cong \operatorname{Cl}(7)$. Counting dimensions yields

$$(14.54) \qquad\qquad \operatorname{Cl}(7)^{\text{even}} \cong \operatorname{End}_{\mathbf{R}}(\mathbf{O}).$$

Because of Corollary 10.50, $g \in \operatorname{Cl}(7)^{\text{even}} \cong \operatorname{End}_{\mathbf{R}}(\mathbf{O})$ belongs to $\operatorname{Spin}(7)$ if and only if

$$(14.55a) \qquad\qquad g\,\hat{g} = 1 \quad \left(\text{i.e., } g \in O(\mathbf{O})\right)$$

and

$$(14.55b) \qquad \text{given } u \in \operatorname{Im}\mathbf{O} \cong V(7), \text{ there exists } w \in \operatorname{Im}\mathbf{O} = V(7)$$

such that

$$(14.56) \qquad\qquad g \circ R_u \circ g^{-1} = R_w.$$

The only possibility for w (apply both sides of (14.56) to $1 \in \mathbf{O}$) is

$$(14.57) \qquad\qquad w = g(g^{-1}(1)u).$$

Therefore, if $g \in \operatorname{Spin}(7) \subset O(\mathbf{O})$, then

$$(14.58) \qquad\qquad \chi_g(u) \equiv g(g^{-1}(1)u) \quad \text{for all } u \in \operatorname{Im}\mathbf{O}$$

is equal to the vector representation

$$(14.59) \qquad\qquad \chi = \operatorname{Ad} : \operatorname{Spin}(7) \to SO(7).$$

Moreover, (apply (14.56) to $g(v)$)

$$(14.60) \qquad g(vu) = g(v)\chi_g(u) \quad \text{for all } u \in \operatorname{Im}\mathbf{O} \text{ and } v \in \mathbf{O}.$$

(Note that (14.58) can be used to extend χ_g to $\mathbf{O}$ with $\chi_g(1) = 1$, and then (14.60) is valid for all $u \in \mathbf{O}$.)

Conversely, if (14.60) is satisfied by $g \in O(\mathbf{O})$, with χ_g defined by (14.58), then (14.56) is valid with $w \equiv \chi_g(u)$ (take $v = g^{-1}(z)$ and apply both sides of (14.56) to z). This proves the following lemma.

Lemma 14.61. *An element* $g \in O(\mathbf{O})$ *belongs to* $\mathrm{Spin}(7)$ *if and only if*

$$(14.62) \qquad g(uv) = g(u)\chi_g(v) \quad \text{for all } u, v \in \mathbf{O},$$

where

$$(14.63) \qquad \chi_g(v) \equiv g(g^{-1}(1)v) \quad \text{for all } v \in \mathbf{O}$$

defines the vector representation of $\mathrm{Spin}(7)$ *on* $V(7) = \mathrm{Im}\,\mathbf{O}$.

Now because of the Triality Theorem 14.19

Corollary 14.64.

$$(14.65) \qquad \begin{aligned} \mathrm{Spin}(7) &= \{g = (g_+, g_-) \in \mathrm{Spin}(8) : g_+ = g_-\} \\ &= \{g = (g_0, g_+, g_-) \in \mathrm{Spin}(8) : g_0(1) = 1\}. \end{aligned}$$

To complete our list of characterizations of $\mathrm{Spin}(7)$, note that if

$$\begin{pmatrix} R_u & 0 \\ 0 & -R_u \end{pmatrix} \in G(1, \mathrm{Im}\,\mathbf{O}),$$

then

$$\lambda \begin{pmatrix} R_u & 0 \\ 0 & -R_u \end{pmatrix} = \begin{pmatrix} R_u & 0 \\ 0 & R_u \end{pmatrix} \in G(6, \mathrm{Im}\,\mathbf{O}).$$

Lemma 14.66. $\mathrm{Spin}(7)$ *is generated by* $\{R_u : u \in S^6 \subset \mathrm{Im}\,\mathbf{O}\}$.

Proof: By Proposition 10.23, $\mathrm{Spin}(7)$ is generated by

$$\lambda G(1, \mathrm{Im}\,\mathbf{O}) \cong G(6, \mathrm{Im}\,\mathbf{O}) \subset \mathrm{Cl}(7)^{\mathrm{even}} \cong \mathrm{End}_{\mathbf{R}}(\mathbf{O}). \quad \blacksquare$$

Theorem 14.67. *Consider the spin representation of* $\mathrm{Spin}(7)$ *on* $\mathbf{S} \equiv R^8$. *Then* $\mathrm{Spin}(7)$ *acts transitively on the 7-sphere*

$$S^7 \equiv \{x \in \mathbf{R}^8 : \|x\| = 1\},$$

and the isotropy subgroup at a point is G_2:

$$(14.68) \qquad \mathrm{Spin}(7)/G_2 \cong S^7.$$

Proof: To compute the isotropy subgroup of $\mathrm{Spin}(7)$, the octonian model for $\mathrm{Cl}(7)$ is useful. Let K denote this isotropy subgroup of $\mathrm{Spin}(7)$ at $1 \in \mathbf{S} \cong \mathbf{O}$. Suppose $g \in K$. Then, by (14.63), $\chi_g(v) \equiv g(g^{-1}(1)v) = g(v)$. Therefore, by (14.62), $g(uv) = g(u)g(v)$, so that $g \in G_2$. Conversely, if

$g \in G_2$, then, by Lemma 6.67, $g \in O(\mathbf{O})$. Since $g(1) = 1$ it follows that $\chi_g = g$, and (14.62) is valid. Therefore, conditions (14.55a and b) are satisfied and $g \in \mathrm{Spin}(7)$. This proves that $K = G_2$. We give two proofs that $\mathrm{Spin}(7)$ acts transitively on S^7.

First, recall that $\mathrm{Spin}(6) \cong \mathrm{SU}(4)$. The spin representation of $\mathrm{Spin}(7)$ on $\mathbf{R}^8$, when restricted to $\mathrm{SU}(4) \cong \mathrm{Spin}(6) \subset \mathrm{Spin}(7)$, is the standard representation of $\mathrm{SU}(4)$ on $\mathbf{C}^4$, where the volume element λ for $\mathbf{R}(6,0)$ provides the complex structure on $\mathbf{R}^8 \cong \mathbf{C}^4$. Note that $\mathrm{Cl}(6)^{\mathrm{even}} \cong \mathrm{End}_{\mathbf{C}}(\mathbf{C}^4)$ and $\mathrm{Cl}(6)^{\mathrm{even}} \subset \mathrm{Cl}(6) \cong \mathrm{Cl}(7)^{\mathrm{even}} \cong \mathrm{End}_{\mathbf{R}}(\mathbf{R}^8)$. Since $\mathrm{SU}(4)$ acts transitively on $S^7 \subset \mathbf{C}^4$, this proves that the larger group $\mathrm{Spin}(7)$ also acts transitively on S^7.

Second, the orbit of $\mathrm{Spin}(7)$ through $1 \in S^7 \subset \mathbf{O}$ is $\mathrm{Spin}(7)/G_2$ which has dimension $7 = 21 - 14$. Since S^7 is connected and $\mathrm{Spin}(7)$ is compact, the orbit must be all of S^7. ∎

Theorem 14.69. *Consider the spin representation $\rho = \rho_+ \oplus \rho_-$ of* $\mathrm{Spin}(7)$ *on* $\mathbf{S}_+ \oplus \mathbf{S}_- = \mathbf{R}^8 \oplus \mathbf{R}^8$. *The orbit through a point* $(a, b) \in \mathbf{S}_+ \oplus \mathbf{S}_-$ *is*

$$\{(x, y) \in \mathbf{S}_+ \oplus \mathbf{S}_- : |x| = |a| \text{ and } |y| = |b|\}.$$

(a) *If $a, b \neq 0$, the isotropy subgroup is G_2 and*

$$(14.70) \qquad\qquad \mathrm{Spin}(8)/G_2 \cong S^7 \times S^7.$$

(b) *If exactly one of a, b vanishes, then the isotropy subgroup is a copy of* $\mathrm{Spin}(7)$ *and*

$$(14.71) \qquad\qquad \mathrm{Spin}(8)/\mathrm{Spin}(7) \cong S^7.$$

Proof of (a): $\mathrm{Spin}(7)$ is diagonally embedded in $\mathrm{Spin}(8)$ by Corollary 14.64. Thus, an element $g \in \mathrm{Spin}(8)$ can be chosen that sends a to $|a|$ because $\mathrm{Spin}(7)$ acts transitively on the 7-sphere. Consequently, we may assume that $a \in \mathbf{O} \cong \mathbf{S}_+$ is nonzero and real.

Consider the subgroup

$$(14.72) \qquad H \equiv \{g \equiv (g_0, g_+, g_-) \in \mathrm{Spin}(8) : g_+(1) = 1\}$$

of $\mathrm{Spin}(8)$. Applying the triality automorphism $\tau(g_0, g_+, g_-) = (g'_+, g'_-, g_0)$ shows that

$$\tau H = \{(g_0, g_+, g_-) \in \mathrm{Spin}(8) : g_0(1) = 1\}.$$

Now $g_0(1) = 1$ if and only if $g_+ = g_-$ (see Corollary 14.64). Therefore

(14.73) $$\tau H = \text{Spin}(7).$$

Thus, $H = \tau^2(\text{Spin}(7))$ is an embedding of $\text{Spin}(7)$ into $\text{Spin}(8)$ sending $h \in \text{Spin}(7)$ to the triple $(h, \chi_h, h') \in \text{Spin}(8)$. In particular, $h' \in \text{Spin}(7)$ can be chosen to send b to $\|b\|$ since $\text{Spin}(7)$ acts transitively on spheres. This proves that $\text{Spin}(8)$ acts transitively on

$$S^7 \times S^7 = \{(x, y) \in \mathbf{S}_+ \oplus \mathbf{S}_- : |x| = |a| \text{ and } |y| = |b|\}.$$

Let K denote the isotropy subgroup of $\text{Spin}(8)$ at

$$(a, b) \equiv (r \cos \theta, r \sin \theta) \in S^7 \times S^7 \subset \mathbf{O} \oplus \mathbf{O},$$

with $r > 0$ and $0 < \theta < \pi/2$. Suppose $g = (g_+, g_-) \in K$. Recall that $g_+(uv) = g_-(u)g_0(v)$ for all $u, v \in \mathbf{O}$. Since $g_-(1) = 1$, $g_+(v) = g_0(v)$. Therefore, $g_0(1) = g_+(1) = 1$ also. Consequently, $g_+(u) = g_-(u)$. This proves $g_0 = g_+ = g_- \in G_2$. Conversely, if $g_0 = g_+ = g_- \in G_2$, then $g \equiv (g_+, g_-)$ fixes (a, b) since a, b are real. Therefore, $K = G_2$. ∎

Proof of (b): Since $\rho_+ : \text{Spin}(8) \to SO(8)$ is surjective, $\text{Spin}(8)$ certainly acts transitively on S^7.

Suppose $(a, 0) \in \mathbf{S}_+ \times \{0\}$, with a real and nonzero. The isotropy H of $\text{Spin}(8)$ at $(a, 0) \in \mathbf{S}_+ \oplus \mathbf{S}_-$ is given by (14.59) so that $H = \tau^2(\text{Spin}(7))$ as desired.

The proof for $(0, a) \in \{0\} \times \mathbf{S}_-$, with a real and nonzero, is similar and so omitted. ∎

A Model for $\text{Cl}(9)$

Let $\mathbf{S} \equiv \mathbf{O} \oplus \mathbf{O}$ and $\mathbf{P} \equiv \mathbf{S} \otimes_{\mathbf{R}} \mathbf{C}$. Choose

(14.74) $$V(9) \equiv \left\{ i \begin{pmatrix} r & R_u \\ R_{\overline{u}} & -r \end{pmatrix} : r \in \mathbf{R}, u \in \mathbf{O} \right\} \subset \text{End}_{\mathbf{C}}(\mathbf{P}).$$

If

$$A(r, u) \equiv i \begin{pmatrix} r & R_u \\ R_{\overline{u}} & -r \end{pmatrix},$$

then $A(r, u)A(r, u) = -(r^2 + |u|^2)Id$. Therefore, by the Fundamental Lemma of Clifford Algebras

(14.75) $$\text{Cl}(9) \cong \text{End}_{\mathbf{C}}(\mathbf{P}).$$

Also note that

(14.76) $\mathrm{Cl}(8)^{\mathrm{even}} \cong \mathrm{End}_{\mathbf{R}}(\mathbf{S})$,

and the unit volume element for $V(9)$ may be chosen to be $i \in \mathrm{End}_{\mathbf{C}}(\mathbf{P})$, the complex structure on $\mathbf{P} \equiv \mathbf{S} \otimes_{\mathbf{R}} \mathbf{C}$.

This model can be developed further, yielding a characterization of Spin(9) via Corollary 10.50 (see Problem 5). However, what we shall need below is a characterization of Spin(9) using Proposition 10.23. This Proposition says Spin(9) is generated by $\lambda G(1, V(9)) = G(8, V(9))$, where $\lambda \equiv i$ is the unit volume element. That is,

Lemma 14.77. *The group* Spin(9) *is generated by the 8-sphere*

(14.78) $\left\{ \begin{pmatrix} r & R_u \\ R_{\bar{u}} & -r \end{pmatrix} : r \in \mathbf{R}, u \in \mathbf{O}, \text{ and } r^2 + |u|^2 = 1 \right\}$

in $\mathrm{End}_{\mathbf{R}}(\mathbf{O} \oplus \mathbf{O}) \cong \mathrm{Cl}(9)^{\mathrm{even}}$.

This lemma can be used to prove that Spin(9) acts transitively on S^{15}.

Theorem 14.79. *Consider the spin representation*

(14.80) $\rho: \mathrm{Spin}(9) \to \mathrm{SO}(16)$,

of Spin(9) *on* $\mathbf{S} \cong \mathbf{R}^{16}$. *Then* Spin(9) *acts transitively on the 15-sphere*

$$S^{15} \equiv \{x \in \mathbf{R}^{16} : \|x\| = 1\},$$

and the isotropy subgroup at a point is Spin(7):

(14.81) $\mathrm{Spin}(9)/\mathrm{Spin}(7) \cong S^{15}$.

Proof: Because Spin(8) $\subset$ Spin(9), Theorem 14.69 says that a unit vector in $\mathbf{O} \oplus \mathbf{O}$ can be mapped to the arc $(\cos\theta, \sin\theta)$, with $0 \le \theta \le \pi/2$ by an element of Spin(8). Lemma 14.77 says, in particular, that

$$\begin{pmatrix} \cos\theta & \sin\theta \\ \sin\theta & -\cos\theta \end{pmatrix} \in \mathrm{Spin}(9) \quad \text{for each } \theta.$$

This implies that each point on the arc $(\cos\theta, \sin\theta), 0 \le \theta \le \pi/2$, can be mapped to the point $(1,0)$ by an element of Spin(9). This proves Spin(9) acts transitively on the unit sphere S^{15} on $\mathbf{R}^{16}$.

Let K denote the isotropy subgroup of Spin(9) at $(1,0) \in \mathbf{S} \equiv \mathbf{O} \oplus \mathbf{O}$. Since Spin(8) $\subset$ Spin(9), part (b) of Theorem 14.69 says that

(14.82) $K \cap \text{Spin}(8) = \tau^2(\text{Spin}(7))$.

This proves that

(14.83) $\tau^2(\text{Spin}(7)) \subset K$.

Therefore,

$$\pi : \text{Spin}(9)/\tau^2(\text{Spin}(7)) \to S^{15}$$

is a covering map, since dim Spin(9) $-$ dim Spin(7) $=$ dim S^{15}. Since S^{15} is simply connected, π must be one to one. ∎

THE CAYLEY PLANE AND THE EXCEPTIONAL GROUP F_4

Consider the relation $\sim$ on $F^n - \{0\}$ defined by

(14.84) $a \sim b$ if $a\lambda = b$ for some scalar $\lambda \in F$.

If $F \equiv \mathbf{R}, \mathbf{C}$, or $\mathbf{H}$,then $\sim$ is an equivalence relation and $\mathbf{P}^{n-1}(F)$ is the projective space of all F-lines through the origin in F^n. However, since $\mathbf{O}$ is not associative the "Cayley projective spaces $\mathbf{P}^{n-1}(\mathbf{O})$" are not well-defined. Alternate (equivalent) definitions of $\mathbf{P}^{n-1}(F)$, for $F \equiv \mathbf{R}, \mathbf{C}$, or $\mathbf{H}$, are available (cf. Problem 4.15).

Instead of considering a line $L \equiv [a]$ in F^n, consider the orthogonal projection A from F^n onto L. (Equip F^n with the standard F-hermitian symmetric inner product.) Then A satisfies

(14.85) $\overline{A}^t = A, \quad A^2 = A, \quad \text{and} \quad \text{trace}_F \, A = 1.$

Recall the notation Herm$(n, F) = \{A \in M_n(F) : \overline{A}^t = A\}$. Now projective space can be described as a subset of the real vector space Herm(n, F).

(14.86) $\mathbf{P}^{n-1}(F) = \{A \in \text{Herm}(n, F) : A^2 = A \text{ and trace}_F \, A = 1\}$.

To complete the proof of (14.86), note that if $A^2 = A$ with $A \in \text{Herm}(n, F)$, then A has eigenvalues 0 and 1, so that A is orthogonal projection from F^n onto $L \equiv \{x \in F^n : Ax = x\}$. In particular, trace$_F \, A = \dim_F L$.

If $A \in M_n(F)$ represents orthogonal projection onto L and $\dim_F L = 1$, then, choosing any unit vector $a \in L$,

(14.87) $A = a\,\overline{a}^t \quad \text{or} \quad Ax = a\langle a, x\rangle \text{ for all } x \in F^n.$

Here $a \in F^n$ represents a column vector.

Now we turn to the Cayley plane.

Definition 14.88. *The Cayley plane is defined by*

(14.89) $\mathbf{P}^2(\mathbf{O}) \equiv \{A \in \mathrm{Herm}(3, \mathbf{O}) : A^2 = A \text{ and trace}_{\mathbf{O}} A = 1\}.$

Lemma 14.90.

(14.91) $$\mathbf{P}^2(\mathbf{O}) = \{a\,\bar{a}^t : a^t = (a_1, a_2, a_3) \in \mathbf{O}^3 \text{ with}$$
$$\|a\| = 1 \text{ and } [a_1, a_2, a_3] = 0\},$$

which is a 16-dimensional compact submanifold of $\mathrm{Herm}(3, \mathbf{O})$.

Proof: Suppose

$$A \equiv \begin{pmatrix} r_1 & \bar{x}_3 & \bar{x}_2 \\ x_3 & r_2 & x_1 \\ x_2 & \bar{x}_1 & r_3 \end{pmatrix} \in \mathrm{Herm}(3, \mathbf{O}).$$

Then

$$A^2 = \begin{pmatrix} r_1^2 + \|x_2\| + \|x_3\| & (r_1 + r_2)\bar{x}_3 + \bar{x}_2\bar{x}_1 & (r_1 + r_3)\bar{x}_2 + \bar{x}_3 x_1 \\ (r_1 + r_2)x_3 + x_1 x_2 & r_2^2 + \|x_1\| + \|x_3\| & (r_2 + r_3)x_1 + x_3\bar{x}_2 \\ (r_1 + r_3)x_2 + \bar{x}_1 x_3 & (r_2 + r_3)\bar{x}_1 + x_2\bar{x}_3 & r_3^2 + \|x_1\| + \|x_2\| \end{pmatrix}.$$

Now suppose $A^2 = A$ and trace$_{\mathbf{O}} A = 1$. Then $r_1 x_1 = x_3\bar{x}_2$, $r_2 x_2 = \bar{x}_1 x_3$, and $r_3 x_3 = x_1 x_2$. Since not all of r_1, r_2, r_3 can vanish, Artin's Theorem implies that for $A \in \mathbf{P}^2(\mathbf{O})$ the entries x_1, x_2, x_3 belong to a quaternion subalgebra $\mathbf{H} \subset \mathbf{O}$. Thus, $A \in \mathrm{Herm}(3, \mathbf{H}) \subset \mathrm{Herm}(3, \mathbf{O})$. Because of this fact, (14.87) is applicable. Choose $a^t \in \mathbf{H}^3$ with $A = a\,\bar{a}^t$ and $\|a\| = 1$. This proves that $\mathbf{P}^2(\mathbf{O})$ is contained in the right hand side of (14.91).

Finally, if $a^t \equiv (a_1, a_2, a_3)$, $\|a\| = 1$, and $[a_1, a_2, a_3] = 0$, then $A \equiv a\,\bar{a}^t$ satisfies trace$A = 1$ and $A^2 = A$ so that $A \in \mathbf{P}^2(\mathbf{O})$.

To prove that $\mathbf{P}^2(\mathbf{O})$ is a 16-dimensional manifold, note that it is covered by the three charts

$$U_1 \equiv \left\{ \frac{a\,\bar{a}^t}{\|a\|} : a^t \equiv (1, a_2, a_3) \in \mathbf{O}^3 \right\} \cong \mathbf{O}^2,$$

$$U_2 \equiv \left\{ \frac{a\,\bar{a}^t}{\|a\|} : a^t \equiv (a_1, 1, a_3) \in \mathbf{O}^3 \right\} \cong \mathbf{O}^2,$$

$$U_3 \equiv \left\{ \frac{a\,\bar{a}^t}{\|a\|} : a^t \equiv (a_1, a_2, 1) \in \mathbf{O}^3 \right\} \cong \mathbf{O}^2. \quad \blacksquare$$

Definition 14.92. *The group F_4 is defined to be the automorphism group of the (Jordan) algebra* Herm$(3, \mathbf{O})$*, equipped with the symmetric (or Jordan) product $A \circ B \equiv \frac{1}{2}(AB + BA)$.*

A linear map $g \in \text{GL}(\text{Herm}(3, \mathbf{O}))$ is an automorphism if

$$(14.93) \qquad g(A \circ B) = g(A) \circ g(B) \quad \text{for all } A, B \in \text{Herm}(3, \mathbf{O}).$$

Note that $A \circ A = AA$, so that the notation A^2 is unambiguous.
In particular, (14.93) implies that

$$(14.93') \qquad\qquad g(A^2) = g(A)^2.$$

This simpler condition of "preserving squares" suffices to guarantee that $g \in F_4$, because polarization of (14.93') yields (14.93).
Given

$$A \equiv \begin{pmatrix} r_1 & \bar{x}_3 & \bar{x}_2 \\ x_3 & r_2 & x_1 \\ x_2 & \bar{x}_1 & r_3 \end{pmatrix} \in \text{Herm}(3, \mathbf{O}),$$

define

$$(14.94) \quad \begin{aligned} & \text{trace} A \equiv r_1 + r_2 + r_3, \\ & \|A\| \equiv r_1^2 + r_2^2 + r_3^2 + 2|x_1|^2 + 2|x_2|^2 + 2|x_3|^2, \\ & \sigma_2(A) \equiv (r_1 r_2 - |x_3|^2) + (r_1 r_3 - |x_2|^2) + (r_2 r_3 - |x_1|^2), \\ & \det\ A \equiv r_1 r_2 r_3 + 2\langle x_1, x_2 x_3 \rangle - r_1|x_1|^2 - r_2|x_2|^2 - r_3|x_3|^2, \end{aligned}$$

in analogy with Herm$(3, \mathbf{R})$.
Note that (i.e., calculate directly)

$$(14.95) \qquad \|A\| = \text{trace}\ (A^2), \quad \sigma_2(A) = \frac{1}{2}((\text{trace}\ A)^2 - \text{trace}\ A^2)$$

and

$$(14.95') \qquad \det A = \frac{1}{6}\ (\text{trace}\ A)^3 - \frac{1}{2}(\text{trace}\ A^2)(\text{trace}\ A) + \frac{1}{3}\ \text{trace}\ A^3,$$

for all $A \in \text{Herm}(3, \mathbf{O})$. Here and in the following $A^3 \equiv A \circ A \circ A$ denotes the Jordan cube, so that $A^3 = \frac{1}{2}(A^2 \cdot A + A \cdot A^2)$ in terms of ordinary matrix multiplication.

Lemma 14.96. *If $g \in F_4$, then*

$$\text{trace}\, g(A) = \text{trace}\, A, \quad \text{for all } A \in \text{Herm}(3, \mathbf{O}).$$

Consequently, F_4 fixes $\| \cdot \|$, σ_2, and det.

Proof: It suffices to show that F_4 fixes trace because of (14.95) and (14.95′). Since each $g \in F_4$ maps 1 to 1, it suffices to show that if trace $A = 0$, then trace $g(A) = 0$. One can show by direct calculation that each $A \in \text{Herm}(3, \mathbf{O})$ satisfies

$$(14.97) \qquad A^3 - (\text{trace}\, A)A^2 + \sigma_2(A)A - \det A = 0.$$

Let

$$C \equiv \{A \in \text{Herm}(3, \mathbf{O}) : A^3 = aA + b \text{ for some } a, b \in \mathbf{R}\},$$

and

$$Q \equiv \{A \in \text{Herm}(3, \mathbf{O}) : A^2 = aA + b \text{ for some } a, b \in \mathbf{R}\}.$$

Note both C and Q are invariant under F_4. Now assume that trace $A = 0$. Because of (14.97) this implies $A \in C$. Therefore $B \equiv g(A) \in C$ so that

$$(14.98) \qquad B^3 = aB + b \quad \text{for some } a, b \in \mathbf{R}.$$

Subtracting (14.97) (with A replaced by B) from (14.98) yields

$$(\text{trace}\, B)B^2 - (\sigma_2(B) + a)B + (\det B - b) = 0.$$

Therefore, either trace $B = 0$ or $B \in Q$. If $B \in Q$, then $A = g^{-1}(B) \in Q$. This proves that g maps $\{A : \text{trace}\, A = 0\} \sim Q$ into $\{A : \text{trace}\, A = 0\}$. Finally, note that $\{A : \text{trace}\, A = 0\} \sim Q$ is a nonempty open subset of $\{A : \text{trace}\, A = 0\}$ (choose A diagonal with trace $A = 0$ and distinct eigenvalues). ■

Theorem 14.99. *F_4 acts transitively on the Cayley plane $\mathbf{P}^2(\mathbf{O})$ with isotropy subgroup equal to (an isomorphic copy of) $\text{Spin}(9)$ at the point*

$$E_1 \equiv \begin{pmatrix} 1 & 0 & 0 \\ 0 & 0 & 0 \\ 0 & 0 & 0 \end{pmatrix} \in \mathbf{P}^2(\mathbf{O}),$$

$$(14.100) \qquad F_4/\text{Spin}(9) \cong \mathbf{P}^2(\mathbf{O}).$$

Proof: Spin(9) is generated by

$$\left\{ \begin{pmatrix} r & R_{\overline{u}} \\ R_u & -r \end{pmatrix} : r \in \mathbf{R},\ u \in \mathbf{O},\ \text{and } r^2 + |u|^2 = 1 \right\}$$

because of Lemma 14.77. Let C denote conjugation on both factors of $\mathbf{O} \oplus \mathbf{O}$. It is more convenient to consider the copy $C \cdot \text{Spin}(9) \cdot C$ of Spin(9). Since

$$C \cdot \begin{pmatrix} r & R_{\overline{u}} \\ R_u & -r \end{pmatrix} \cdot C = \begin{pmatrix} r & L_u \\ L_{\overline{u}} & -r \end{pmatrix},$$

the group $C \cdot \text{Spin}(9) \cdot C$ is generated by

$$\left\{ \begin{pmatrix} r & L_u \\ L_{\overline{u}} & -r \end{pmatrix} : r \in \mathbf{R},\ u \in \mathbf{O},\ \text{and } r^2 + |u|^2 = 1 \right\}.$$

Suppose

$$g = \begin{pmatrix} r & L_u \\ L_{\overline{u}} & -r \end{pmatrix}$$

is a generator for $C \cdot \text{Spin}(9) \cdot C$. Then

$$g(x) = \begin{pmatrix} rx_1 + ux_2 \\ \overline{u}x_1 - rx_2 \end{pmatrix} \quad \text{for} \quad \begin{pmatrix} x_1 \\ x_2 \end{pmatrix} \in \mathbf{O} \oplus \mathbf{O},$$

determines the spinor representation of $C \cdot \text{Spin}(9) \cdot C$. Also,
(14.101)
$$\chi_g \begin{pmatrix} \rho & L_v \\ L_{\overline{v}} & -\rho \end{pmatrix} = \begin{pmatrix} (r^2 + |u|^2)\rho + 2r\langle u, v \rangle & L_{2r\rho u - r^2 v + u\overline{v}u} \\ L_{2r\rho\overline{u} - r^2\overline{v} - \overline{u}v\overline{u}} & -\left(|u|^2 + r^2\right)\rho - 2r\langle u, v \rangle \end{pmatrix}$$

determines the vector representation of $C \cdot \text{Spin}(9) \cdot C$ on

$$\left\{ \begin{pmatrix} \rho & L_v \\ L_{\overline{v}} & -\rho \end{pmatrix} : \rho \in \mathbf{R}, v \in \mathbf{O} \right\}.$$

Extend χ_g to act on

$$\left\{ \begin{pmatrix} r_2 & L_v \\ L_{\overline{v}} & r_3 \end{pmatrix} : r_2, r_3 \in \mathbf{R}, v \in \mathbf{O} \right\}$$

by defining

$$\chi_g \begin{pmatrix} 1 & 0 \\ 0 & 1 \end{pmatrix} = \begin{pmatrix} 1 & 0 \\ 0 & 1 \end{pmatrix}.$$

The action of $C \cdot \text{Spin}(9) \cdot C$ on $\text{Herm}(3, \mathbf{O})$ is defined as follows. Given

(14.102) $$A \equiv \begin{pmatrix} r_1 & \overline{x}_1 & \overline{x}_2 \\ x_1 & r_2 & v \\ x_2 & \overline{v} & r_3 \end{pmatrix} \in \text{Herm}(3, \mathbf{O})$$

and

(14.103) $$g \equiv \begin{pmatrix} r & L_u \\ L_{\overline{u}} & -r \end{pmatrix} \quad \text{a generator for } C \cdot \text{Spin}(9) \cdot C,$$

define

$$g(A) \equiv \begin{pmatrix} r_1 & \overline{g(x)}^t \\ g(x) & \chi_g(a) \end{pmatrix}, \quad \text{where } a \equiv \begin{pmatrix} r_2 & L_v \\ L_{\overline{v}} & r_3 \end{pmatrix} \text{ and } x \equiv \begin{pmatrix} x_1 \\ x_2 \end{pmatrix}.$$

Now we can prove that

Lemma 14.104. $C \cdot \text{Spin}(9) \cdot C \subset F_4$.

Note that this implies that $\text{Spin}(9) \subset \text{Isot}_{E_1}(F_4)$.

Proof: Given $A \in \text{Herm}(3, \mathbf{O})$ and g a generator for $C \cdot \text{Spin}(9) \cdot C$ as in (14.102) and (14.103), let

(14.105) $$G \equiv \begin{pmatrix} 1 & 0 & 0 \\ 0 & r & u \\ 0 & \overline{u} & -r \end{pmatrix}.$$

Then, using (14.101) for $\chi_g(a)$, a direct calculation shows that

(14.106) $$GAG = \begin{pmatrix} r_1 & \overline{g(x)}^t \\ g(x) & \chi_g(a) \end{pmatrix} = g(A).$$

Suppose $U \equiv (u_{ij}) \in M_n(\mathbf{O})$, with $\text{span}\{u_{ij}, 1\}$ contained in a subalgebra isomorphic to $\mathbf{C} \subset \mathbf{O}$. (For example, $U \equiv G$ defined by (14.105).) Then for any element $z \in \mathbf{O}$, $\text{span}\{u_{ij}, 1, z\}$ is contained in an associative subalgebra of $\mathbf{O}$. In particular, $u_1 z u_2 u_3 z u_4$ is well-defined for all $u_1, u_2, u_3, u_4 \in \text{span}\{u_{ij}, 1\}$, so that

(14.107) $$(u_1 z u_2)(u_3 z u_4) = u_1[z(u_2 u_3)z]u_4.$$

Polarization yields

(14.107′) $$\begin{aligned} (u_1 z u_2)(u_3 w u_4) &+ (u_1 w u_2)(u_3 z u_4) \\ &= u_1[z(u_2 u_3)w]u_4 + u_1[w(u_2 u_3)z]u_4 \end{aligned}$$

This implies that

(14.108) $(UAU)(UBU) + (UBU)(UAU) = U(AU^2B)U + U(BU^2A)U,$

for all $A, B \in \text{Herm}(n, \mathbf{O})$. Applying this identity, with $U \equiv G$ defined by (14.104), yields

$$g(A)g(A) = (GAG)(GAG) = GA^2G = g(A^2),$$

since $G^2 = 1$. Thus $g \in F_4$. This proves that $C \cdot \text{Spin}(9) \cdot C$ is a subgroup of F_4. ∎

Since F_4 preserves the trace, both the conditions, $A^2 = A$ and trace $A = 1$, defining $\mathbf{P}^2(\mathbf{O})$ are preserved by F_4. Next we prove that F_4 acts transitively on $\mathbf{P}^2(\mathbf{O})$. Suppose $A \in \mathbf{P}^2(\mathbf{O})$:

$$A \equiv \begin{pmatrix} r_1 & \overline{x}^t \\ x & a \end{pmatrix}, \quad x \equiv \begin{pmatrix} x_1 \\ x_2 \end{pmatrix}, \quad \text{and} \quad a \equiv \begin{pmatrix} r_2 & v \\ \overline{v} & r_3 \end{pmatrix}.$$

Choose $g \in C \cdot \text{Spin}(9) \cdot C \subset F_4$ so that $\chi_g(a) = \begin{pmatrix} r_2' & 0 \\ 0 & r_3' \end{pmatrix}$. This is possible because $\chi : \text{Spin}(9) \to \text{SO}(9)$ acts transitively on the 8-sphere in each 9-plane $\{a \in \text{Herm}(2, \mathbf{O}) : \text{trace}\, a = \text{constant}\}$. The isotropy of $C \cdot \text{Spin}(9) \cdot C$ at $\begin{pmatrix} r_2' & 0 \\ 0 & r_3' \end{pmatrix}$ is the same as the isotropy at $\begin{pmatrix} 1 & 0 \\ 0 & -1 \end{pmatrix}$, which contains $C \cdot \text{Spin}(8) \cdot C$. As proven earlier, $\text{Spin}(8) \subset \text{End}(\mathbf{O} \oplus \mathbf{O})$ has orbits $S^7 \times \{0\}$, $S^7 \times S^7$, and $\{0\} \times S^7$. In all cases, there exists $h \in \text{Spin}(8)$ so that $h(x_1, x_2) = (y_1, y_2)$ with $y_1, y_2 \in \mathbf{R}$. This proves that the orbit of F_4 (in fact, $C \cdot \text{Spin}(9) \cdot C \subset F_4$) through any point $A \in \mathbf{P}^2(\mathbf{O})$ contains a point $B \in \mathbf{P}^2(\mathbf{O})$ with each entry $b_{ij} \in \mathbf{R}$, i.e., $B \in \text{Herm}(3, \mathbf{R})$.

Given $g \in \text{SO}(3) \subset M_3(\mathbf{R})$, define

(14.109) $\rho_g(A) = gAg^t$ for all $A \in \text{Herm}(3, \mathbf{O})$.

This defines $\text{SO}(3)$ as a subgroup of F_4. If A is real, then we may choose g so that $B \equiv \rho_g(A)$ is diagonal. The conditions $B^2 = B$ and trace $B = 1$ imply that B has eigenvalue 1 with multiplicity one and eigenvalue 0 with multiplicity 2. Finally, by applying a permutation $\pi \in \text{SO}(3)$ to B, we obtain E_1, proving that F_4 acts transitively on $\mathbf{P}^2(\mathbf{O})$.

It remains to show that each $g \in F_4$, which fixes E_1, belongs to $C \cdot \text{Spin}(9) \cdot C$.

Given $A \equiv \begin{pmatrix} r_1 & \overline{x}^t \\ x & a \end{pmatrix} \in \text{Herm}(3, \mathbf{O})$, note that

(14.110) $2A \circ E_1 - A = \begin{pmatrix} r_1 & 0 \\ 0 & -a \end{pmatrix}.$

Thus, if $g \in F_4$ fixes E_1, then g maps the subspace of $\mathrm{Herm}(3, \mathbf{O})$ defined by $a = 0$ into itself. Therefore,

$$(14.111) \qquad\qquad g(A) = \begin{pmatrix} r & \overline{f(x)}^t \\ f(x) & h(a) \end{pmatrix},$$

where $h \in SO(9)$, acting on $\mathrm{Herm}(2, \mathbf{O})$ while leaving the identity fixed. Now by utilizing an element of $\mathrm{Spin}(9)$, we may assume that $h = 1$. It remains to show that f is the identity.

Suppose A is of the special form

$$\begin{pmatrix} 0 & \overline{x}_1 & \overline{x}_2 \\ x_1 & 0 & 0 \\ x_2 & 0 & 0 \end{pmatrix}.$$

Since $h \equiv 1$, g fixes the basic diagonal matricies E_2 and E_3. Note that

$$2A \circ E_2 = \begin{pmatrix} 0 & \overline{x}_1 & 0 \\ x_1 & 0 & 0 \\ 0 & 0 & 0 \end{pmatrix}.$$

Thus, $2A \circ E_3 = A$ if and only if $x_2 = 0$. Consequently if $x_2 = 0$, then $y_2 = 0$, where $y = f(x)$. Similarly, if $x_1 = 0$, then $y_1 = 0$. This proves that $y_1 = f_1(x_1)$ and $y_2 = f_2(x_2)$. Since

$$A^2 = \begin{pmatrix} |x|^2 & 0 & 0 \\ 0 & 0 & x_1\overline{x}_2 \\ 0 & x_2\overline{x}_1 & 0 \end{pmatrix},$$

the condition $g(A^2) = g(A)g(A)$ becomes

$$x_1 x_2 = f_1(x_1)\overline{f_2(x_2)} \quad \text{for all } x_1, x_2 \in \mathbf{O}.$$

If $u \equiv f_1(1)$, then $u^{-1} = \overline{f_2(1)}$, and it follows that $f_1(x_1) = x_1 u$ and $f_2(x_2) = x_2 \overline{u}^{-1}$. The above condition on f becomes $(x_1 u)(u^{-1}\overline{x}_2) = x_1 \overline{x}_2$ for all $x_1, x_2 \in \mathbf{O}$, which implies that $u = 1$. Therefore $f(x) = x$. This completes the proof of Theorem 14.99. ∎

The complex Lie group $F_4^{\mathbf{C}}$ can be defined to be the automorphism group of the complexified Jordan algebra $\mathrm{Herm}(3, \mathbf{O}) \otimes_{\mathbf{R}} \mathbf{C} \cong \mathrm{Herm}(3, \mathbf{O_C})$. Here $\mathbf{O_C} \equiv \mathbf{O_C} \otimes_{\mathbf{R}} \mathbf{O}$ denotes the algebra of complexified octonians. Note that $\mathbf{O_C}$ contains the split octonians $\tilde{\mathbf{O}}$ as a subalgebra. The noncompact or split case of F_4, denoted $\tilde{F}_4$ is defined to be the automorphism group of the Jordan algebra $\text{Herm}(3, \tilde{\mathbf{O}})$.

In addition to F_4 and $\tilde{F}_4$, there is a third real form of $F_4^{\mathbf{C}}$, which we denote by F_4'. Let

$$
\text{Herm}'(3, \mathbf{O}) \equiv \left\{ A \in M_3(\mathbf{O_C}) : A \equiv \begin{pmatrix} r_1 & -i\overline{x}_1 & -i\overline{x}_2 \\ ix_1 & r_2 & x_3 \\ ix_2 & \overline{x}_3 & r_3 \end{pmatrix} \right.
$$

$$
\left. \text{and } x_1, x_2, x_3 \in \mathbf{O}, r_1, r_2, r_3 \in \mathbf{R} \right\}.
$$

(Here i denotes the $i \in \mathbf{C}$ in the complexification $\mathbf{O_C} \equiv \mathbf{O} \otimes_{\mathbf{R}} \mathbf{C}$ of $\mathbf{O}$.) Define F_4' to be the automorphisms of the Jordan algebra $\text{Herm}'(3, \mathbf{O})$. Note that all three Jordan algebras:

$$
\text{Herm}(3, \mathbf{O}), \quad \text{Herm}(3, \tilde{\mathbf{O}}), \quad \text{and} \quad \text{Herm}'(3, \mathbf{O})
$$

have the same complexification $\text{Herm}(3, \mathbf{O_C})$.

The results of this section have analogues for $\tilde{F}_4$ and F_4'. For example, define

$$
\mathbf{P}^2(\tilde{\mathbf{O}}) \equiv \left\{ A \in \text{Herm}(3, \tilde{\mathbf{O}}) : A^2 = A \text{ and } \text{trace}_{\mathbf{O}} A = 1 \right\}.
$$

Then the orbit of $\tilde{F}_4$ through

$$
\begin{pmatrix} 1 & 0 & 0 \\ 0 & 0 & 0 \\ 0 & 0 & 0 \end{pmatrix}
$$

is $\mathbf{P}^2(\tilde{\mathbf{O}})$ with isotropy subgroup $\text{Spin}^0(5, 4)$:

(14.112) $$\mathbf{P}^2(\mathbf{O}) \cong \tilde{F}_4/\text{Spin}^0(5, 4).$$

Also,

(14.112)′ $\quad F_4'/\text{Spin}(9) \cong \left\{ A \in \text{Herm}'(3, \mathbf{O}) : A^2 = A \text{ and } \text{trace}\, A = 1 \right\}.$

The proofs of (14.112) and (14.112)′ can be adapted from the proof of $\mathbf{P}^2(\mathbf{O}) \cong F_4/\text{Spin}(9)$, with suitable modifications and additions.

CLIFFORD ALGEBRAS IN LOW DIMENSIONS

The models presented in the previous sections for $\mathrm{Cl}(7), \mathrm{Cl}(8), \mathrm{Cl}(3,4)$, $\mathrm{Cl}(4,4)$, and $\mathrm{Cl}(9)$ are quite useful. Similar models for the other low dimensional Clifford algebras are available. Some of these models are collected in this section (dimensions 2, 3, and 4 and the various signatures). In these low dimensions, the algebraic structure of $\mathrm{Cl}(r,s)$ along with the canonical involutions uniquely determines the choice of vectors $\Lambda^1 \mathbf{R}(r,s)$, and hence uniquely determines the full Clifford structure. This is because the hat involution is minus one on $\Lambda^1 \mathbf{R}(r,s)$, and plus one on $\Lambda^3 \mathbf{R}(r,s)$. Thus, in dimensions $n \leq 4$, $\Lambda^1 \mathbf{R}(r,s)$ is uniquely defined as the subspace of $\mathrm{Cl}^{\mathrm{odd}}(r,s)$ fixed by the hat involution, sending a to $\hat{a}$.

Dimension 2 and Signature $2,0$ or $0,2$

Spinors.

$$\mathrm{Cl}(2,0)^{\mathrm{even}} \cong \mathrm{Cl}(0,2)^{\mathrm{even}} \cong \mathbf{C},$$

where a choice λ of unit volume element corresponds to $i \in \mathbf{C}$. A canonical choice for the spinor space $\mathbf{S}$ (a complex 1-dimensional vector space with $\mathrm{Cl}(2,0)^{\mathrm{even}} \cong \mathrm{End}_{\mathbf{C}}(\mathbf{S})$) is possible once an orientation has been selected. Just take $\mathbf{S} \equiv \mathrm{Cl}(2,0)^{\mathrm{even}}$ with complex structure λ, and let $\mathrm{Cl}(2,0)^{\mathrm{even}}$ act on $\mathbf{S}$ by left multiplication. Thus

$$\mathrm{Spin}(2) = \{a \in \mathbf{C} : \|a\| = 1\} \equiv S^1$$

and the canonical involution (hat = check) on $\mathrm{Cl}^{\mathrm{even}} \cong \mathbf{C}$ is just conjugation on $\mathbf{C}$.

Pinors. The two larger algebras $\mathrm{Cl}(2,0)$ and $\mathrm{Cl}(0,2)$ containing $\mathrm{Cl}^{\mathrm{even}} \cong \mathbf{C}$ are the two Cayley–Dickson doubles of $\mathbf{C}$.

First,

$$\mathrm{Cl}(2,0) \cong \mathrm{Cl}(2,0)^{\mathrm{even}} \oplus \mathrm{Cl}(2,0)^{\mathrm{odd}},$$

can be rewritten as:

$$\mathbf{H} \quad \cong \quad \mathbf{C} \quad \oplus \quad \mathbf{C}^{\perp}$$

where $\mathbf{C}^{\perp} \equiv \mathrm{span}\{j,k\}$. In particular, the space of vectors $\mathbf{R}(2,0) = \Lambda^1 \mathbf{R}(2,0) \subset \mathrm{Cl}(2,0)$ is just $\mathbf{C}^{\perp}$. One can take $\mathbf{P} \equiv \mathrm{Cl}(2,0)$ with right $\mathbf{H}$-structure $\mathbf{H} \cong \mathrm{Cl}(2,0)$.

Second, the pinor space $\mathbf{P}(0,2)$ is a real 2-dimensional vector space with $\mathrm{Cl}(0,2) \cong \mathrm{End}_{\mathbf{R}}(\mathbf{P})$. Now

$$\mathrm{Cl}(2,0) \cong \mathrm{Cl}(2,0)^{\mathrm{even}} \oplus \mathrm{Cl}(2,0)^{\mathrm{odd}},$$

can be rewritten as:

$$M_2(\mathbf{R}) \quad \cong \quad \mathbf{C} \quad \oplus \quad \mathbf{C}^\perp$$

where $\mathbf{C}^\perp \equiv \mathrm{span}\{C, R\}$, $C \equiv \left(\begin{smallmatrix} 1 & 0 \\ 0 & -1 \end{smallmatrix}\right)$ conjugation, $R \equiv \left(\begin{smallmatrix} 0 & 1 \\ 1 & 0 \end{smallmatrix}\right)$ reflection, and $i \equiv J \equiv \left(\begin{smallmatrix} 0 & -1 \\ 1 & 0 \end{smallmatrix}\right)$. In particular, $\mathbf{C}^\perp \equiv \mathrm{span}\{C, R\} = \Lambda^1 \mathbf{R}(0, 2)$, i.e., the set of complex antilinear maps in $M_2(\mathbf{R})$ is the space of vectors.

In both cases, the vector representation of $\mathrm{Spin}(2) \cong S^1$ is equivalent to the representation $\chi : \mathrm{Spin}(2) \to \mathrm{End}_\mathbf{R}(\mathbf{C})$ defined by $\chi_a(z) = a^2 z$.

Remark. Suppose M is a 2-dimensional (positive definite) Riemannian manifold. Then, via the canonical isomorphism $\mathrm{Cl}(2, 0) \cong \Lambda(\mathbf{R}(2, 0))$, the exterior bundle ΛTM is naturally a bundle of **H**-algebras, with the subbundle $\Lambda^{\mathrm{even}} TM = \Lambda^0 TM \oplus \Lambda^2 TM$ a bundle of **C**-algebras. Since in this dimension $\Lambda^{\mathrm{even}} TM$ acts on TM by Clifford multiplication, $\Lambda^{\mathrm{even}} TM$ is naturally a subbundle of $\mathrm{End}(TM)$. Note M is oriented if and only if $\Lambda^{\mathrm{even}} TM = M \times \mathbf{C}$ is trivial. (Similarly, ΛTM is naturally a bundle of $M_2(\mathbf{R})$-algebras.) In this dimension (if M is oriented), $\Lambda^{\mathrm{even}} TM = M \times \mathbf{C}$ can be taken as the bundle of spinors $M \times \mathbf{S}$ with the bundle isomorphism $\mathrm{Cl}^{\mathrm{even}}(2) \cong \mathrm{End}_\mathbf{C}(\mathbf{S})$.

Dimension 2 and Signature 1, 1

Spinors. $\mathrm{Cl}(1, 1)^{\mathrm{even}} \cong \mathbf{R} \oplus \mathbf{R} \cong \mathbf{L}$ (double numbers of Lorentz numbers), where a choice of unit volume element corresponds to $\tau \in \mathbf{L}$. A canonical choice for the spinor space $\mathbf{S} \equiv \mathbf{S}_+ \oplus \mathbf{S}_-$ (two real vector spaces $\mathbf{S}_+$ and $\mathbf{S}_-$ with $\mathrm{Cl}(1, 1)^{\mathrm{even}} \cong \mathrm{End}_\mathbf{R}(\mathbf{S}_+) \oplus \mathrm{End}_\mathbf{F}(\mathbf{S}_-)$) is possible once an orientation has been selected. Just take $\mathbf{S}_\pm \equiv \mathrm{Cl}(1, 1)^{\mathrm{even}}_\pm = \{a \in \mathrm{Cl}(1, 1)^{\mathrm{even}} : \lambda a = \pm a\}$, and let $\mathrm{Cl}(1, 1)^{\mathrm{even}}$ act on $\mathbf{S}$ by left multiplication. The canonical involution on $\mathrm{Cl}^{\mathrm{even}} \cong \mathbf{L}$ is conjugation on $\mathbf{L}$. Now

$$\mathrm{Spin}^0(1, 1) = \left\{ \pm e^{\tau \theta} : \theta \in \mathbf{R} \right\} \cong \mathbf{R} \otimes \mathbf{Z}_2$$
$$\mathrm{Spin}(1, 1) = \left\{ \pm e^{\tau \theta} : \theta \in \mathbf{R} \right\} \cup \left\{ \pm \tau e^{\tau \theta} : \theta \in \mathbf{R} \right\}.$$

Pinors. The pinor space $\mathbf{P}$ is just $\mathbf{S}$ and $\mathrm{Cl}(1, 1) \cong M_2(\mathbf{R}) \cong \mathrm{End}_\mathbf{R}(\mathbf{P})$, with

$$\mathrm{Cl}(1, 1) \cong \mathrm{Cl}(1, 1)^{\mathrm{even}} \oplus \mathrm{Cl}(1, 1)^{\mathrm{odd}}$$

corresponding to

$$M_2(\mathbf{R}) = \mathbf{L} \quad \oplus \quad \mathbf{L}^\perp$$

where $\tau \equiv \left(\begin{smallmatrix} 1 & 0 \\ 0 & -1 \end{smallmatrix}\right)$ and $\mathbf{L}^\perp = \mathrm{span}\{R, J\}$ with $R \equiv \left(\begin{smallmatrix} 0 & 1 \\ 1 & 0 \end{smallmatrix}\right)$ and $J \equiv \left(\begin{smallmatrix} 0 & -1 \\ 1 & 0 \end{smallmatrix}\right)$. That is, $\mathbf{L} \subset M_2(\mathbf{R})$ consists of the **L**-linear maps and $\mathbf{L}^\perp \subset M_2(\mathbf{R})$ consists of the **L**-antilinear maps.

Dimension 3 and Signature 3, 0 or 0, 3

Signature 3, 0.

(14.114) $$\text{Cl}(3,0) \cong M_1(\mathbf{H}) \oplus M_1(\mathbf{H}),$$

with $\tau = \begin{pmatrix} 1 & 0 \\ 0 & -1 \end{pmatrix}$ a choice of unit volume element. Also $\mathbf{S} \equiv \mathbf{H}$ with $\mathbf{P} \equiv \mathbf{H} \oplus \mathbf{H}$. The hat inner product $\hat{\varepsilon}$ on $\mathbf{P} \equiv \mathbf{H} \oplus \mathbf{H}$ is the standard $\mathbf{H}$-hermitian symmetric inner product. Thus, the hat anti-automorphism applied to $\begin{pmatrix} u & 0 \\ 0 & v \end{pmatrix}$ equals $\begin{pmatrix} \overline{u} & 0 \\ 0 & \overline{v} \end{pmatrix}$. Now $\text{Cl}(3,0)^{\text{even}} \cong M_1(\mathbf{H})$ considered to be diagonally embedded in $M_1(\mathbf{H}) \oplus M_1(\mathbf{H}) \cong \text{Cl}(3,0)$. Thus,

$$\text{Spin}(3) \cong \{a \in \mathbf{H} : \|a\| = 1\} \equiv S^3 \subset \mathbf{H} \cong \text{Cl}(3,0)^{\text{even}}.$$

That is, $\text{Spin}(3) \cong \mathrm{HU}(1)$. The vectors $\Lambda^1\mathbf{R}(3,0)$ can be distinguished as a subspace of $\Lambda\mathbf{R}(3,0) \cong \text{Cl}(3,0)$ by noting that $A \in \Lambda^1\mathbf{R}(3,0)$ if and only if a is odd and a is orthogonal to the unit volume element $\Lambda = \begin{pmatrix} 1 & 0 \\ 0 & -1 \end{pmatrix}$. Thus,

$$\Lambda^1\mathbf{R}(3, \ 0) = \left\{ \begin{pmatrix} u & 0 \\ 0 & -u \end{pmatrix} : u \in \text{Im}\,\mathbf{H} \right\} \subset M_1(\mathbf{H}) \oplus M_1(\mathbf{H}) \cong \text{Cl}(3,0).$$

Conversely, adopting this as the definition of $\Lambda^1\mathbf{R}(3,0) \subset M_1(\mathbf{H}) \oplus M_1(\mathbf{H})$, it is easy to deduce (from the Fundamental Lemma for Clifford Algebras) the isomorphism (14.114) and all of the rest of the information above about $\text{Cl}(3,0)$.

The vector representation $\text{Ad} : \text{Spin}(3) \to \text{SO}(3)$ is given by $\text{Ad}_a u = au\overline{a}$, for all $u \in \text{Im}\,\mathbf{H}$, where $a \in \text{Spin}(3) \cong S^3 \cong \mathrm{HU}(1) \subset \mathbf{H}$.

Remark. If M is a 3-dimensional (positive definite) Riemannanian manifold, then the vector bundle $E \equiv \Lambda^{\text{even}}TM = \text{Cl}^{\text{even}}(TM)$ is naturally a bundle of $\mathbf{H}$-algebras. Letting E act on itself on the left embeds E as a subbundle of $\text{End}_{\mathbf{R}}(E)$. Let C denote the centralizer of E in $\text{End}_{\mathbf{R}}(E)$. Then C is a bundle of $\mathbf{H}$-algebras that acts on E on the right, giving E the structure of a right $\mathbf{H}$-vector bundle with coefficient (or scalar) bundle C. Because of dimensions, $\text{Cl}^{\text{even}}(TM) \cong \text{End}_{\mathbf{H}}(E)$, so that E may be taken as the global bundle of spinors for M.

Signature 0, 3.
$$\text{Cl}(0,3) \cong M_2(\mathbf{C}),$$

where $\lambda \equiv i$ is a choice of unit volume element. Thus $\mathbf{P} \equiv \mathbf{C}^2$. The check inner product $\check{\varepsilon}$ is the standard $\mathbf{C}$-hermitian symmetric positive definite inner product on $\mathbf{C}^2$; thus $\check{A} = A^* = \overline{A}^t$. Therefore, consulting the Table in Lemma 9.27, we see that $\Lambda^1\mathbf{R}(0,3)$ must consist of those $A \in M_2(\mathbf{C})$ that satisfy

$$A \text{ is orthogonal to } 1 \quad (\text{i.e., } \mathrm{trace}_{\mathbf{C}}\, A = 0)$$

and

$$A = A^*, \quad \text{i.e., } A \text{ is hermitian symmetric}).$$

This proves

$$\Lambda^1\mathbf{R}(0,3) = \mathrm{Herm}_0(\mathbf{C}^2),$$

the space of trace free hermitian symmetric 2×2 matrices.

Note that the Pauli spin matrices

$$\sigma_1 \equiv \begin{pmatrix} 0 & 1 \\ 1 & 0 \end{pmatrix}, \quad \sigma_2 \equiv \begin{pmatrix} 0 & -i \\ i & 0 \end{pmatrix}, \quad \sigma_3 \equiv \begin{pmatrix} 1 & 0 \\ 0 & -1 \end{pmatrix}$$

form a basis for $\Lambda^1\mathbf{R}(0,3) \subset M_2(\mathbf{C}) \cong \mathrm{Cl}(0,3)$, with $\lambda = i = \sigma_1\sigma_2\sigma_3$ the volume element. Since $\Lambda^1\mathbf{R}(0,3) = \mathrm{Herm}_0(\mathbf{C}^2)$,

$$\Lambda^2\mathbf{R}(0,3) \cong i\,\mathrm{Herm}_0(\mathbf{C}^2)$$

so that

$$\mathrm{Cl}(0,3)^{\mathrm{even}} \cong \left\{ \begin{pmatrix} z & -\overline{w} \\ w & \overline{z} \end{pmatrix} : z, w \in \mathbf{C} \right\}.$$

Therefore,

$$\mathrm{Spin}(3) = \left\{ \begin{pmatrix} z & -\overline{w} \\ w & \overline{z} \end{pmatrix} : z, w \in \mathbf{C} \text{ with } |z|^2 + |w|^2 = 1 \right\} \cong \mathrm{SU}(2).$$

The vector representation $\mathrm{Ad} : \mathrm{Spin}(3) \to \mathrm{SO}(3)$ is given by

$$\mathrm{Ad}_A h \equiv AhA^* \quad \text{for all } h \in \mathrm{Herm}_0(\mathbf{C}^2),$$

where $A \in \mathrm{Spin}(3) = \mathrm{SU}(2)$.

Remark 14.113. The isomorphism $\mathrm{Cl}(3,0)^{\mathrm{even}} \cong \mathrm{Cl}(0,3)^{\mathrm{even}}$ induces an isomorphism of the two realizations of $\mathrm{Spin}(3)$, namely,

$$\mathrm{HU}(1) \cong \mathrm{SU}(2).$$

Thus, the Clifford algebra point of view leads naturally to an isomorphism, which was first discussed in Proposition 1.40; the isomorphism between

$$\mathrm{Cl}^{\mathrm{even}}(0,3) = \left\{ \begin{pmatrix} z & -\overline{w} \\ w & \overline{z} \end{pmatrix} : z, w \in \mathbf{C} \right\} \subset M_2(\mathbf{C}) = \mathrm{Cl}(0,3),$$

and

$$\mathrm{Cl}^{\mathrm{even}}(3,0) \cong M_1(H).$$

Remark. One can show that

(a) $\mathrm{Pin}(3,0) \cong \mathbf{Z}_2 \times S^3 \subset \mathbf{H} \oplus \mathbf{H}$ with $\mathbf{Z}_2 = \{(1,1),(1,-1)\}$,

(b) $\mathrm{Pin}(0,3) \cong \{A \in U(2) : \det_{\mathbf{C}} A = \pm 1\} \subset M_2(\mathbf{C}),$,

(c) Center $\mathrm{Pin}(3,0) \cong \mathbf{Z}_2$,

(d) Center $\mathrm{Pin}(0,3) \cong \{\pm 1, \pm i\} = \mathbf{Z}_4$.

Thus by (c) and (d), $\mathrm{Pin}(3,0)$ and $\mathrm{Pin}(0,3)$ are *not* isomorphic as abstract groups.

Dimension 3 and Signature 2,1 or 1,2

Spinors.
$$\mathrm{Cl}(2,1)^{\mathrm{even}} \cong \mathrm{Cl}(1,2)^{\mathrm{even}} \cong M_2(\mathbf{R}).$$

The space of spinors $\mathbf{S} \equiv \mathbf{R}^2$, and the spinor inner product ε is the standard volume form ($\mathbf{R}$ skew) on $\mathbf{S} \equiv \mathbf{R}^2$. Thus, the canonical involution on $\mathrm{Cl}^{\mathrm{even}} \cong M_2(\mathbf{R})$ is the cofactor transpose, and

$$\mathrm{Spin}^0(2,1) \cong \mathrm{SL}(2,\mathbf{R})$$
$$\mathrm{Spin}\ (2,1) \cong \{A \in M_2(\mathbf{R}) : \det A = \pm 1\}.$$

Pinors for Signature 2, 1.

$$\mathrm{Cl}(2,1) \cong M_2(\mathbf{C}) \cong \mathrm{Cl}(2,1)^{\mathrm{even}} \otimes_{\mathbf{R}} \mathbf{C}.$$

The pinor space $\mathbf{P}$ is $\mathbf{C}^2$ and the unit volume element λ equals $i \in M_2(\mathbf{C})$. The hat inner product $\hat{\varepsilon}$ on $\mathbf{C}^2$ is the standard complex volume element $\hat{\varepsilon} \equiv dz_1 \wedge dz_2$ on $\mathbf{C}^2$. Thus, the hat involution on $\mathrm{Cl}(2,1) \cong M_2(\mathbf{C})$ is the cofactor transpose. Consequently, $\Lambda^1 \mathbf{R}(2,1) \subset M_2(\mathbf{C}) \cong \mathrm{Cl}(2,1)$ is distinguished as

$$\{iA : A \in M_2(\mathbf{R}) \text{ and trace } A = 0\} = \left\{ \begin{pmatrix} ix & iy \\ iz & -ix \end{pmatrix} : x,y,z \in \mathbf{R} \right\}.$$

Pinors for Signature 1, 2.

$$\mathrm{Cl}(1,2) \cong M_2(\mathbf{R}) \oplus M_2(\mathbf{R}) \cong \mathrm{Cl}(1,2)^{\mathrm{even}} \otimes_{\mathbf{R}} \mathbf{L}$$

The pinor space is given by $\mathbf{P} = \mathbf{P}_+ \oplus \mathbf{P}_-$ with $\mathbf{P}_\pm \equiv \mathbf{R}^2$, and the hat inner product $\hat{\varepsilon}$ on $\mathbf{P}$ equals $\varepsilon \oplus \varepsilon$, where $\varepsilon \equiv dx_1 \wedge dx_2$ is the standard volume element on $\mathbf{R}^2$. Thus, the hat involution is the cofactor transpose applied to both 2×2 real matrices in $\mathrm{Cl}(1,2) \cong M_2(\mathbf{R}) \oplus M_2(\mathbf{R})$. The space $\Lambda^1 \mathbf{R}(1,2)$ of vectors can be distinguished in $\mathrm{Cl}(1,2)$ and equals

$$\Lambda^1 \mathbf{R}(1,2) = \left\{ \begin{pmatrix} A & 0 \\ 0 & -A \end{pmatrix} : A \in M_2(\mathbf{R}) \text{ and trace } A = 0 \right\}.$$

Remark. Thus, in both cases, signature $2, 1$ and $1, 2$, the vector representation

$$\mathrm{Ad} : \mathrm{Spin}(2,1) \to \mathrm{SO}(2,1)$$

is given by

$$\mathrm{Ad}_A B \equiv ABA^{-1} \quad \text{for all } B \in M_2(\mathbf{R}) \text{ such that trace } B = 0,$$

where $A \in \mathrm{Spin}(2,1) \equiv \mathrm{SL}(2, \mathbf{R})$.

Dimension 4, Signature 4, 0 and 0, 4

Spinors.

$$\mathrm{Cl}(4,0)^{\mathrm{even}} \cong \mathrm{Cl}(0,4)^{\mathrm{even}} \cong M_1(\mathbf{H}) \oplus M_1(\mathbf{H}).$$

The space of spinors $\mathbf{S} \equiv \mathbf{S}_+ \oplus \mathbf{S}_-$ with $S_\pm \equiv \mathbf{H}$, so that $\mathrm{Cl}^{\mathrm{even}} \cong \mathrm{End}_{\mathbf{H}}(\mathbf{S}_+) \oplus \mathrm{End}_{\mathbf{H}}(\mathbf{S}_-)$. The spinor inner products $\varepsilon_\pm$ are the standard $\mathbf{H}$-hermitian symmetric (positive definite) inner product on $\mathbf{S}_\pm \equiv \mathbf{H}$. Thus, the canonical involution on $\mathrm{Cl}^{\mathrm{even}} \cong M_1(\mathbf{H}) \oplus M_1(\mathbf{H})$ is conjugation of both factors. Therefore,

$$\mathrm{Spin}(4) = S^3 \times S^3 \equiv \left\{ \begin{pmatrix} a & 0 \\ 0 & b \end{pmatrix} : \|a\| = \|b\| = 1 \right\} \subset M_1(\mathbf{H}) \oplus M_1(\mathbf{H}).$$

The unit volume element is $\lambda \equiv \begin{pmatrix} 1 & 0 \\ 0 & -1 \end{pmatrix}$.

Pinors.
As algebras both

$$\mathrm{Cl}(4,0) \cong M_2(\mathbf{H}), \text{ and } \mathrm{Cl}(0,4) \cong M_2(\mathbf{H}).$$

Thus, $\mathbf{P}(4,0)$ and $\mathbf{P}(0,4)$ are both $\mathbf{H}^2$. The standard $\mathbf{H}$-hermitian symmetric positive definite inner product on $\mathbf{H}^2$ corresponds to the hat inner product $\hat{\varepsilon}$ on $\mathbf{P}(4,0) \cong \mathbf{H}^2$ while it corresponds to the check inner product

$\tilde{\varepsilon}$ on $\mathbf{P}(0,4) \cong \mathbf{H}^2$. Consequently, $V(4,0) = \Lambda^1 \mathbf{R}(4,0)$ is distinguished as the subspace

$$(14.115) \qquad V(4,0) = \left\{ \begin{pmatrix} 0 & L_u \\ -L_{\bar{u}} & 0 \end{pmatrix} : u \in \mathbf{H} \right\} \subset \text{Cl}(4,0) \cong M_2(\mathbf{H}),$$

i.e., those odd elements that are $\mathbf{H}$-hermitian skew; while $V(0,4) \equiv \Lambda^1 \mathbf{R}(0,4)$ is distinguished as the subspace

$$(14.116) \qquad V(0,4) \equiv \left\{ \begin{pmatrix} 0 & L_u \\ L_{\bar{u}} & 0 \end{pmatrix} : u \in \mathbf{H} \right\} \subset \text{Cl}(0,4) \cong M_2(\mathbf{H}).$$

Conversely, adopting (14.115) as the definition of $V(4,0) \subset M_2(\mathbf{H})$ and (14.116) as the definition of $V(0,4) \subset M_2(\mathbf{H})$, it is easy to deduce (from the Fundamental Lemma of Clifford Algebras) the isomorphisms $\text{Cl}(4,0) \cong M_2(\mathbf{H})$ and $\text{Cl}(0,4) \cong M_2(\mathbf{H})$ and all the rest of the information above about $\text{Cl}(4,0)$ and $\text{Cl}(0,4)$.

Dimension 4, Signature $1,3$ and $3,1$ (Special Relativity)

Spinors.

$$\text{Cl}(3,1)^{\text{even}} \cong \text{Cl}(1,3)^{\text{even}} \cong M_2(\mathbf{C}),$$

and the unit volume element λ for $\mathbf{R}(3,1)$ or $\mathbf{R}(1,3)$ can be chosen to be $\lambda \equiv i \in M_2(\mathbf{C})$. The space of spinors $\mathbf{S}$ is a 2-dimensional complex vector space with isomorphisms

$$\text{Cl}(3,1)^{\text{even}} \cong \text{Cl}(1,3)^{\text{even}} \cong \text{End}_{\mathbf{C}}(\mathbf{S}).$$

The spinor inner product ε on $\mathbf{S}$ is $\mathbf{C}$-skew. (Hence, we may take $\mathbf{S} \equiv \mathbf{C}^2$ and $\varepsilon \equiv dz_1 \wedge dz_2$, the standard complex volume form on $\mathbf{C}^2$.) Therefore, the canonical involution on $\text{Cl}^{\text{even}} \cong M_2(\mathbf{C})$ is the cofactor transpose sending

$$A = \begin{pmatrix} a & b \\ c & d \end{pmatrix} \quad \text{to} \quad A^* \equiv \begin{pmatrix} d & -b \\ -c & a \end{pmatrix}.$$

Thus,

$$\text{Spin}^0(3,1) \cong \text{Spin}^0(1,3) \cong \text{SL}(2,\mathbf{C})$$

and

$$\text{Spin}(3,1) \cong \text{Spin}(1,3) \cong \{A \in M_2(\mathbf{C}) : \det_{\mathbf{C}} A = \pm 1\}.$$

The reduced spin group $\text{Spin}^0(3,1) \cong \text{SL}(2,\mathbf{C})$ acts on the space $\mathbf{S}, \varepsilon$ of spinors preserving the $\mathbf{C}$-skew inner product ε. The conjugate representation of $\text{SL}(2,\mathbf{C})$ is on the space $\overline{\mathbf{S}}$ of conjugate spinors. ($\mathbf{P}_{\mathbf{C}} = \mathbf{S} \oplus \overline{\mathbf{S}}$ is the space of complexified pinors.)

In this important case, the spinor representations yield a classification of all representations of $SL(2, \mathbf{C})$ with important implications in general relativity. A brief description is included.

Start with S, J a complex vector space with complex structure J. As usual, consider the complexification $S_{\mathbf{C}} = S \otimes_{\mathbf{R}} \mathbf{C} = S^{1,0} \oplus S^{0,1}$ decomposed into the $+i$ eigenspace $S^{1,0}$ and the $-i$ eigenspace $S^{0,1}$ for J (extended to be complex linear on S), along with the canonical conjugation on $S_{\mathbf{C}}$. Now assume $\dim_{\mathbf{C}} S = 2$, let $\mathbf{S} \equiv S^{1,0}$, $\overline{\mathbf{S}} \equiv S^{0,1}$, and assume that $\mathbf{S}$ is equipped with a complex volume element (a $\mathbf{C}$-skew inner product ε). Representations of the group $SL(\mathbf{S}) \cong SL(2, \mathbf{C})$ will be denoted by the vector space alone, i.e., the action is assumed to be obvious.

For example, $SL(2, \mathbf{C})$ has a natural induced action on $\mathbf{S}^p$ the pth symmetric tensor product of $\mathbf{S}$, and on $\overline{\mathbf{S}}^q$ the qth symmetric tensor product of $\overline{\mathbf{S}}$. It is a standard fact in representation theory that

(1) $\mathbf{S}^p \otimes \overline{\mathbf{S}}^q$ (the *space of p, q spinors*) is irreducible, $p, q \geq 0$,
(2) there are no other irreducible representations of $SL(2, \mathbf{C})$.

For example, ε defines an isomorphism $\flat$ of $\mathbf{S}$ with dual space $\mathbf{S}^*$ that is the intertwining operator for $\mathbf{S} \cong \mathbf{S}^*$. Also, ε provides an isomorphism between $\Lambda^2 \mathbf{S}$ and $\mathbf{C}$, the trivial representation. The vector representation of $\text{Spin}^0(3, 1) \cong SL(2, \mathbf{C})$ is obtained as follows. Let $H \equiv (\mathbf{S} \otimes \overline{\mathbf{S}})_{\mathbf{R}}$ denote the subspace of $\mathbf{S} \otimes \overline{\mathbf{S}} \equiv \Lambda^{1,1} \mathbf{S}_{\mathbf{C}}$ fixed by conjugation. Let

$$\frac{1}{2}(\operatorname{Im} h) \wedge (\operatorname{Im} h) = \langle h, h \rangle \varepsilon \wedge \overline{\varepsilon} \in \Lambda^{2,2} \mathbf{S}_{\mathbf{C}}$$

define a bilinear form $\langle \ , \ \rangle$ on H. One can identify H with the space of hermitian symmetric 2×2 matrices with square norm equal to the determinant. In particular, $H \cong \mathbf{R}(1, 3)$ is Minkowski space. Note that replacing ε by $e^{i\theta}$ dos not change the metric on H.

Using ε, the space of p, q spinors (with the $SL(2, \mathbf{C})$ action) can be identified with the space $\mathbf{C}_{p,q}[z, \overline{z}]$ of polynomials which are homogeneous of degree p in $z \in \mathbf{S}$ and homogeneous of degree q in $\overline{z} \in \overline{\mathbf{S}}$ (with the natural action of $SL(2, \mathbf{C})$ on $\mathbf{C}[z, \overline{z}]$).

Remark. On a Lorentzian manifold M equipped with a spinor bundle $\mathbf{S}, \varepsilon$, note that: (i) $T^* M \otimes \mathbf{C} \cong \mathbf{S} \otimes \overline{\mathbf{S}}$, and $\mathbf{S} \otimes \mathbf{S} \cong \mathbf{S}^2 \oplus \Lambda^2 \mathbf{S} \cong \mathbf{S}^2 \oplus \mathbf{C} \cong \mathbf{S}^{2,0} \oplus \mathbf{S}^{0,0}$. More generally, one can show that: (ii) $\mathbf{S} \otimes \mathbf{S}^p = \mathbf{S}^{p+1} \oplus \mathbf{S}^{p-1}$. These facts, (i) and (ii), enable one to define four different (if $p \geq 1$ and $q \geq 1$) differential operators on p, q spinor fields s based on decomposing the covarient deviative Ds of s as follows:

$$Ds \in T_{\mathbf{C}}^* M \otimes \mathbf{S}^{p,q} = \mathbf{S}^{p-1,q-1} \oplus \mathbf{S}^{p+1,q-1} \oplus \mathbf{S}^{p+1,q+1} \oplus \mathbf{S}^{p-1,q+1}.$$

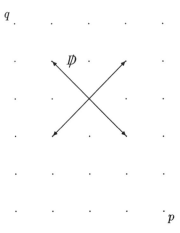

Figure 14.117

If $q = 0$, then there are just two differential operators. the one taking a $p, 0$ spinor field to a $p - 1, 1$ spinor field is called the *Dirac operator* and denoted by $\not{D}$. The other, taking a $p, 0$ spinor field to a $p + 1, 1$ spinor field is called the *twistor operator*. The Dirac equation

$$\not{D}f = 0, \quad \text{where } f \text{ is a } p, 0 \text{ spinor field,}$$

is called the equation for a *massless particle of* spin $s = p/2$. The case $p = 2$ $(s = 1)$ is that of an electron.

Pinors for Signature 1, 3: The pinor space $\mathbf{P}$ is a real 4-dimensional vector space. A choice of unit volume element λ provides a complex structure $I \equiv \lambda$ on $\mathbf{P}$. The spinor space $\mathbf{S}$ can be taken to be $\mathbf{P}$ with the complex structure I. The spinor inner product ε is complex skew and

$$\varepsilon \equiv \hat{\varepsilon} + i\breve{\varepsilon}, \quad \text{or } \hat{\varepsilon} \equiv \operatorname{Re}\varepsilon.$$

It is possible to choose a conjugation c on $\mathbf{P}$ so that

$$\hat{\varepsilon}(cx, cy) = \hat{\varepsilon}(x, y) \quad \text{for all } x, y \in \mathbf{P}.$$

Note that

$$\Lambda^{\mathrm{odd}}\mathbf{R}(1, 3) \cong \overline{\operatorname{End}_{\mathbf{C}}}(\mathbf{P})$$

(the space of all complex antilinear maps of $\mathbf{P}$) can be expressed as

$$\overline{\operatorname{End}_{\mathbf{C}}}(\mathbf{P}) = \{Ac : A \in \operatorname{End}_{\mathbf{C}}(\mathbf{P})\}.$$

Now $c^2 = 1$ and c is complex antilinear. Therefore, given $A \in \mathrm{End}_{\mathbf{C}}(\mathbf{P})$, $\hat{\varepsilon}(Acx, y) = \mathrm{Re}\,\varepsilon(Acx, y) = \mathrm{Re}\,\varepsilon(cx, A^*y) = \hat{\varepsilon}(cx, A^*y) = \hat{\varepsilon}(x, cA^*c^2y) = \hat{\varepsilon}(x, \overline{A}^*cy)$. This proves that

$$(Ac)^{\wedge} = \overline{A}^*c \quad \text{for all } A \in \mathrm{End}_{\mathbf{C}}(\mathbf{P}).$$

Therefore,

$$\Lambda^1 \mathbf{R}(1, 3) \cong \left\{ Ac : A \in \mathrm{End}_{\mathbf{C}}(\mathbf{P}) \text{ and } \overline{A}^* = -A \right\}.$$

That is

$$V(1, 3) \equiv \Lambda^1 \mathbf{R}(1, 3) \text{ must be the subspace}$$

$$(14.118) \qquad V(1, 3) = \left\{ \begin{pmatrix} z & x \\ y & -\overline{z} \end{pmatrix} c : z \in \mathbf{C}, \text{ and } x, y \in \mathbf{R} \right\}$$

of $\mathrm{End}_{\mathbf{R}}(\mathbf{P})$.

The square norm $\| \; \|$ on $V(1, 3)$ is given by

$$(14.119) \qquad \|Ac\| = (Ac)(Ac)^{\wedge} = Ac\overline{A}^*c = AA^* = \det_{\mathbf{C}} A,$$

which equals $-\|z\| - xy$ if

$$A \equiv \begin{pmatrix} z & x \\ y & -\overline{y} \end{pmatrix}.$$

Conversely, adopting (14.118) as the definition of $V(1, 3) \subset \mathrm{End}_{\mathbf{R}}(\mathbf{P})$ $\cong M_4(\mathbf{R})$, where $V(1, 3)$ is equipped with the square norm given by (14.119), it is easy to deduce the isomorphism $\mathrm{Cl}(1, 3) \cong \mathrm{End}_{\mathbf{R}}(\mathbf{P})$ and all the rest of the information presented above about $\mathrm{Cl}(1, 3)$.

Pinors for Signature 3, 1: The pinor space $\mathbf{P}$ is a 2-dimensional (right) quaternionic vector space; so that $\mathrm{Cl}(3, 1) \cong \mathrm{End}_{\mathbf{H}}(\mathbf{P})$. One can show that, under an isomorphism $\mathrm{End}_{\mathbf{H}}(\mathbf{P}) \cong M_2(\mathbf{H})$, the space of vectors $\Lambda^1 \mathbf{R}(3, 1)$ is isomorphic to

$$\left\{ j \begin{pmatrix} w & x \\ y & -\overline{w} \end{pmatrix} : x, y \in \mathbf{R} \text{ and } w \in \mathbf{C} \right\}.$$

Left multiplication by i, denoted L_i, corresponds to a choice of unit volume element.

SQUARES OF SPINORS AND CALIBRATIONS

The *pinor multiplication* introduced in Chapter 13 is a useful tool for constructing calibrations. The Clifford element $\phi \equiv x \hat{\mathrm{o}} y \in \mathrm{Cl}(r, s)$ obtained from multiplying two pinors x and y may also be considered as a differential form $\phi \in \Lambda \mathbf{R}(r, s)^*$; under the natural identifications of $\mathrm{Cl}(r, s)$ with $\Lambda \mathbf{R}(r, s)$, and $\Lambda \mathbf{R}(r, s)$ with $\Lambda \mathbf{R}(r, s)^*$. In this section, we consider the positive definite case.

The special case $n \equiv 8p$ is particularly interesting. Recall

$$\mathrm{Cl}(8k) \cong \mathrm{End}_{\mathbf{R}}(\mathbf{P}),$$

with $\mathbf{P} \equiv \mathbf{S}_+ \oplus \mathbf{S}_-$ decomposing into the real vector spaces $\mathbf{S}_\pm$, of positive and negative spinors. The spinor inner product ε on $\mathbf{S}_+$ and $\mathbf{S}_-$ is positive definite and real. Given a differential form $\phi \in \Lambda \mathbf{R}(n)^*$, let $\phi_k \in \Lambda^k \mathbf{R}(n)^*$ denote the degree k part of ϕ.

Theorem 14.120. *Consider* $\mathbf{R}(n)$ *with* $n \equiv 8p$. *Given a unit positive spinor* $x \in \mathbf{S}_+$, *let*

$$(14.120) \qquad \phi \equiv 16^p x \circ x \in \mathrm{Cl}(8p) \cong \Lambda \mathbf{R}(n)^*.$$

Then

(a) $\phi \in \sum \Lambda^{4m} \mathbf{R}(n)^*$ *only has components of degree* $4m$.
(b) *Each* ϕ_k *is a calibration, i.e.,* $\phi_k(\xi) \leq 1$ *for all* $\xi \in G(k, \mathbf{R}(n))$.
(c) *Equality* $\phi_k(\xi) = 1$ *occurs if and only if* $\xi x = x$,
(d) *The isotropy subgroup of* $O(n)$ *that fixes* ϕ *is isomorphic to the subgroup* K_x *of* $\mathrm{Spin}(n)$ *that fixes the spinor* x.

Corollary 14.122. *Under the hypothesis of Theorem 14.120, the degree* k-*form* ϕ_k *is a nontrivial calibration* $(k = 4m)$ *if and only if there exist an* $\xi \in G(k, \mathbf{R}(n))$ *such that* $\xi x = x$. *The contact set*

$$G(\phi) \equiv \{\xi \in G(k, \mathbf{R}(n)) : \phi(\xi) = 1\}$$

is equal to $\{\xi \in G(k, \mathbf{R}(n)) \subset \mathrm{Spin}(n) : \xi x = x\}$, *which via the map* χ *is isomorphic to* $\{\mathrm{Ref}_{\mathrm{span}\,\xi} : \xi \in G(k, \mathbf{R}(n)) \text{ and } \mathrm{Ref}^*_{\mathrm{span}\,\xi}(\phi) = \phi\} \subset \mathrm{SO}(n)$.

Proof: By Theorem 13.73 ($\dim_{\mathbf{R}} \mathbf{P} = 16^p$),

$$(14.123) \qquad \phi(\xi) = 16^p \langle x \hat{\mathrm{o}} x, \xi \rangle = \varepsilon(\xi x, x) \quad \text{for all } \xi \in \mathrm{Cl}(n).$$

This formula is very useful.

(a) If $\xi \in \mathrm{Cl}(n)^{\mathrm{odd}}$ then $x \in \mathbf{S}_+$ implies $\xi \in \mathbf{S}_-$ so that $\phi(\xi) = \varepsilon(\xi x, x) = 0$. If ξ has degree $= 2 \bmod 4$, then $\hat{\xi} = -\xi$ and hence

$$\varepsilon(\xi x, x) = \varepsilon(x, \hat{\xi} x) = -\varepsilon(\xi x, x),$$

so that again $\phi(\xi) = 0$.
(b) By the Cauchy–Schwarz inequality, Equation (14.123) implies that

(14.124) $\qquad \phi(\xi) \leq |\xi x| \, |x| = 1 \quad \text{for all } \xi \in G\left(k, \mathbf{R}(n)\right) \subset \mathrm{Cl}(n)$

where $|y|^2 \equiv \varepsilon(y, y)$ is the spinor norm on $\mathbf{S}_+$.
(c) Also,

(14.125) $\qquad\qquad \phi(\xi) = 1$ if and only if $\xi x = x$.

Note that $|\xi x| = 1$ in (14.124) because $\xi \in G(k, \mathbf{R}(n)) \subset \mathrm{Spin}(n)$ for $k = 0 \bmod 4$, which implies that $\xi \hat{\xi} = 1$.
(d) Consider the vector representation $\chi : \mathrm{Pin}(n) \to O(n)$. Given $a \in \mathrm{Pin}(n)$ and $v \in \mathbf{R}(n)$ then $\chi_a(v) \equiv \tilde{a} v a^{-1}$. More generally, given $a \in \mathrm{Pin}(n)$, $\chi_a \in O(n)$ acts on $\Lambda \mathbf{R}(n)^* \cong \Lambda \mathbf{R}(n)$ by $\chi_a(\phi) = \tilde{a} \phi a^{-1} = 16^p \tilde{a}(x \circ x) \hat{a}$. By Lemma 8.44(a), this equals $16(\tilde{a} x) \circ (a x)$. Therefore, if $\chi_a(\phi) = \phi$ then $(\tilde{a} x) \circ (a x) = x \circ x$. By the definition of spinor multiplication, this implies that $\tilde{a} x = c s$ for some scalar $c \in \mathbf{R}$. Since $\tilde{a}$ preserves the pinor norm, $c = \pm 1$. The element $a \in \mathrm{Pin}(n)$ must be either even or odd. If a is odd, then $\tilde{a} x \in \mathbf{S}_-$ and $\tilde{a} x = \pm x$ is impossible. Thus $a x = \pm x$ and

$$\{a \in \mathrm{Pin}(n) : \chi_a(\phi) = \phi\} = \{a \in \mathrm{Spin}(n) : a x = \pm x\}.$$

Since, under the vector representation χ, both a and $-a$ have the same image, this proves that

(14.126) $\qquad \chi : K_x \equiv \{a \in \mathrm{Spin}(n) : a x = x\} \to \{g \in O(n) : g^* \phi = \phi\}$

is surjective. Since $-1 \notin K_x$, this is an isomorphism. ∎

Using the octonian model for $\mathrm{Cl}(8) \cong \mathrm{End}_{\mathbf{R}}(\mathbf{O} \oplus \mathbf{O})$ with $\mathbf{S}_+ \equiv \mathbf{O}$, the square of a positive spinor is easily calculated. First note that $\mathrm{Spin}(8)$ acts transitively on the unit sphere in $\mathbf{S}_+$, so that any two squares are $SO(8)$-equivalent in $\Lambda \mathbf{R}(8)^*$. Therefore, we may choose the multiplicative unit in $\mathbf{S}_+ \equiv \mathbf{O}$, denoted 1_+, as a unit positive spinor without loss of generality.

Theorem 14.127.

(a) $16\ 1_+ \circ 1_+ = 1 + \Phi + \lambda$, where $\Phi \in \Lambda^4 \mathbf{O}^*$ is defined by

(14.127) $\Phi(u_1 \wedge u_2 \wedge u_3 \wedge u_4) \equiv \langle u_1, u_2 \times u_3 \times u_4 \rangle,$

and λ is the unit volume element determined by the orientation on $\mathbf{O}$.
(b) The 4-form Φ is a calibration (called the *Cayley calibration*) on $\mathbf{R}^8 \cong$
$\mathbf{O}$, that is fixed by the group $\mathrm{Spin}(7) \cong \{g \in \mathrm{Spin}(8) : g_+(1) = 1\}$.

Proof: By (14.123),

$$16(1_+ \ \hat{\circ}\ 1_+)(u_1 \wedge u_2 \wedge u_3 \wedge u_4) = \varepsilon(R_{u_1} R_{\overline{u}_2} R_{u_3} R_{\overline{u}_4} 1_+, 1_+),$$

which equals

$$\langle ((\overline{u}_4 u_3)\, \overline{u}_2)\, u_1, 1 \rangle = \langle (\overline{u}_4 u_3)\, \overline{u}_2, \overline{u}_1 \rangle$$
$$= \langle u_1, u_2\, (\overline{u}_3 u_4) \rangle = \Phi\, (u_1 \wedge u_2 \wedge u_3 \wedge u_4)$$

if u_1, u_2, u_3, u_4 are orthogonal. Therefore, the degree 4 component of
$16\ 1_+\ \hat{\circ}\ 1_+$ is Φ. Similarly, one shows that the degree 8 component is
λ. ∎

See Bryant and Harvey [6] for other examples of pinor multiplication
in dimension 8. In the next multiple of 8, dimension 16, all unit positive
spinors are not the same under the action of $\mathrm{Spin}(16)$. The orbit structure
of $\mathrm{Spin}(16)$ acting on $\mathbf{S}_+$ is very interesting. Each orbit can be used to
construct calibrations by squaring a spinor in that orbit. See Dadok and
Harvey [7] for the details.

The case dimension $n = 0$ mod 8 can be used to understand the case
dimension $m = 7$ mod 8 (see the next Theorem and Problem 9).

Recall that if a unit vector $e_0 \in V(n)$ is chosen so that $V(n-1) = e_0^\perp$,
then sending a vector $u \in V(n-1)$ to $e_0 \cdot u \in \mathrm{Cl}(n)$ induces an isomorphism
of $\mathrm{Cl}(n-1) \cong \mathrm{Cl}(n)^{\mathrm{even}}$. Furthermore, if $\mathrm{Cl}(n) \cong \mathrm{End}_{\mathbf{R}}(\mathbf{P})$ is a pinor
representation with $\mathbf{P} \equiv \mathbf{S}_+ \oplus \mathbf{S}_-$, then $\mathrm{Cl}(n)^{\mathrm{even}} \cong \mathrm{End}_{\mathbf{R}}(\mathbf{S}_+) \oplus \mathrm{End}_{\mathbf{R}}(\mathbf{S}_-)$.
Therefore, an identification $\mathbf{S} \equiv \mathbf{S}_+ \equiv \mathbf{S}_-$ induces an isomorphism

$$\mathrm{Cl}(n-1) \cong \mathrm{End}_{\mathbf{R}}(\mathbf{S}) \oplus \mathrm{End}_{\mathbf{R}}(\mathbf{S}),$$

which is a pinor representation of $\mathrm{Cl}(n-1)$.

Now, a unit positive spinor $x \in \mathbf{S}_+(n)$ can also be considered a unit
pinor $y \equiv (x, 0) \in \mathbf{P}(n-1) \equiv \mathbf{S} \oplus \mathbf{S}$.

Theorem 14.129. *Let*

$$(14.130) \qquad \phi \equiv 16^p \ x \circ x \in \text{Cl}(n), \quad n \equiv 8p,$$

and

$$(14.131) \qquad \psi \equiv 16^p \ y \ \hat{\circ} \ y \in \text{Cl}(n-1).$$

Let the orthogonal decomposition

$$\phi \equiv e_0^* \wedge \alpha + \beta$$

define α, β *in* $\Lambda \mathbf{R}(n-1)^*$. *Then*

$$(14.132) \qquad \psi \equiv \alpha + \beta.$$

Proof: By Theorem 13.73

$$(14.131) \qquad \psi(\xi) = \hat{\varepsilon}(\xi y, y) = \varepsilon(\xi x, x) \quad \text{for all } \xi \in \text{Cl}(n-1).$$

If ξ has degree 1 or 2 mod 4, then $\hat{\xi} = -\xi$ so that $\psi(\xi) = 0$. We must show that

$$(14.134) \qquad \psi(\xi) = \beta(\xi) \quad \text{if degree } \xi = 0 \text{ mod } 4,$$

and

$$(14.135) \qquad \psi(\xi) = \alpha(e_0 \wedge \xi) \quad \text{if degree } \xi = 3 \text{ mod } 4.$$

Let $u_1, \ldots, u_{n-1}$ denote an orthonormal basis for $V(n-1)$ and $e_0, e_1 \equiv u_1, \ldots, e_{n-1} \equiv u_{n-1}$ an orthonormal basis for $V(n)$. Then $\text{Cl}(n-1)$ embeds in $\text{Cl}(n)$ by sending $u_j \in \text{Cl}(n-1)$ to $e_0 \cdot e_j \in \text{Cl}(n)$. If $\xi \equiv u_{i_1} \cdot u_{i_2} \cdot u_{i_3} \cdot u_{i_4} \in \text{Cl}(n-1) \cong \Lambda \mathbf{R}(n-1)$, then the corresponding element, also denoted ξ, in $\text{Cl}(n) \cong \Lambda \mathbf{R}(n)$ is equal to

$$e_0 u_{i_1} \cdot e_0 u_{i_2} \cdot e_0 u_{i_3} \cdot e_0 u_{i_4} = u_{i_1} \cdot u_{i_2} \cdot u_{i_3} \cdot u_{i_4}.$$

Therefore, $\psi(\xi) = \varepsilon(\xi x, x) = \beta(\xi)$. If $\xi = u_{i_1} \cdot u_{i_2} \cdot u_{i_3} \in \text{Cl}(n-1) \cong \Lambda \mathbf{R}(n-1)$, then the corresponding element of $\text{Cl}(n) \cong \Lambda \mathbf{R}(n)$ is $e_0 u_{i_1} \cdot e_0 u_{i_2} \cdot e_0 u_{i_3} = e_0 u_{i_1} \cdot u_{i_2} \cdot u_{i_3}$. Therefore, $\psi(\xi) = \phi(e_0 \cdot \xi) = \alpha(\xi)$. ∎

312   *Problems*

PROBLEMS

1. A complex volume element λ on $\mathbf{C}^4$ determines a nondegenerate complex inner product $\langle \, , \rangle$ on $\Lambda^2 \mathbf{C}^4$ by

$$\alpha \wedge \beta = \langle \alpha, \beta \rangle \lambda.$$

Show that the induced action χ of $\mathrm{SL}(4, \mathbf{C})$ on $\Lambda^2 \mathbf{C}^4$ provides a double cover

$$\mathrm{SL}(4, \mathbf{C}) \xrightarrow{\chi} O(6, \mathbf{C}).$$

2. Verify the isomorphisms in Theorem 14.5.

3. (a) Show that the (*triality automorphism*) τ defined by (14.27) is an automorphism of $\mathrm{Spin}(8)$ of order three that permutes the representations χ, $\overline{p}_+$, and ρ_-.

   (b) Show that $(g_0, g_+, g_-) \in \mathrm{Spin}(8)$ if and only if one (all) of the following triples belongs to $\mathrm{Spin}(8)$:

$$\left(g_0, g_+, g_-\right), \quad \left(g_-', g_+', g_0'\right),$$

$$\left(g_+', g_-', g_0\right), \quad \left(g_0', g_-, g_+\right),$$

$$\left(g_-, g_0', g_+'\right), \quad \left(g_+, g_0, g_-'\right).$$

4. (a) Show that $\mathrm{Spin}(3, 4)$ acts transitively on $S^7 = S_+^7 \cup S_-^7$,

$$S_\pm^7 \equiv \{x \in \tilde{\mathbf{O}} : \|x\| = \pm 1\} \subset \tilde{\mathbf{O}} \cong \mathbf{S},$$

with isotropy subgroup at $1 \in \tilde{\mathbf{O}} \cong \mathbf{S}$ equal to split $\tilde{G}_2$.

   (b) Show that $\mathrm{Spin}^0(3, 4)/G_2 \cong S_+^7$.

5. Show that, under the spinor representation $\rho = \rho_+ \oplus \rho_-$ of $\mathrm{Spin}(4, 4)$ on $\mathbf{S} \equiv \mathbf{S}_+ \oplus \mathbf{S}_- \cong \tilde{\mathbf{O}} \oplus \tilde{\mathbf{O}}$,

   (a) $\mathrm{Spin}(4, 4)/\tilde{G}_2 \cong \{(x, y) \in \tilde{\mathbf{O}} \times \tilde{\mathbf{O}} : \|x\| = \pm 1 \text{ and } \|y\| = \pm 1\}$.

   (b) $\mathrm{Spin}(4, 4)/\mathrm{Spin}(3, 4) \cong \{x \in \tilde{\mathbf{O}} : \|x\| = \pm 1\}$.

6. Show that

$$\mathrm{Spin}(5, 4)/\mathrm{Spin}(3, 4) \cong \{x \in \mathbf{R}(8, 8) : \|x\| = \pm 1\}.$$

7. Use Corollary 10.50 to characterize $\mathrm{Spin}(9)$ in the model $\mathrm{Cl}(9)^{\mathrm{even}} \cong \mathrm{End}_{\mathbf{R}}(\mathbf{O} \oplus \mathbf{O})$ defined by (14.74).

8. (a) Show that, under the action of F_4 on $\mathrm{Herm}(3, \mathbf{O})$, each $A \in$ $\mathrm{Herm}(3, \mathbf{O})$ may be put in the canonical form

$$A = \begin{pmatrix} r_1 & 0 & 0 \\ 0 & r_2 & 0 \\ 0 & 0 & r_3 \end{pmatrix}, \quad \text{with } r_1, r_2, r_3 \in \mathbf{R} \text{ and } r_1 \leq r_2 \leq r_3.$$

(b) If two (but not all three) of the eigenvalues r_1, r_2, r_3 are equal, then the orbit through A is

$$F_4/\mathrm{Spin}(9) \cong \mathbf{P}^2(\mathbf{O}).$$

(c) If all three of the eigenvalues r_1, r_2, r_3 of A are distinct, then the orbit through A is

$$F_4/\mathrm{Spin}(8) \cong \text{An } S^8 \text{ bundle over } \mathbf{P}^2(\mathbf{O}).$$

9. Prove that $16 \, 1_+ \, \hat{\mathrm{o}} \, 1_+$ considered as an element of $\mathrm{Cl}(7) \cong \Lambda \mathbf{R}(7)^*$ is equal to $1 + \phi + \psi + \lambda$, where ϕ is the associative calibration, ψ is the coassociative calibration, and λ is the unit volume form on $\mathrm{Im}\,\mathbf{O}$.

REFERENCES

1. Arnold, V. I., *Mathematical Methods of Classical Mechanics*, Springer–Verlag, New York, 1978.
2. Atiyah, M. F., R. Bott, and A. Shapiro, Clifford modules, *Topology*, 3 (1964), 3–38.
3. Besse, A. L., *Einstein Manifolds*, Springer–Verlag, Berlin, 1987.
4. Bryant, R. and R. Harvey, Submanifolds in hyperkähler geometry, *Jour. Amer. Math. Soc.*, 2 (1989), 1–31.
5. Bryant, R. and R. Harvey, Stabilizers of calibrations, Rice University preprint, 1989.
6. Bryant, R. and R. Harvey, Geometry of G_2 and Spin(7) structures, Rice University preprint, 1989.
7. Dadok, J. and R. Harvey, Calibrations and spinors, Rice University preprint, 1989.
8. Gureirch, G. B., *Foundations of the Theory of Algebraic Invariants*, P. Noordhoff LTD., Groningen, The Netherlands, 1964.
9. Harvey, R. and Lawson, H. B., Jr., Calibrated geometries, *Acta Math.* 148 (1982), 47–157.
10. Helgason, S., *Differential Geometry, Lie Groups, and Symmetric Spaces*, Academic Press, New York, 1978.
11. Lawlor, G., The Angle Criterion, *Invent. Math.*, 95 (1989), 437–446.
12. Lawson, H. B., Jr. and M. Michelsohn, *Spin Geometry*, Princeton University Press, Princeton, New Jersey, 1989.
13. Nance, D., Sufficient conditions for a pair of n-planes to be area-minimizing, Math. Ann. 279 (1987), 161–164.
14. O'Niell, B., *Semi-Riemannian Geometry*, Academic Press, New York, 1983.
15. Penrose, R. and W. Rindler, *Spinors and Space-Time* (2 vols.), Cambridge University Press, Cambridge, 1986.
16. Salamon, S., *Riemannian Geometry and Holonomy Groups*, Wiley, New York, 1989.

Subject Index

Perspectives in Mathematics